Advanced adhesives in electronics

Related titles:

Silicon–germanium (SiGe) nanostructures
(ISBN 978-1-84569-689-4)
Nanostructured silicon–germanium (SiGe) provides the prospect of novel and enhanced electronic device performance. This book reviews the materials science and technology of SiGe nanostructures, including crystal growth, fabrication of nanostructures, material properties and applications in electronics.

High-temperature superconductors
(ISBN 978-1-84569-578-1)
High-temperature superconductors reviews growth techniques, properties and applications of key high-temperature superconducting films. The first part of the book provides an overview of high-temperature superconductor properties such as optical conductivity and transport properties. Part II reviews the growth and properties of particular types of superconducting film, while Part III describes how they can be applied in practice in various areas of electronics.

Optical switches: Materials and design
(ISBN 978-1-84569-579-8)
Optical communication using optical fibres as the transmission medium is essential to handling the massive growth of both telecom and datacom traffic. Different technologies which can be applied to switching optical signals are addressed. The book features electro-optical, thermo-optical, micro-electro-mechanical (MEMS)-based and semiconductor optical amplifier (SOA)-based optical switches. *Optical switches* also covers switching based on optical nonlinear effects, liquid and photonic crystal optical switches as well as fibre, holographic, quantum optical and other types of optical switches.

Details of these and other Woodhead Publishing materials books can be obtained by:

- visiting our web site at www.woodheadpublishing.com
- contacting Customer Services (e-mail: sales@woodheadpublishing.com; fax: +44 (0) 1223 832819; tel.: +44 (0) 1223 499140 ext. 130; address: Woodhead Publishing Limited, 80 High Street, Sawston, Cambridge CB22 3HJ, UK)

If you would like to receive information on forthcoming titles, please send your address details to: Francis Dodds (address, tel. and fax as above; e-mail: francis.dodds@woodheadpublishing.com). Please confirm which subject areas you are interested in.

Advanced adhesives in electronics

Materials, properties and applications

Edited by

M. O. Alam and C. Bailey

Oxford Cambridge Philadelphia New Delhi

Published by Woodhead Publishing Limited,
80 High Street, Sawston, Cambridge CB22 3HJ, UK
www.woodheadpublishing.com

Woodhead Publishing, 1518 Walnut Street, Suite 100, Philadelphia, PA 19102-3406, USA

Woodhead Publishing India Private Limited, G-2, Vardaan House, 7/28 Ansari Road,
Daryaganj, New Delhi – 110002, India
www.woodheadpublishingindia.com

First published 2011, Woodhead Publishing Limited

British Library Cataloguing in Publication Data
A catalogue record for this book is available from the British Library.

Library of Congress Control Number: 2011927290

ISBN 978-1-84569-576-7 (print)
ISBN 978-0-85709-289-2 (online)

The publisher's policy is to use permanent paper from mills that operate a sustainable forestry policy, and which has been manufactured from pulp which is processed using acid-free and elemental chlorine-free practices. Furthermore, the publisher ensures that the text paper and cover board used have met acceptable environmental accreditation standards.

Typeset by Replika Press Pvt Ltd, India
Printed by TJI Digital, Padstow, Cornwall, UK

Contents

Contributor contact details

Editors and Chapter 1

M. O. Alam and C. Bailey
School of Computing and Mathematical Science
University of Greenwich
Old Royal Naval College
Park Road
London
SE10 9LS
UK

E-mail: M.O.Alam@gre.ac.uk; C.Bailey@gre.ac.uk

Chapter 2

J. Felba
Faculty of Microsystem Electronics and Photonics
Wroclaw University of Technology
ul. Janiszewskiego 11/17
50-372 Wrocław
Poland

E-mail: jan.felba@pwr.wroc.pl

Chapter 3

J. W. C. de Vries and J. F. J. M. Caers
Philips Applied Technologies
High Tech Campus 7
5656AE Eindhoven
The Netherlands

E-mail: j.w.c.de.vries@philips.com; j.f.j.caers@philips.com

Chapter 4

J. E. Morris
Department of Electrical & Computer Engineering
Portland State University
P.O. Box 751
Portland
OR 97207-0751
USA

E-mail: j.e.morris@ieee.org

Chapter 5

Q. K. Tong
Henkel Corporation
10 Finderne Avenue
Bridgewater
NJ 08807
USA

E-mail: quinn.tong@us.henkel.com

Chapter 6

M. Inoue
The Institute of Scientific and Industrial Research
Osaka University
8-1 Mihogaoka
Ibaraki
Osaka 567-0047
Japan

E-mail: inoue@sanken.osaka-u.ac.jp

Chapter 7

C. Bailey
School of Computing and Mathematical Science
University of Greenwich
Old Royal Naval College
Park Road
London
SE10 9LS
UK

E-mail: C.Bailey@gre.ac.uk

Chapter 8

M. A. Uddin and H. P. Chan
Department of Electronic Engineering
City University of Hong Kong
Kowloon
Hong Kong

E-mail: afsarhk@gmail.com

1
Introduction to adhesives joining technology for electronics

M. O. ALAM and C. BAILEY, University of Greenwich, UK

Abstract: Polymer adhesives are now considered as invaluable materials for modern-day, low-cost miniaturized consumer electronic products because of their low cost and high production throughput. This introductory chapter describes the different types of adhesives that are used in electronic packaging along with a brief introduction to electronic assemblies and the uses of adhesives in electronics.

Key words: integrated circuit, miniaturized consumer electronic products, electronic packaging, through-hole packages, surface mount packages.

1.1 Introduction

The assembly of modern electronic components started in the early 1960s, after the emergence of the integrated circuit in the 1950s. However, all the early electronic components were connected by metallic systems such as solder (e.g. tin–lead), and wire bonding (gold and aluminium), and were mounted on ceramic substrates up until the 1980s. In fact, most of the applications for electronic systems at this time were for high-reliability applications such as defence and were not driven by cost. With the introduction of electronics into consumer products, polymeric substrates and adhesives started to attract attention because of their low cost and high production throughput. In fact, polymer adhesives are now considered as invaluable materials for modern-day, low-cost miniaturized consumer electronic products. However, the introduction of polymers and adhesives in the packaging of semiconductors results in a number of manufacturing and reliability problems. Outgassing, ion contamination, void/air entrapment, moisture absorption, CTE (coefficient of thermal expansion) mismatch and related thermal stress are some of the notorious problems that have had to be addressed by polymer experts and electronic packaging engineers over the last few decades. Therefore, polymer adhesive technology (both materials and their processing) has improved significantly during that period. Because of the tremendous growth of the consumer electronics, adhesive technology is now a multi-billion pound business, fuelling continuous R&D efforts from both industry and academia.

Adhesives are used as both functional and structural materials in electronic packaging. They provide electrical connections, thermal paths for extracting heat away from a semiconductor, and can be used to provide structural integrity of the package and system. They are applied as either paste or a solid film. Bare chips, single-chip packages, and multi modules can all be attached on the plastic printed circuit board (PCB) by adhesives without any soldering or wiring. Some of the advanced adhesives are used in flexible PCBs, optoelectronics, and sensors, as well as in smart cards. This book is dedicated to advanced adhesives that have attracted attention from the electronic packaging community. This chapter aims to introduce the various types of adhesives used in electronic packaging, along with a brief introduction to electronic assemblies and the uses of adhesives in electronics.

1.2 Classification of adhesives used in electronic packaging

Adhesives used to package electronic components are classified based on (i) their *physical form* such as pastes or films, (ii) *chemical type* such as epoxies, acrylics or polyimides, (iii) *molecular structure* such as thermoplastic or thermosetting, (iv) *curing method* such as heat curable, uv-curable, (v) *function* such as electrical adhesives, thermal adhesives, and (vi) *application* such as die attach, underfill adhesive.

Interestingly, the same adhesive can be termed differently, based on its various classifications, and sometimes it is desirable to mention all kinds of classification while describing an adhesive. For example, Anisotropic Conductive Adhesive (ACA) is a film-type (physical form), epoxy-based (chemical formulation), thermosetting (molecular structure), heat-curable (curing method) adhesive used in electronics.

While this book is dedicated to such advanced adhesives, a very brief classification of all adhesives used in electronics is given in here.

1.2.1 Classification based on physical form

All adhesives used in electronics can be applied either as a paste or a film. Pastes are semi-solid materials that are usually screen printed or stencil printed. Sometimes, they can be dispensed, even through a needle. Film adhesives are solid tapes that can be cut as required for the bonding area. Isotropic Conductive Adhesives (ICA) are generally used as pastes while Anisotropic Conductive Adhesives (ACA) are applied as films.

1.2.2 Classification based on chemical formulation

Adhesives are much more complicated than metals or ceramics. They are commonly referred to by their polymer type. The most common polymer type used in electronic adhesives is epoxy. The next is probably acrylic, then polyimide and then polyurethane, and so on. There are hundreds of formulations on each generic polymer in the market, with some proprietary additives.

1.2.3 Classification based on molecular structure

Polymer adhesives are broadly classified as thermoplastic, or thermosetting, depending on the molecular structure. *Thermoplastics* are linear polymers with either straight chains or branched chains (see Fig. 1.1a,b). They melt (liquefy) at a specific temperature and then resolidify on cooling; therefore they are easy to process and rework. Examples of thermoplastics are polyurethanes

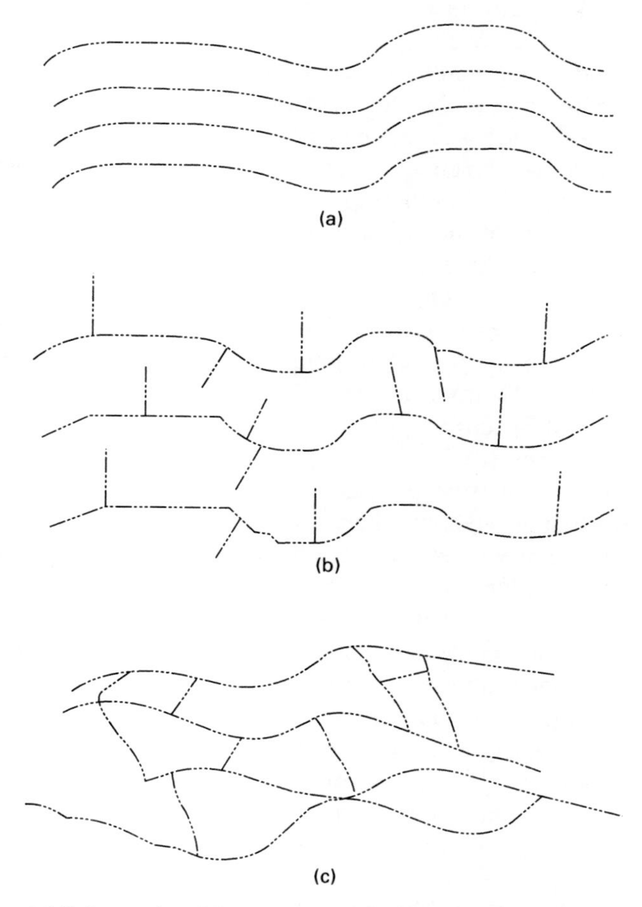

1.1 Schematic of linear (thermoplastic) and cross-linked (thermosetting) polymer. (a) Straight chain structure; (b) branched chain structure; (c) cross-linked structure.

and polyamides. However, some linear polymer adhesives with aromatic or heterocyclic components in their structures possess high thermal stability and do not melt; instead, they decompose on further heating. *Thermosetting* adhesives become highly cross-linked polymers on curing (see Fig. 1.1c), and unlike thermoplastics, they will not then melt at any temperature. Therefore, they are not reworkable once they are cross-linked. Examples are epoxies, cyanate esters and phenolics.

1.2.4 Classification based on curing process

There are various curing processes to give workable strength to adhesives. While dispensing the adhesive in its initial state, the molecular weight is low; therefore it is usually in a liquid state. However, during the curing process, molecules join together to form a three-dimensional network, leading to high molecular weight which increases the viscosity and finally forms a solid. Heat-curing is the most widely used and simplest method of polymerization. Curing kinetics are temperature- and time-dependent.[1] For any polymer system, the kinetics' parameters dictate the rate of curing at different temperatures. Final properties of the adhesive are directly related to the curing percentage of the polymer components. The curing profile is measured by monitoring the desired adhesive properties such as electrical resistivity, hardness, bond strength or dielectric constant, as a function of temperature and time. The degree of polymerization, i.e. curing percentage, can be measured by experimental techniques such as DSC (Differential Scanning Calorimetry), Thermomechanical Analysis (TMA) or FTIR (Fourier Transformed Infrared Spectroscopy).

Curing kinetics are mostly controlled by the added hardener (often called the catalyst) used in adhesive formulations. Short curing is desirable to increase the industrial throughput. Therefore, selecting a suitable hardener is important to cope with the mass production required for consumer electronics. Recently, snap-cured adhesives have been developed, which can be cured within a few seconds at relatively low temperatures.

Some adhesives polymerize in the presence of ultraviolet or even visible light, which makes them attractive for wider applications in heat-sensitive parts, especially in photonics packaging. They also show rapid curing. In certain cases, however, slight heat is required to complete the curing process. Chapter 8 describes this photo curing process and their uses in photonics.

Microwave curing is another rapid curing method that is attractive for its advantage in providing a uniform heating capability over a large section. However, unlike conventional microwaves, it uses Variable Frequency Microwaves (VFM) that rapidly vary the operating frequency. Swept frequencies generate a uniform energy distribution as well as reduced arcing of metal fillers and semiconductor dies.[2,3]

1.2.5 Classification based on functions

Electronic adhesives are used to serve for mechanical connections, electrical connections, thermal dissipation and stress mitigations. As the name implies, the main function of an electronic adhesive is to mechanically bond dissimilar materials with the required strength and maintain that strength over the life-time of the product. Although polymer adhesives are not intrinsically electrically/thermally conductive, their monomers can be wetted by metallic particles and, after curing, those metallic particles are impregnated in their microstructure. The common polymer matrix is epoxy based and silver and nickel particles of 5 to 20 microns in diameter are typically used as the conductive fillers.[4] Depending on the metal particle content in the adhesive, they are classified either as anisotropic conductive adhesive/film (ACA/ACF) or isotropic conductive adhesive (ICA). ACA or ACF types contain less filler metal (<5 % by volume) and therefore can conduct electricity in only one direction. For example, they can be used to provide electrical connections for flip-chip components where the metal particles provide electrical connection between the pads on the die and those on the substrate. The die is assembled onto the substrate by applying the ACA or ACF onto the substrate and then placing the die over the substrate/adhesive assembly and applying pressure and temperature until good electrical contact is made and the adhesive is cured. At elevated temperature, the adhesive is cured or set and the pressure helps to entrap the conductive particles between the bump and the pad; therefore, electrical contact is made only in the z-direction from the substrate to the die. Hence the name anisotropic. Nickel- or gold-coated polymer particles are also used as conductive fillers in ACF.

ICA contains a much higher volume percentage of metallic filler material (such as silver) above the percolation limit, so when they are set by curing, they can conduct electricity in all directions. Hence the name isotropic. Silver-filled epoxies can also be used as thermally conductive adhesives. However, special additives such as ceramics having high thermal conductivity but very low electrical conductivity (e.g. AlN, BN, Al_2O_3) are used as filler materials where only thermal conductivity is desired and there is a risk of short circuiting. Recently, carbon nano tubes have been used as filler materials because of their high thermal conductivity.[5] Thermally conductive adhesives are extensively used in power electronic devices and heat sinks. Chapter 2 illustrates the materials, and the conduction modelling and processing of thermally conductive adhesives.

Flip chip on plastic PCB is receiving significant attention in consumer electronics; however, the low CTE of silicon and the high CTE of a PCB pose a significant threat to the thermal fatigue reliability of solder joints. Underfill materials provide an example of how adhesives are used to reduce thermal expansion related stress (see Chapter 5 for the details) and hence thermal fatigue of solder joints.

Modelling is now playing a very important role in deciding what adhesives to use in electronic packaging. Modelling techniques, both at the micro and macro scale, can be used to identify the electrical, thermal and mechanical performance of adhesives. In addition, modelling can help identify optimal properties for adhesives during the packaging process and subsequently during the life of the package in different environments. Chapter 7 gives some details on modelling techniques that can be used for adhesives in electronic packaging.

1.3 A brief overview of electronic assemblies

It is assumed that readers of this book are aware of the basic construction of electronic assemblies. However, a brief overview of such assemblies might assist in recapping where and how advanced adhesives are used in electronics. Basically, electronic products starts from the semiconductor materials such as silicon. Sequential processing steps (such as doping, oxidation and metallization) on a silicon wafer constitute a device, where devices with a specific design can be interconnected to form a functional circuit. The integration of many such circuits or components on a single chip is called an integrated circuit (IC) or IC chip (or die). By this process, thousands of dies can be produced from a single silicon wafer. Of course, there are many types of ICs whose size can vary from 1 mm to 30 mm. ICs are classified by (i) materials and composition, (ii) degree of integration or number of transistor elements, (iii) principles of operation, (iv) manufacturing method, (v) device type, etc. An IC can be part of a single component, such as a power amplifier or a power transistor, or be a fully integrated microprocessor as used in modern PCs and high performance servers or workstations. Theoretically, there is no limit to integrating multiple functions on a single chip, giving rise to concepts such as System-on-Chip (SOC). However, adding more functionality to a single piece of silicon by decreasing down beyond 22 nm node technology is very costly and time-consuming, due to design and fabrication difficulties.[6,7] Thus, in addition to continued developments in SOC, System-in-Package (SiP) is gaining interest to help fulfil the requirements of a number of applications. SiP is characterized by any combination of more than one active electronic components of different functionality plus (optionally) passives and other devices such as MEMS or optical components assembled into a single standard package that provides multiple functions. SiP requires expertise in packaging to ensure that it meets its objectives.

Before the advent of ICs, discrete components such as transistors, diodes, capacitors, resistors, inductors, etc. were mounted on a PCB to form a circuit block, which was then connected to other circuit blocks to build a complete functional unit. ICs have enabled the monolithic integration of most of those blocks onto a single chip, resulting in miniaturized products that have become low in cost and high in reliability. However, every individual IC has to be packaged before it can be used. Packaging starts where the IC chip stops. The typical parameters important for IC packaging include the number and distribution of Input/Outputs (I/Os), materials used, power size of the IC chip, and the number of chips to be assembled into a single package. The constraint for any packaging engineer is that the final package needs to be reliable and cost effective.

Packaging a single IC does not generally lead to a complete system for a piece of electronic equipment, since a typical electronic system requires a number of active and passive components. An IC is known to be an active component in an electronic system. Passive components such as resistors, capacitors and inductors are also integrated with the packaged IC to build a system-level package (e.g. the whole 'electronic package'). Thus, the term 'electronic packaging' is used in a broader spectrum than that of IC packaging. The functions of an electronic package are to protect, power and cool the ICs and/or components, as well as to provide electrical and mechanical connection between the ICs and other components, and to communicate with the outside world.

Figure 1.2 shows schematically the hierarchy of an electronic package.[6,8]

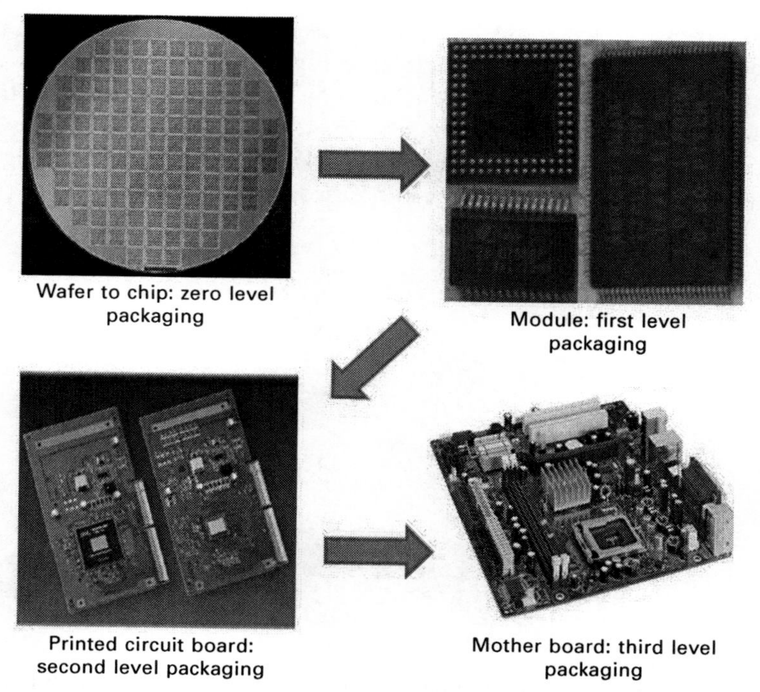

1.2 Hierarchy of the electronic packaging: optical photographs of the real wafer/package/assembly in different levels.

Typically, there are several levels of packaging, starting from chip level (zero level) to module (first level packaging), card or PCB (second level packaging) and mother board (third level packaging). The number of levels may vary, depending on the degree of integration and the totality of the packaging needs. For example, high performance servers might contain a large number of levels, whereas, consumer electronic products may consist of only one or two levels of packaging. However, among all the levels, the first level packaging, i.e. IC packaging is the most critical step for packaging engineers. Most of the adhesives discussed in different chapters of this book are related to IC packaging.

Since, there are so many types of ICs and their packaging requirements vary over a wide spectrum, it is impractical to have one packaging solution for all ICs. To resolve this problem, many types of IC packaging technologies have been developed that vary in their structures, materials, fabrication methodology, building technologies, size, thickness, number of I/O connections, heat removal capability, electrical performance, reliability and cost. In general, based on the methodology used in assembling the IC packages to the PCB, these are classified into two categories:

(i) through-hole packages, and
(ii) surface mount packages.

If the packages have pins that can be inserted into holes in the PCB, they are called through-hole packages (see Fig. 1.3a). If the packages are not inserted into the PCB, but are mounted on the surface of the PCB, they are called surface mount packages (see Fig. 1.3b).

The advantage of the surface mount package, as compared to through-hole, is that both sides of the PCB can be used, and therefore a higher packing

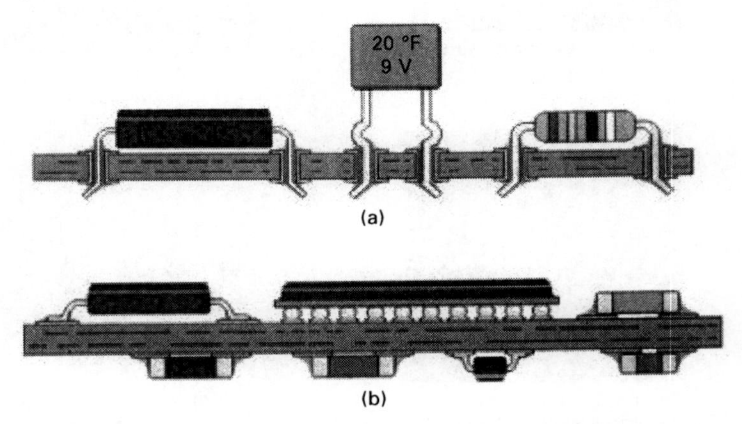

1.3 Board level integration of the packages for (a) through-hole packages and (b) surface mount packages.[6]

density can be achieved on the board. Dual-in-line packages (DIPs) and pin grid arrays (PGA) are two common through-hole packages (see Fig. 1.4). In DIPs, the I/Os, or the pins, are distributed along the sides of the package. To achieve higher I/O connections, PGAs are used where the pins are distributed in an area-array fashion underneath the package surface. However, most of the advances and varieties are observed in the surface mount packages. The small outline package (SOP, see Fig. 1.5a) is the most widely used package in modern memories for low I/O applications because of its extremely low cost. The quad flat package (QFP) is an extension of the SOP with larger I/O connections (Fig. 1.5c). Both the SOP and QFP have leads that can be attached to the surface of the PCB.

In the late 1980s, packages with solder balls were developed as an alternative to packages with leads.[6,8] The solder balls can be placed underneath the surface of the package in an area array and significantly increase the I/O count of surface mount packages. A ball grid array (BGA) package is an example of this technology (see Fig. 1.5c). Smaller, thinner and lighter packages are required in this modern age of portable and hand-held products.[7] Micro BGA and/or chip scale packages (CSPs) have been developed to address these demands of modern electronics.[6] For example, a DIP of the 1960s was roughly 100 times the size of the IC die. Since then, the evolution in packaging technology has reduced the ratio of package area to die area by a factor of 4 to 5. It is important to note that the package sizes are progressively approaching the size of the chip. The CSP, by definition, is a package whose area is less than 1.2 times the area of the IC it packages.

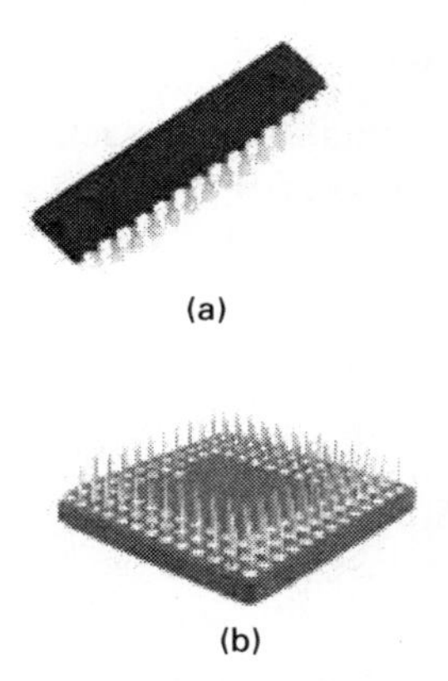

1.4 Optical images of two common through-hole packages. (a) Dual-in-line packages (DIP) and (b) pin grid arrays (PGA). (Courtesy of Motorola.)

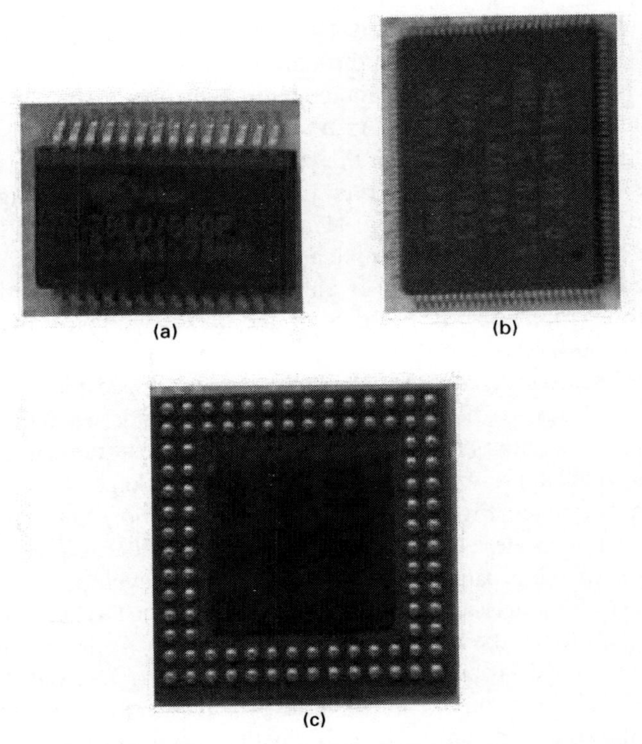

1.5 Optical images of three common surface mount packages. (a) Small outline package (SOP); (b) quad flat package (QFP); (c) ball grid array (BGA) package.

1.4 Typical uses of advanced adhesives in electronics

Figure 1.6 shows a schematic cross-sectional view of a wire bond BGA package with Cu heat spreader. Three adhesive materials have been used in this package, while molding resin has been used to encapsulate the chip and wire bonding. Among these three adhesives, the thermally conductive adhesive that bonds the chip to the heat sink is considered as an advanced adhesive. Details of this adhesive can be found in Chapter 2. In some applications, wire bonding is replaced by ACF or ACA; however, a different design to that shown in Fig. 1.6 is used.

Figure 1.7 shows two examples of direct bonded chip-on-board. The first one is bonded by flip-chip solder joints (Fig. 1.7a) while the second one is bonded by anisotropic conductive adhesives (Fig. 1.7b). To reduce the thermal expansion related mismatch between the chip (3ppm) and the substrate (18ppm), the space between the solder joints is filled with an

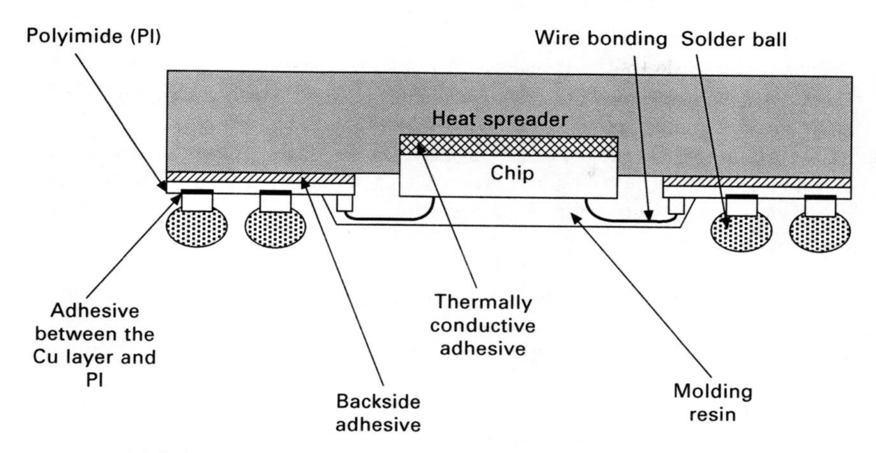

1.6 Schematic cross-sectional view of a wire bond BGA package.

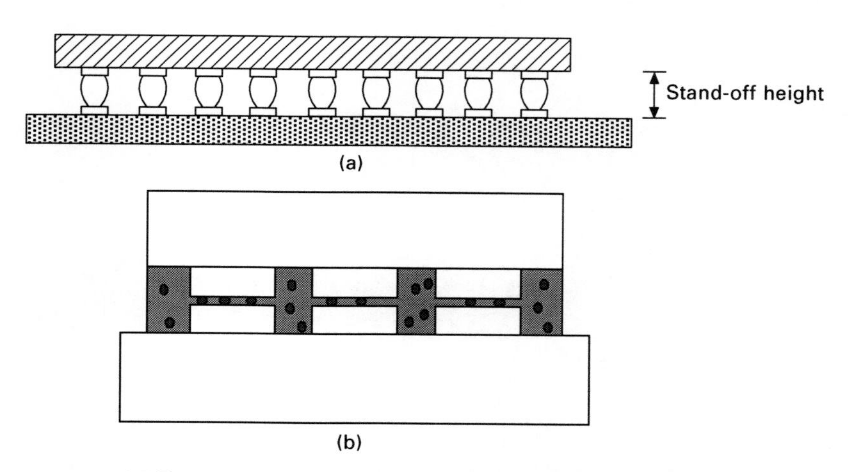

1.7 Two examples of direct bonded chip on board. (a) Flip-chip solder joints bonding; (b) flip-chip ACF joints bonding.

adhesive known as an underfill material. Chapter 5 describes the underfill process for flip-chip applications. On the other hand, it is clear from Fig. 1.7b that ACA itself occupies the room between chip to substrate.

1.5 References

1. Y. C. Chan, M. A. Uddin, M. O. Alam and H. P. Chan, 'Curing Kinetics of Anisotropic Conductive Adhesive Film', *Journal of Electronic Materials*, **32**(3), 131–136, March 2003.
2. C. Bailey, T. Tilford, H. Lu and M. P. Y. Desmulliez, 'Design for Manufacture and Reliability of Polymer Based Electronics', *2nd International Conference on Polymers in Defence and Aerospace Applications*, Hamburg, Germany, 2010.

3. T. Tilford, S. Pavuluri, C. Bailey and M. P. Y. Desmulliez, 'On Variable Frequency Microwave Processing of Heterogeneous Chip-on-Board Assemblies', *Proc. International Conference on Electronic Packaging Technology & High Density Packaging (ICEPT-HDP 2009)*, Beijing, China, ISBN: 978-1-4244-4659-9, pp. 927–931.
4. M. A. Uddin, M. O. Alam, Y. C. Chan and H. P. Chan, 'Adhesion strength and contact resistance of Flip Chip On Flex (FCOF) packages – Effect of curing degree of anisotropic conductive film', *Microelectronics Reliability*, **44**(3), pp. 505–514, March 2004.
5. J. Felba, T. Fałat and A. Wymysłowski, 'Influence of thermo-mechanical properties of polymer matrices on the thermal conductivity of adhesives for microelectronic packaging' *Materials Science – Poland*, Vol. **25**(1), 45–55, 2007.
6. R. R. Tummala, *Fundamentals of Microsystems Packaging*, McGraw-Hill: New York, 2001.
7. R. R. Tummala and V. K. Madisetti, 'System on chip or system on package?', *Design & Test of Computers, IEEE*, **16**(2), 48–56, 1999.
8. J. H. Lau, *Ball Grid Array Technology*, McGraw-Hill: New York, 1995.

Part I

Types of adhesives

2
Thermally conductive adhesives in electronics

J. FELBA, Wroclaw University of Technology, Poland

Abstract: The heat dissipation problem is becoming a crucial barrier in the process of electronic devices and systems' miniaturization. This chapter discusses heat transport in adhesives used for electronic packaging consisting of a polymer base material matrix and a thermally conducting filler. Practically, only the material of the filler influences the adhesive thermal conductivity, which is limited mainly by the thermal contact resistance between filler particles. The result is that the thermal conductivity of composites saturated with micro- or nanometer-sized particles of very well thermally conducting materials – such as silver, diamond, or even carbon nanotubes – does not exceed a level of several W/m·K.

Key words: microelectronics, heat conductance, thermal resistance, thermal conductivity, polymer matrix, composite fillers, thermally conductive adhesives, nanocomposites, thermal conductivity measurement.

2.1 Introduction

Miniaturization is a steady process that results in increasing the number of electronic elements in a volume unit. All electronic devices require a source of power, partially converted into heat which has to be transferred by the surrounding structures to the ambient air outside the system and/or to a heat sink. Thermal conduction, convection and radiation, as well as phase-changing processes play a role in electronics cooling and have to be efficient enough to keep systems at a not too high temperature. Generally, failure models indicate a direct link between component reliability and operating temperature. Thus, a rise in temperature from 75 °C to 125 °C can be expected to result in a five-fold increase in failure rate.[1] This shows that the heat dissipation problem is becoming a crucial barrier in miniaturization processes, and successful thermal packaging depends on a combination of proper materials and heat transfer mechanisms to stabilize the component temperature at an acceptable level.

The first level of packaging (single chip or multichip modules) is primarily concerned with conducting the heat from the chip to the package surface and then into the printed circuit board. At this packaging level, improvement of heat flow between the silicon die and the outer surface of the package is the most effective way to lower the chip temperature. At the second level of packaging, heat removal typically occurs both by conduction in the printed circuit board (which connects the electronic elements making systems) and by convection to air. The convection seems to be important for assembled packages while the radiation is rather out of the temperature work-range of electronic products.[1]

The problem of heat transport and heat dissipation, especially in the first but also in the second packaging level, may be solved by conductive adhesives. There are two main requirements for such materials, namely sufficient mechanical strength of the joined components and high thermal conductivity.

Thermally conductive adhesives formulated as composites consist of a polymer base material matrix and a thermally conducting filler. All polymeric materials (epoxy or other types, thermoset or thermoplastic) have very low thermal conductivity, in the range of 0.2 to 0.3 W/m·K. In practice therfore, only the filler material is responsible for heat transport. Conduction is provided by conductive additives, since high conductivity requires high filler content, considerably above the percolation threshold. It is believed that at this concentration, all conductive particles contact each other and form a three-dimensional network. The size, shape and content of the filler particles significantly change the viscosity and rheology of the composite in comparison with the pure polymer. The expected conductivity value is obtained after the curing process due to better contacts between the filler particles resulting from shrinkage of the polymer matrix. This means that the polymer base material, which is responsible for components joining, in spite of its low conductivity, may play a crucial role in heat transfer in the composite. Thermally conducting fillers in the form of micro- or nanometer-sized balls, flakes, wires, fibers, etc. are dispersed randomly in a matrix. The material of filler influences the thermal conductivity of the adhesive, which is limited mainly by the so-called 'thermal contact resistance' between filler particles. The contact thermal resistance depends on both the properties of the materials and the geometric parameters of the contact areas between the particles. The geometric parameters are related to the contact pressure within the contact area. Therefore, the thermal conductivity of composites saturated with very well thermally conducting materials such as silver, diamond or even carbon nanotubes, does not exceed a level of several W/m·K. Reported higher values need information about the method of thermal conductivity measurement used and its accuracy.

2.2 Model of heat conductance

The basic principle of heat conductance through a thermally conductive adhesive layer dx (Fig. 2.1), also known as Fourier's law, states that the rate of heat transfer through a material is proportional to the negative gradient in the temperature and to the area at right angles to that gradient

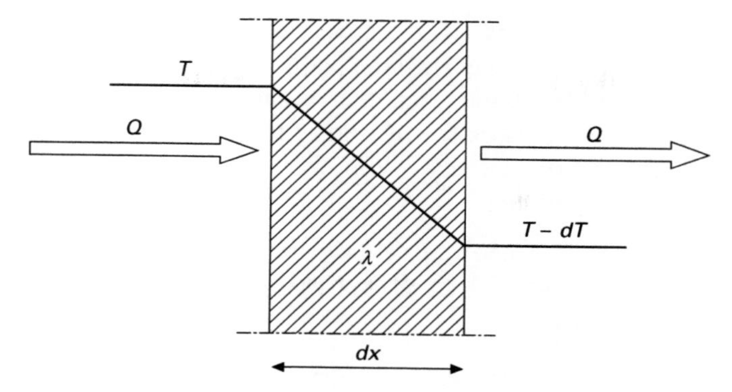

2.1 Heat transfer through a layer.

through which the heat flows and can be written as follows:

$$\frac{dQ}{dt} = -\lambda A \frac{dT}{dx} \qquad [2.1]$$

where Q is the quantity of heat energy (J), t is time (s), λ is a thermal conductivity (W/m·K), dT/dx is the temperature gradient in the heat flow direction (K/m), x is the distance along the direction of heat flow (m), A is the area of the cross-section (m^2). The thermal conductivity is given by

$$\lambda = \frac{1}{3}(c_e v_e L_e + c_{ph} v_{ph} L_{ph}) = \lambda_e + \lambda_{ph} \qquad [2.2]$$

where c_e and c_{ph} are the heat capacities per unit volume (J/m^{3}K) of electrons and phonons, respectively, v_e and v_{ph} are their root-mean-square velocities and L_e, L_{ph} are their mean free paths.

The thermal conductivity of electron type λ_e is dominant for metals, and one can roughly assume that

$$\lambda \cong \lambda_e \qquad [2.3]$$

In such a case, the ratio of thermal conductivity λ_e and electrical conductivity σ_e, according to Wiedemann–Franz's law, can be expressed as

$$\frac{\lambda_e}{\sigma_e} = LT \qquad [2.4]$$

where T is temperature (K) and L is the Lorenz constant, the theoretical value of which is 2.44×10^{-8} W·Ω/K^2.

The transport of heat in nonmetals occurs mainly by phonons. For insulators, thermal conductivity linearly increases with temperature, and, additionally, phonon heat conduction efficiency is dependent on filler material size. The thermal conductivity of a phonon type λ_{ph} in nanostructures may differ significantly from that in macrostructures.[2] As the size of a nanostructure becomes comparable or smaller than the phonon's mean-free-path, phonons collide with the boundary more often than in bulk materials. This additional collision mechanism increases the resistance to heat flow and thus reduces the effective thermal conductivity of thin films, wires, nanotubes and other forms of nano-sized particles.

2.3 Heat transport in thermally conductive adhesives

It is assumed that at the macro scale, the formulation of thermally conductive adhesives consisting of a polymer base material matrix, filler and some special additives, is uniform and isotropic, although slight heterogeneity has also been measured.[3] As the polymer matrix has very low thermal conductivity (below 0.3 W/m·K), heat transport between joining surfaces is possible when particles of filler form conductive paths (Fig. 2.2).

2.3.1 Models of heat transport

There are many analytical models for predicting heat transport between filler particles used in thermally conductive adhesives. According to literature reviews,[4,5] the most common and important analytical models, mostly taking into account spherical particles, can be listed as follows:

- *The Maxwell–Garnett effective medium model*,[5–8] which can be considered as a good model for low volume fractions, up to 40 % (also when carbon nanotubes are used as the fillers[9,10]). Therefore, this model is not used in modeling of heat transport for most common composites, in which conductive particle content exceeds 60 %.[4]
- *The Bruggeman symmetric and asymmetric models*,[4] which can include thermal contact resistance between particle and matrix. These models are capable of predicting thermal conductivity of spherical particles for larger volume fractions.
- *Percolation models*. Percolation is a geometrical phenomenon, which means that above some volume fraction (called the percolation threshold) there is a continuous path for heat conduction through the particles because the conducting particles start to touch each other, as shown in Fig. 2.2. The classic percolation theory assumes a statistical distribution of the filler particles in a matrix and some network models of interaction between particles with random size distribution are proposed.[11] Nevertheless, especially when nano-sized particles are used, the distribution of filler in a polymer matrix cannot be completely random because of the aggregation of particles. In the case of nano-Ag with epoxy resin, it is observed that

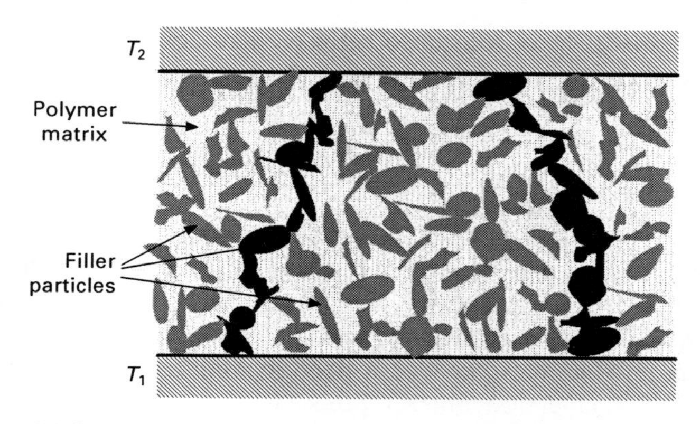

2.2 Conductive paths (black shapes of filler) between joining surface at different temperatures.

a conductive network can be formed even when the content of particles is lower than the percolation threshold estimated conventionally.[12] With increasing temperature, self-organization of particles helps in the formation of a conducting path throughout the matrix.

Apart from more general models taking into account the filler particles in the form of spheres or ovals, there are analytical models predicting thermal conductivity of composites with carbon nanotubes, even with their various orientation distributions in a matrix.[13]

2.3.2 Thermal resistance of adhesives

If free electrons accomplish conduction (Equation 2.3), then the formulas for thermal and electrical conduction are the same. Assuming that electrical and thermal flow lines are geometrically equal, the concept of a thermal resistance can be introduced

$$\Theta_\lambda = \frac{l}{\lambda A} \qquad [2.5]$$

where Θ_λ is the thermal resistance of a thermal conductor (K/W), l is its length (m) and A is the area of the conductor cross-section (m^2). The total thermal resistance in 'a three-dimensional network' of conducting filler particles can be summed up using the concepts of serial and parallel resistances, as in an electrical circuit. The thermal resistance Θ_p of the singular contact path between substrate and component in the steady state conditions can be expressed in a simple equation

$$\Theta_p = \Theta_{\lambda 1} + \Theta_{\lambda 2} + \cdots + \Theta_{\lambda \mathrm{n}} + \Theta_{TC1} + \Theta_{TC2} + \cdots + \Theta_{TC(n-1)} \qquad [2.6]$$

where Θ_λ is the thermal resistance of a filler particle (bulk resistance) and

Θ_{TC} is the thermal contact resistance between particles making a chain in the direction of the temperature gradient. The total thermal resistance of the adhesive Θ_{TA} is

$$\Theta_{TA} = \left(\frac{1}{\Theta_{p1}} + \frac{1}{\Theta_{p2}} + \cdots + \frac{1}{\Theta_{pn}} \right)^{-1} \quad [2.7]$$

Equation 2.6 points out that the less contact between particles, the greater the thermal conductivity of the adhesive. For the same distance between joining surfaces, it requires larger particles of a filler. Practically, the experiments show[14] that at an alumina filler with a fixed loading level of 50 wt%, the thermal conductivity increases about 1.7 times when filler of <10 μm average particle size is replaced by a much larger one – not higher than 149 μm (– 100 mesh). This way of thermal conductivity improvement (by enlarging filler particles) is not good because the miniaturized packaging scale, and the technology of adhesive dispensing, generally require that the dimension of filler particles should not be higher than a few micrometers. According to Equation 2.7 it is natural that the more conducting paths there are, the lower is the thermal resistance of the composite. This can be ensured by higher filler loadings. In Fig. 2.3 the thermal conductivity of the epoxy adhesives as a function of the volume loading of AlN filler is presented. The non-linear conductivity increases with increase of the filler content, suggesting that additional paths for heat transport are being formed not only by new

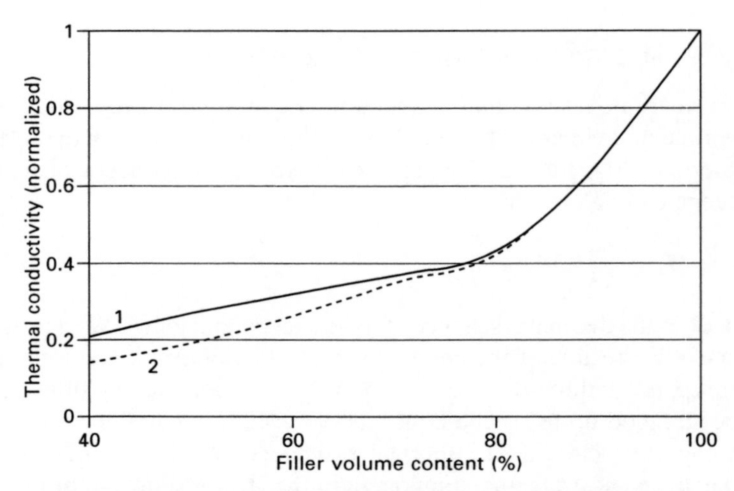

2.3 Normalized thermal conductivity of adhesive *vs* filler volume content (in relation to maximum content). Line 1 – mix of AlN whiskers and AlN particles with maximum content of 60 vol% (according to Ref. 15). Line 2 – AlN particles with maximum content of 62 vol% (according to Ref. 16).

chains, but also by multiplication of thermal contacts between filler particles in the whole network of particles. This method of conductivity improvement is limited because of a need for usable viscosity of the formulation and for sufficient mechanical strength of adhesion of the adhesive.

For a single particle of filler, its bulk conductivity is determined by the material used. From this point of view, diamond, with a conductivity of 2000 W/m·K, is a better material than alumina (slightly better than 30 W/m·K). However, with adhesives formulated in a similar way to electrically conductive adhesives, which contain silver flakes with average particle dimensions of several micrometers and a thermosetting epoxy resin as polymer matrix, the thermal conductivity value does not exceed 3 W/m·K.[17] This enormous difference between the thermal conductivity of pure silver (about 420 W/m·K) and a thermally conductive adhesive with this filler is due to constraint by the existing thermal contact resistance between filler particles. It is the main 'bottleneck' for heat transport inside composite formulations.

The problem of thermal constriction resistance is well-known for the region where two solid contact members touch each other. The real contact area is only a small fraction of the nominal or apparent area. This is mainly due to the non-flat shape of the contacting surfaces and surface roughness (typical for filler particles of an adhesive), the hardness of the contacting materials, or additional materials between the contacting surfaces. It would be difficult to analyze the accidental contact area between two flake-shape filler particles of an adhesive. Thus, for understanding the problem of thermal constriction resistance, let us assume two cylindrical metallic units P_1 and P_2, 'representing filler particles' contacting flat surfaces (Fig. 2.4). When the contact areas are nominally flat and hard, then they touch each other in maximum of three points. In fact, these points become the small areas

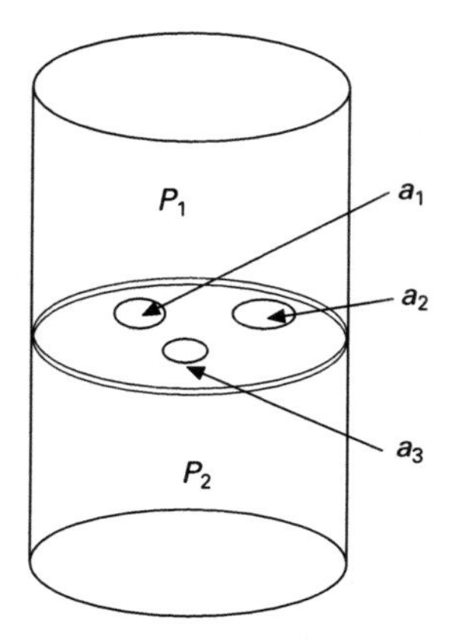

2.4 Two cylindrical filler particles P_1 and P_2 forming the apparent contact area $A_c = a_1 + a_2 + a_3$.

(a-spots) since contact member materials are deformable.[18] When the contact surfaces are perfectly clean, then the total metallic contact area A_c is the sum of all a-spot areas. When cylindrical unit P_1 has a higher temperature (T_1) then unit P_2 (temperature T_2), the steady state heat flow between them is constricted through small conducting a-spots.

In the case of contacting adhesive filler particles, the number of a-spots is unknown. The total thermal resistance of such units is the sum of the bulk thermal resistances $\Theta_{\lambda 1}$ and $\Theta_{\lambda 2}$ of both contact members, the constriction resistances Θ_{c1} and Θ_{c2} of contact members, and the real thermal resistance of the a-spot (or A_c – the sum of all a-spot areas) bump Θ_a (if a-spot is treated as a 3D structure).

The thermal constriction resistance depends significantly on both the area of the a-spot and the gap between contact members. Increasing the contact area and lowering the distance between filler particles cause a decrease of the thermal constriction resistance. As an example, the results of calculating the gap distance influence on this resistance are presented in Fig. 2.5. The thermal constriction resistances of two cylindrical copper particles, with radius of 10 mm and 40 mm length each (Fig. 2.4), were calculated using the finite element method.[19] The gap between both contact members ranged from 0 up to 10 μm. Due to symmetry, only one quarter of the system was analyzed. The computer calculations were performed assuming that the thermal conductivity of the polymer matrix could be neglected and that the contact surface was perfectly clean, without oxides and alien films. In fact, the thermal constriction resistances can be treated as the thermal contact resistance Θ_{TC} from Equation 2.6.

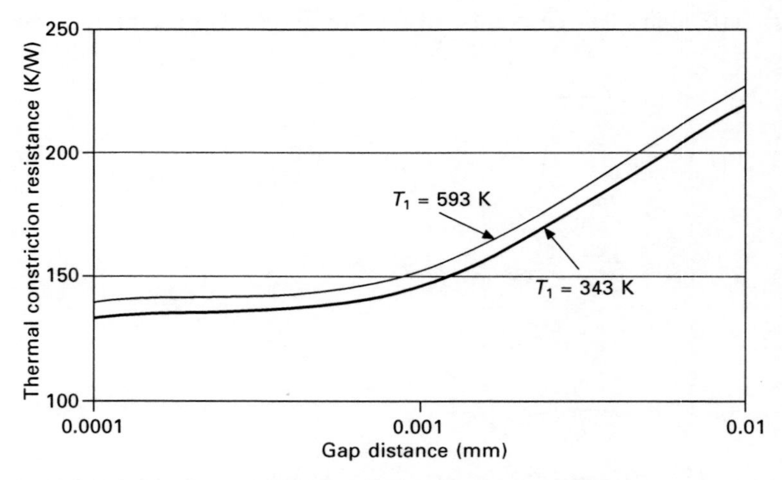

2.5 The thermal constriction resistance of two-particle unit *vs* different gap distance between filler particles; the radius of the a-spot is 0.01mm, T_2 = 293 K (according to Ref. 19).

For the analyzed two-particle unit at the temperature T_1 of 393 K (T_2 = 293 K), the bulk resistance of both filler particles is about 0.7 K/W, while the thermal contact resistance Θ_{TC} = 221 K/W for both the radius of a-spot and the gap equal to 10 μm. This means that in such a configuration, the total resistance of a thermally conductive adhesive depends mainly on the values of thermal contact resistance and numbers of contacts between filler particles making a chain in the direction of temperature gradient (Θ_{TC} and n in Equation 2.6). The bulk resistance plays a minimal role.

2.3.3 Bond thermal resistance

Thermal conductive adhesives usually are applied between two surfaces at different temperatures, e.g. between a heat generating component and the eventual heat sink. In such a case of heat transport, the thermal resistance Θ_T can be expressed by the formula

$$\Theta_T = \Theta_{BR'} + \Theta_{TA} + \Theta_{BR''} \qquad [2.8]$$

where $\Theta_{BR'}$ and $\Theta_{BR''}$ stand for the bond thermal resistance of contacts between the adhesive layer and the joined elements, while Θ_{TA} represents the total thermal resistance of the adhesive (Equation 2.7). In such a case, the bond thermal resistance influence on Θ_T cannot be treated as the thermal contact resistance between solid surfaces analyzed previously. The surface of the joined element is glued by a polymer composite with a conductive filler, and macroscopic irregularities such as flatness deviations and waviness are of no importance; however, surface roughness may influence the bond thermal resistance to a large extent.

The value of the bond thermal resistance depends on the shapes and dimensions of both the surface irregularities and of the particles of the adhesive filler. As this contact is difficult to specify, an imaginary 3D material layer representing the bond thermal contact with thickness of l_{BR} (including all these irregularities) and an unknown thermal conductivity λ_{BR}, can be introduced.[20] With such an assumption, the heat flow Q causes the temperature gradient ΔT on every layer between the joined surfaces; two bond thermal contacts and adhesive (Fig. 2.6).

The value of thermal conductivity λ_{BR} can be extracted from the result of an experiment when the temperature gradient ΔT_T can be measured.[17,21] For the steady state of the heat flow, it is possible to measure and calculate[17,22] the thermal resistance of the adhesive Θ_{TA} and bond thermal resistance Θ_{BR} with the assumption that

$$\Theta_{BR} = \Theta_{BR'} = \Theta_{BR''} \qquad [2.9]$$

Knowing the contact area and thickness of layers from Equation 2.5, the

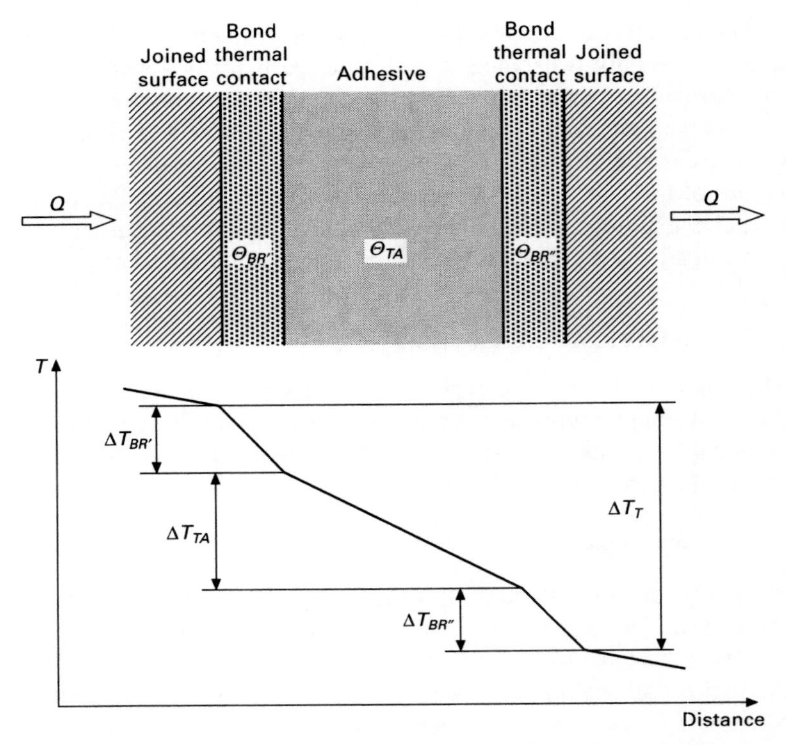

2.6 Temperature gradient on whole joint (ΔT_T) and on individual layers.

conductivity of the adhesive λ_{TA}, as well as the thermal conductivity of the imaginary 'bond' layer λ_{BR} can be established.

In fact, the thickness of the 'bond' layer is unknown and it would be much more valuable to define a new parameter describing the bond thermal conductivity, which will not depend on the l_{BR} thickness. This parameter can be referenced as the relative thermal conductivity of the thermal contact and can be defined as the thermal conductivity on a unit thickness λ'_{BR} and then

$$\lambda_{BR} = \lambda'_{BR} \cdot l_{BR} \qquad [2.10]$$

It has been calculated[22] that the participation of the adhesive bond resistance Θ_{BR} in the total resistance depends on the adhesive layer thickness l_{TA}. Figure 2.7 shows such dependence, plotted according to equation

$$\Theta_{BR} = \frac{2\lambda_{TA}}{\lambda'_{BC} \cdot l_{TA} + 2\lambda_{TA}} 100 \qquad [2.11]$$

with the following values resulting from measurement:[22]

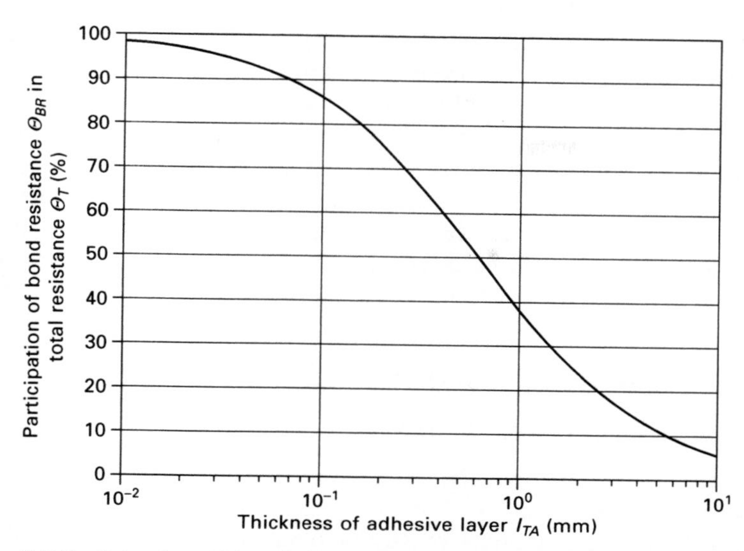

2.7 Participation of bond thermal resistance Θ_{BR} in the total thermal resistance Θ_T of the adhesive joint *vs* thickness of the adhesive layer.[22]

Table 2.1 Participation of bond thermal resistance in the total thermal resistance of a joint; adhesive thickness 40–60 μm[23]

Sign	Type of adhesive formulation	$2\Theta_{BR}/\Theta_T$
A	Silver-loaded reworkable thermoplastic	0.87
B	Silver-loaded interpenetrating network of thermoplastics and thermoset resins	0.76
C	Silver-loaded thermoset matrix	0.76

$\lambda_{TA} = 1.46$ W/m·K, $\lambda_{BR} = 0.0474$, and $\lambda'_{BR} = 4.74 \cdot 10^3$ W/m^2·K (with assumption that l_{BR} = 10 μm).

It can be concluded from calculation that the bond thermal resistance may influence the thermal resistance of the joint significantly. Similar results were observed in experiment.[23] Three thermally conductive adhesives were used to bond together two thick platelets of silicon carbide-reinforced aluminum (AlSiC) composites. The adhesive thickness was kept to a minimum, ca. 40–60 μm. The results of thermal resistance measurements at room temperature are listed in Table 2.1.

Certainly, bond thermal resistance depends strongly on the conductive filler content in adhesive formulations and on the bonded materials. It has been calculated[24] for different silver-filled epoxy adhesives and combination of Si – Al bonded materials, that $2\Theta_{BR}/\Theta_T$ may change from 36 % to over 90 % (with thickness of adhesive layer in the range 26–54 μm).

Participation of bond thermal resistance in total thermal resistance depends on joint temperature. It has been measured[23] that for all types of adhesive formulations described in Table 2.1, an increase of operating temperature from 23 °C to 300 °C causes a decrease of the ratio Θ_{BR}/Θ_T of about 18 %.

Based on calculation and results of experiments, it can be generally concluded that the adhesive bond resistance in comparison to the total resistance of thermally conductive adhesive joints is an important factor, especially in the case of microelectronic packaging where the adhesive layer is in the order of micrometers.

2.4 Thermally conductive fillers

2.4.1 Micrometer-sized fillers

Since the 1990s, isotropically conductive adhesives have been widely used in electronic packaging. Such composites consist of a polymer resin and electrically conductive fillers, mostly silver, although gold, nickel and copper are also used. Silver is unique among all of the cost-effective metals by nature of its oxide, which unlike other metal oxides, is a good conductor. Silver, with thermal conductivity 420 W/m·K, has been considered to be the most suitable material also as the thermally conductive filler. In fact, typically formulated electrically conductive adhesives with fillers of micronized silver particles can be also used as thermally conductive adhesives. The particles usually have the form of very thin flakes (Fig. 2.8a), but producers may also use particles of different shapes (e.g. more spherical, Fig. 2.8b), which influences the thermal contact resistance.

As was stated earlier, larger particles of filler cause higher thermal conductivity of composites because of the lower number of contact points between particles, which generate thermal contact resistances Θ_{TC} (Equation 2.6). This means that with the use of nano-sized silver particles as fillers, the summarized value of Θ_{TC} increases significantly and the thermal resistance of the singular contact path Θ_p between two surfaces also increases.

2.4.2 Nanometer-sized fillers

Nanosilver as a base material of composites for ink-jet technology is used in today's miniaturized electronics, mostly for conductive microstructures and contacts with dimensions in the range of tens of micrometers.[27] Nevertheless, the results of experiments show that nanosilver particles of 3–7 nm diameter added to a micro-sized filler in thermally conductive adhesives may improve the adhesive thermal conductivity, even by 2.55 times.[28] This is explained by making the contact resistance between basic (micro-sized) particles lower, owing to the fusion process of nano-particles at temperature of 200 °C.

2.8 Micro-sized silver particles used as a filler of thermally conductive adhesives: (a) typical flake shape, (b) spherical shape.[25] (Courtesy of Amepox Microelectronics.[26])

Theoretically, according to Equation 2.7, the multiplying of the conductive paths may decrease thermal resistance of adhesive Θ_{TA}. Because of this, there are efforts to enhance the electrical conductance of adhesives containing micro-sized silver particles by adding silver nanowires[29,30] or carbon nanotubes.[31] In a system composed of silver particles and either nanowires or carbon nanotubes, these slender particles are excellent materials to contact the particles together. As an example, Fig. 2.9 shows that the carbon nanotubes are spanning over the gap between two flake silver particles and forming a conductive pathway. Unfortunately, for the heat transport the role of such a single pathway is not important and only a much denser network of such additional ways of heat transport may improve the thermal conductivity of an adhesive.

Carbon nanotubes were discovered in 1991 by Sumio Iijima,[32] firstly as multi-wall structures with outer diameters of 4–30 nm and a length of

2.9 Scanning electron microscopy image of adhesive containing silver flakes and bundles of carbon nanotubes as the filler. (Courtesy of CANOPY Project partners.[31])

up to 1 μm which consisted of two or more seamless graphene cylinders concentrically arranged. Single-wall carbon nanotubes, which are seamless cylinders each made of a single graphene sheet, were reported two years later. Their diameters range from 0.4 to 2–3 nm, and their length is usually of the micrometer order, i.e. with a very high aspect ratio. At present, carbon nanotubes can be obtained using various techniques: arc discharge, pyrolysis of hydrocarbons over catalysts, laser and solar vaporization, and electrolysis. These nanoscale materials can exhibit different morphologies such as straight, curled, hemitoroidal, branched, spiral, and helix-shaped, with different numbers of walls.[33,34]

The high value of thermal conductivity of carbon nanotubes is the most desirable feature that can improve heat transport in adhesives. The thermal conductivity of individual multi-wall carbon nanotubes is estimated to fall in the range of 2000 to over 3000 W/m·K.[35–37] It is also reported that the λ of a carbon nanotube depends on both its chiral vector and length. Non-equilibrium molecular dynamic modeling has clearly shown[38] that the correct chirality may increase the conductivity by tens of percent, and the longer is the nanotube, the higher is the thermal conductivity.

As the measurement of the thermal properties of a single nanotube is extremely difficult, and conventional methods cannot be used for this purpose, the values of λ presented in the literature vary. Additionally, experimental results show that the thermal conductivity of a carbon nanotube depends on both temperature and tube diameter. The λ value increases with decreasing tube diameter and increases with increase in temperature, appearing to have an asymptote near 320 K.[37]

The first composites with carbon nanotubes were proposed in the last decade of the twentieth century. Various materials have been used as matrices, firstly polymers – mostly thermoplastics based on epoxy resin, but also polymethyl methacrylate (PMMA) and some gels.[39] The dispersion process requires a crucial technology to develop a nanocomposite because the specific surface area of carbon nanotubes is too large to well disperse in the matrix resin. According to theoretical calculations,[40] the specific surface area of single-wall nanotubes may reach a value of 1315 m^2/g. For multi-walled tubes the value is lower, but even for higher tube diameters the specific surface area amounts to a few hundreds of m^2/g. If the filler is not properly dispersed, the thermal property of the composite cannot be significantly increased. Several processing methods are used for this reason, many of them based on improving nanotubes/matrix interactions,[41–44] such as an ultrasonic bath, ultrasonic finger, melt-mixing, speed mixer, *in-situ* polymerization or solution processing. To prepare carbon nanotubes for proper dispersion on a macroscopic scale, some preprocessing is required. They can be purified to eliminate non-nanotube material, followed by deagglomeration for dispersing individual nanotubes. As the phonon scattering across the interface between particle fillers can greatly affect the thermal conductivity of a polymer composite, to improve heat transport, functionalization of the carbon nanotubes surfaces is necessary. This is usually done by a chemical treatment, consisting of many steps.[41,45]

In fact, even using complicated processes for improving nanotube surfaces and interactions with matrix, it is impossible to achieve filler content higher than a few weight percent. It is reported that the dispersion of 5 wt% carbon nanotubes in an epoxy resin has been obtained by an ultrasonic treatment. Although the carbon nanotubes were well separated, they remained poorly distributed.[46] At such a filler content, the viscosity of the composite is more then ten times higher in comparison to the pure matrix.[39] It is desirable to disperse this filler as uniformly as possible, because the thermal conductivity of a composite rises almost linearly with rising content of carbon nanotubes in the polymer matrix.[44]

Nevertheless, the thermal conductivity of composites filled with single- or multi-walled carbon nanotubes is lower than for composites with micro-sized fillers. The measured λ value usually reached less then 1 W/m·K.[47–50] This is lower than the calculated value based on the model of the heat transport with neglecting all interactions between the nanotubes.[51] For contents of 'average shape', multi-wall nanotubes at a single weight percent, the model predicts the thermal conductivity of a composite at the level of 1.8 W/m·K. It has been calculated that effective conductivity increases linearly with increasing nanotube loading; the conductivity of the adhesive is not very sensitive to the nanotube length (increases marginally with increase in length) and the conductivity can change drastically with a change in the diameter – a smaller diameter of the nanotubes can significantly increase the overall conductivity.

Adding of carbon black with nano-sized particles may improve the thermal conductivity of epoxy-based composites with 2 wt% carbon nanotubes.[42] by about 20 %. Nevertheless, thermal conductivity of such composites is reported as only ca. 0.6 W/m·K.

2.4.3 Diamond fillers

Adhesives filled with diamond seem also to be very promising for heat transport composites because specially purified synthetic diamonds could have the highest thermal bulk conductivity, 2000 W/m·K or higher. Unfortunately, thermal measurements using micron-sized synthetic diamond powder as filler showed a relatively low thermal conductivity in comparison to one with micron-sized silver. This is believed to be due to the presence of nitrogen and other impurities in the synthetic diamond.[52] Additionally, the thermal conductivity of diamond is strongly dependent on the microstructure of the material[53] and is correlated to grain size, even for the bulk material.[54] Probably, there are effects of structural defects such as stacking faults, twins or dislocations from synthetic processes. Typically, where the grain size is in excess of 30–50 μm, the effects of grain boundaries on thermal conductivity become insignificant, whereas submicron grain sizes can reduce the thermal conductivity by more than two orders of magnitude. As a result, adhesives with diamond filler have lower thermal conductivities than those with micro-sized silver as a filler.[52]

2.4.4 Electrically insulating fillers

Thermally conductive but electrically insulating adhesives are increasingly important for electronic packaging in applications where a composite ought to meet at least three requirements, namely mechanical bonding, heat dissipation and acting as an insulating layer. Usually, composites with insulating fillers are characterized by lower coefficients of thermal expansion than conducting adhesives. Beside diamond, boron nitride, aluminum nitride, alumina and other materials can be used as electrically insulating fillers.

Aluminum nitride (AlN) with its relatively high thermal conductivity of 110–200 W/m·K[36,52,55] is often used as a conducting filler. Generally, the thermal conductivity of an adhesive with this filler increases with increasing filler particle size and its volume fraction.[15,16] The same influence of alumina (Al_2O_3) particle size on the thermal conductivity of adhesives has been observed.[14,56] The possible maximum volume content of such fillers in a polymer, which results in a higher value of composite thermal conductivity, can be obtained when a mixture of particles with different diameters is used.[57] Generally, the thermal conductivity of the polymer composites can be enhanced by applying a hybrid filler.[58]

2.5 Role of polymer base materials

2.5.1 Types of polymer base materials

The main role of the polymer base material of an adhesive is to form mechanical bonds at interconnections. This is done after the polymerization process, when an adhesive has the form of a solid body. Before hardening by polymerization, it should have a suitable viscosity and provide the right characteristics for printing, stenciling, and other industrial dispensing methods.

Polymers are commonly classified as either thermoplastics – typically able to be melted or softened with heat, or thermosets – which resist melting and cannot be re-shaped. Thermoset epoxies are by far the most common adhesive matrices and for electrically conductive adhesives they have found use since the early 1950s. Thermosets are crosslinked polymers and generally have an extensive three-dimensional molecular structure. The polymerization process is strongly accelerated at higher temperatures, although thermosets cured at room temperatures are commercially available when the correct catalyst or hardener is added. Crosslinks are chemical bonds occurring between polymer chains that prevent substantial movement, even at elevated temperature. Silver–epoxy can be considered the base line for isotropic conductive adhesives used for electronic component assembly.[59] The curing of an epoxy is complex and the whole reaction consists of several steps. The chemistry of the cure begins with the formation and linear growth of the polymer chains, which soon begin to branch, and then to crosslink. This sudden and irreversible transformation from a viscous liquid to an elastic gel marks the first appearance of the infinite network. Gelation typically occurs below 80 % conversion, and beyond the gel point the reaction continues with substantial increase in crosslink density towards the formation of one infinite network.[60]

Vitrification is another phenomenon of the growing chains or network. This transformation from a viscous liquid or an elastic gel to a glass begins to occur as the glass transition temperature of these growing chains or networks become coincidental with the cure temperature. Further curing in the glassy state is extremely slow and, for all practical purposes, vitrification brings an abrupt halt to curing.[60]

Polymeric materials have thermal conductivities about 2000 times lower than that of silver and much lower than those of other fillers. Thermal conductivities of all polymeric materials, epoxy or other types, thermoset or thermoplastics, range from 0.2 to 0.3 W/m·K and so these practically do not take part in the heat transport. Nevertheless, the thermal conductivity of resins can be improved by introducing a high-order structure having microscopic anisotropy while maintaining macroscopic isotropy. In the case of diepoxy monomers with a biphenyl group or two phenyl benzoate groups as mesogens, the thermal conductivities are up to five times higher than those of conventional epoxy resins.[61] This is possibly due to these molecular groups, because mesogens form highly ordered, crystal-like structures which suppress phonon scattering.

In commercial applications, the filler is responsible for virtually all the heat transport, but the thermomechanical properties of the polymer matrix may strongly influence the thermal conductivity of adhesives for microelectronic packaging, mostly by shrinkage during polymerization and then relaxation processes.

2.5.2 Role of cure shrinkage of polymer base material

The level of volume shrinkage depends mainly on the type of resin, and may reach a few percent of its initial value. Figure 2.10 presents a typical volume shrinkage chart of an epoxy resin during curing at 150 °C. Of course, the time of curing depends strongly on temperature. In the case presented in Fig. 2.10, the volume shrinkage indicates the high polymerization level of the tested resin after about 25 minutes of heating at 150 °C, while temperature 60 °C needs more than 300 min to achieve the final shrinkage (7.56 %).[63] The role of the shrinkage process during polymerization at high temperature

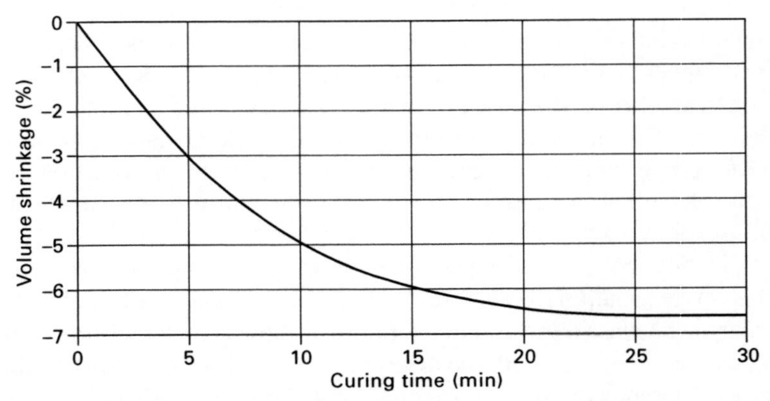

2.10 Volume shrinkage of epoxy resin *vs* curing time (according to Ref. 62).

is easy to observe in the case of electrically conductive adhesives, which usually consist of a similar type of resin and filler (e.g. micrometer-sized silver flakes) as thermally conductive adhesives. For such composites, the curing time depends mostly on the curing temperature, the type of polymer base material and additives. The cure process produces a shrinkage of the polymer matrix, which exerts a pressure on the conductive particles, forcing them into closer contact. Independent of the dominant mechanism of current transport,[64] closer contact between filler particles strongly decreases the electrical resistance of adhesive joints. Examples of such resistance changes for two different composites are presented in Fig. 2.11.

Also in the case of thermally conductive adhesives the shrinkage of the polymer during the polymerization process makes real contact between filler particles. Let us assume that in the beginning of the shrinkage process such particles are in the form of perfect spheres, and that they touch each other at one contact point 'A' (Fig. 2.12). Further shrinkage results in stress between particles which may cause elastic or plastic strain of the contact members. For the epoxy resin with shrinkage characteristics presented on Fig. 2.10, the contact pressure was estimated at more than 0.16 GPa. It has been calculated that if two silver particles of 1 μm diameter ($2R$) were deformed by this shrinkage force contact radius r between them increases from 'point' to more than 0.08 μm[62] (Fig. 2.13).

The change of the contact area influences the thermal contact resistance. To answer the question of how the contact state between particles affects

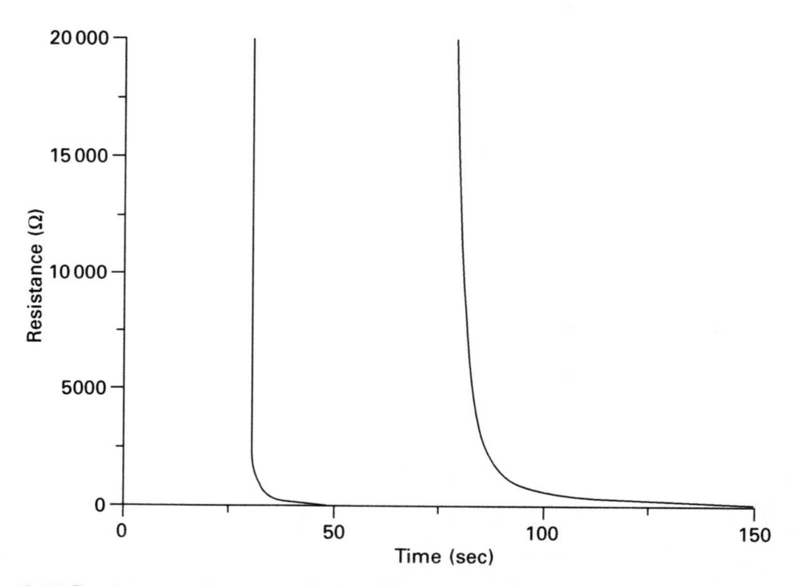

2.11 Resistance changes of electrically conductive adhesives (with different polymers and catalyst) during curing time at 150 °C.[65]

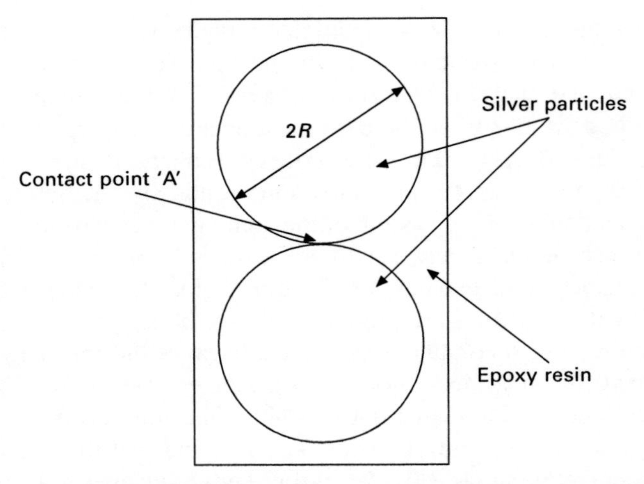

2.12 2D model of thermally conductive adhesive with spherical silver particles.

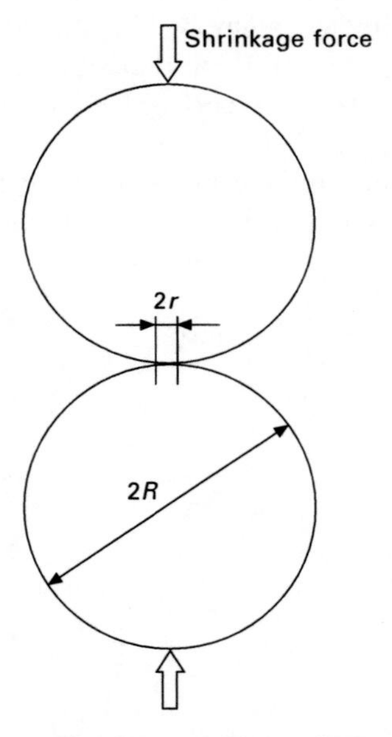

2.13 Influence of polymer shrinkage on contact between filler particles.

the thermal conductivity of the adhesive, numerical modeling of the silver-filled formulation was used. The simulation was performed not only for two balls (Fig. 2.13) but for the more complex 3D structure called the unit cell.[66,67] The unit cell has the maximum possible packing efficiency (V_{filler}/

V_{total} = 74.05 %) in 3-D for spherical particles, i.e. hexagonal close-packed (*HCP)* or face-centered cubic (*FCC)* crystal equivalent. It was assumed that particles contact each other in the ideal pure metallic contact area (the impurities, roughness etc. were neglected).

The unit cell was analyzed using ANSYS 8.0 commercial finite element method software for two different contact area radii (*r*) of balls (Fig. 2.13). The simulation results show that increasing the *r*/*R* ratio from 0.02 to 0.05 causes a change of the thermal conductivity of the cell from 26.45 to 49.12 W/m·K. Such results of calculation show that the role of the polymer matrix may be crucial for the level of heat transfer in composites. But this state may not be stable, as the stresses in the resins may relax with time. This is because thermosetting polymers are characterized by viscoelastic properties. The viscoelastic behavior of the matrix cannot be neglected because of the stress relaxation.

2.5.3 Viscoelasticity of polymer base materials

The general linear viscoelastic equation is the basic equation for modeling the development of viscoelastic stresses (σ_{ij}) as a function of temperature and loading time in fully cured polymer materials[68,69]

$$\sigma_{ij}(t, T) = \int_{-\infty}^{t} \{2G(t-s,T)\dot{\varepsilon}_{ij}^{d} + K(t-s, T)\dot{\varepsilon}_{V}^{eff}\}\, ds \qquad [2.12]$$

where G and K denote the shear and bulk relaxation moduli, respectively (they are time (t)- and temperature (T)-dependent), and $\dot{\varepsilon}_{V}^{eff}$ is the effective strain contribution.

Shear and bulk moduli also depend on the conversion level (degree of cure, α). Therefore, the viscoelastic stress described by Equation 2.12 can be written as

$$\sigma_{ij}(t, T, \alpha) = \int_{-\infty}^{t} \{2G(t-s,T,\alpha)\dot{\varepsilon}_{ij}^{d} + K(t-s,T,\alpha)\dot{\varepsilon}_{V}^{eff}\}ds \qquad [2.13]$$

and $\dot{\varepsilon}_{V}^{eff}$ consists of mechanical, thermal, and cure shrinkage parts[70,71]

$$\varepsilon_{V}^{eff} = \varepsilon_{V}^{mech} - \varepsilon_{V}^{cure} - \varepsilon_{V}^{T}$$
$$\varepsilon_{V}^{T} = 3\beta_{L}^{g,r}(T - T_{ref})$$
$$\varepsilon_{V}^{cure} = -3\gamma_{L}(\alpha - \alpha_{ref}) \qquad [2.14]$$

where β_{L}^{g} and β_{L}^{r} are the linear coefficients of thermal expansion measured below the glass transition temperature T_g (glassy region) and above T_g (rubbery region), respectively, and γ_L is the linear cure shrinkage. The change of the modulus during curing can be explained by the change in the molecular structure of the matrix.

If the time is relatively short ($t \to 0$), shear and bulk relaxation moduli are treated as instantaneous shear and bulk moduli G_0 and K_0, respectively, and they can be determined from the values of the instantaneous (obtained from the high-rate tests) elastic (Youngs) modulus E_0 and Poisson's ratio ν_0

$$G_0 = \frac{E_0}{2(1+\nu_0)} \quad \text{and} \quad K_0 = \frac{E_0}{3(1-2\nu_0)} \qquad [2.15]$$

Poisson's ratio is often assumed to be time-independent in viscoelastic materials; therefore Equation 2.15 can be used for the whole time-scale of the shear, bulk, and elastic relaxations while strains are small. The advantage of such an approach is that while modeling the behavior of viscoelastic materials, only one modulus needs to be measured (e.g. the shear relaxation modulus) and the other can be calculated from Equation 2.15.

The effect of temperature in Equation 2.12 is usually not included as $G(t,T)$, but as $G(t_{red}[T])$, where t_{red} is the so-called reduced time scale and $G(t_{red})$ is referred to as the master curve. The reduced time scale is defined as

$$t_{red} \equiv \int_0^t \alpha_T dt \qquad [2.16]$$

where α_T is the temperature-dependent shift factor, which can be described by the WLF (Williams, Landel, Ferry) equation

$$\alpha_T = \exp\left\{\frac{-C_1(T - T_{ref})}{C_2 + T - T_{ref}}\right\} \qquad [2.17]$$

where T_{ref} is the reference temperature, C_1 and C_2 are material parameters. The reduced time scale t_{red} can be replaced by reduced frequency scale $f_{red} = \alpha_T f$.

The standard way of measuring dynamic changes of the mechanical parameters of shear modulus for polymeric materials is by a continuous monitoring of the thermo-mechanical properties from the liquid to the fully cured state by applying the liquid compound (e.g. resin + hardener) in small gaps between parallel plates of the shear clamps of a dynamic mechanical analyzer (DMA). The samples are subject to a series of sinusoidal strains or stresses at different frequencies (a frequency sweep). The temperature is then increased by 5–10 °C and another frequency sweep is applied. This procedure is repeated from about 80 °C below the glass transition temperature (T_g) to about 80 °C above it. Far below the glass transition temperature the modulus data (stress amplitude divided by strain amplitude) is frequency independent. This modulus is called the glassy modulus. Far above the glass transition temperature, the material either melts, as is typical of thermoplasts, or displays the non-zero, frequency-independent rubbery modulus of thermosets. In between, in the so-called viscoelastic region, the modulus is frequency-dependent and lies between the glassy and the rubbery values. It is customary to shift the individual modulus *vs* frequency curves along the logarithmic frequency axis until they overlap and form a master curve. The shift (α_T) is different for each temperature. The master curve, together with this shift factor, completely describes the temperature and frequency-dependent modulus data and can even be used to predict the mechanical behavior at time scales and temperatures different from those of the test conditions.[72]

When constructing the master curves, the shift factor α_T has to be evaluated (see Equation 2.17) by estimating the values of the WLF equation parameters. As an example, the parameters of the WLF equation for an epoxy with volume shrinkage of $\sim 3{\cdot}10^{-2}$ (the linear reaction cure shrinkage assumed as 1×10^{-2}) are collected in Table 2.2.[69]

2.5.4 Influence of polymer base materials on contact pressure between filler particles

To study the contact pressure occurring between the filler particles due to the cure shrinkage of the polymer matrix (with parameters from Table 2.2) and its influence on thermal conductivity, the 2D model shown in Fig. 2.12 was implemented by ABAQUS FEM software.[73] The model represents two identical spherical silver particles surrounded by cylindrically shaped epoxy resin. The particles touch each other at one point (*A*) at the beginning of the simulation and the contact pressure and its change in time after curing at different temperatures were monitored at this point.

The simulation procedure was divided into two steps. In the first step, a steady state analysis was performed to simulate the shrinkage caused by curing the epoxy and to monitor the initial contact pressure between silver particles. In the second step, transient analysis was performed and various temperatures (T_0 changes from 40 °C to 70 °C) were applied to observe the temperature dependence of the contact pressure relaxation. The results of the simulations, i.e. the contact pressures between particles *vs* time for different temperatures, are shown in Fig. 2.14.[69]

Table 2.2 Estimated parameters for WLF Equation 2.17

Parameter	Estimate
T_{ref}	60 °C
C_1	17.66
C_2	80.58

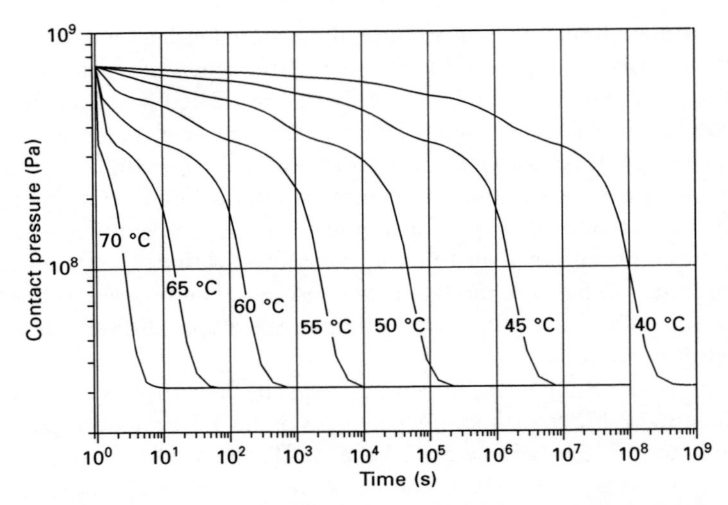

2.14 Simulation results: contact pressure between filler particles versus time for different temperatures.[69]

The contact pressure occurring between filler particles due to the cure shrinkage relaxes with time. The time needed to reach the fully relaxed state strongly depends on the working temperature of the system. In the considered case, the contact pressure is fully relaxed when it decreases from an initial value of 0.73 GPa to 0.03 GPa, as it is shown in Fig. 2.14. When the temperature is 70 °C, the contact pressure becomes fully relaxed after 10 seconds, but when it is lower (e.g. 40 °C), the full relaxation state is reached after about 10^9 seconds (more than 30 years!).

The thermal conductivities of the tested structures have been calculated (using Finite Elements Method) before and after relaxation. After the relaxation of the contact pressure, the thermal conductivity drops to around 50 % (0.484) of its initial value for the unrelaxed structure.[69]

The shrinkage phenomenon of the polymer base seems to be very important in heat transport. On the other hand, shrinkage means lowering of the formulation volume, which may lead to the occurance of some defects. Much more dangerous in effects may be the presence of solvents as polymer base additives.[28] The minimum void ratio was obtained when the solvent vaporized effectively before the resin had been cured.

2.6 Thermal conductivity of adhesives and methods for its measurement

Thermally conductive adhesives reported in the literature are characterized by thermal conductivities not higher than a few W/m·K for formulations with both commonly used filler materials (e.g. silver[17,52]) and special materials (e.g. diamond powder[52] or carbon fiber[74]). But there are also reports about polymer-matrix and silver-filled adhesives having a few times higher thermal conductivity.[28,75,76] Probably, various methods of thermal conductivity measurement and their errors, as well as low accuracy are the reasons for such significant differences in the measured thermal conductivity of similar adhesive formulations. To compare thermal data of different adhesive formulations, full information about the measurement methods used is necessary.

There are two main categories of techniques used to measure thermal conductivity – steady-state techniques and transient techniques. The heat flow method and the guarded hot plate method are the best examples of steady-state techniques. Hot wire and laser flash can be listed as the most popular measurement methods of the transient technique, but also the 3ω, photoacoustic method, the pulsed photothermal displacement technique, and the thermal-wave technique are in use.[77,78]

2.6.1 Steady-state methods of adhesive thermal conductivity measurement

Steady-state methods are based on establishing a steady temperature gradient over a known thickness of a sample, and on controlling the heat flow from one side to the other. The determination of the thermal conductivity of the insulation follows from the basic law of heat flow (Equation 2.1).

In the guarded hot plate method, the specimen is placed between uniformly distributed heaters and a heat sink (or cold plate), both of which contain temperature sensors that measure the temperature drop across the sample after a steady-state heat flow has been established. The method's error is diminished by modifying the base measurement configuration, e.g. by the line heat source applying,[79] more stable operation at liquid nitrogen temperatures[80] or measurement under vacuum conditions.[81] In result, the temperature drop ΔT_T can be measured (Fig. 2.7) and by changing the thickness of the sample, it is possible to extract the thermal conductivity λ_{TA} of the tested material.[17,22] The measured sample can be prepared either as a disc and directly placed between hot and cold plates,[17,22] or additional blocks made from a material with a low value of λ can be joined to the sample.[82] In such a case, the additional temperature drops (Fig. 2.7) have to be taken into consideration.

As an example, the measurement setup is shown in Fig. 2.15. The system consists of a heater, a heat sink, a pair of contact members (with a known thermal conductivity λ, e.g made from iron,[17,82] aluminum,[83] or copper,[84,85]), as well as temperature sensors.[21] The whole setup is placed in a high vacuum environment in order to minimize the effect of heat dissipation through convection. The radiation is out of the temperature work-range, and therefore is neglected, and the heat conduction is a dominating factor in transport heat

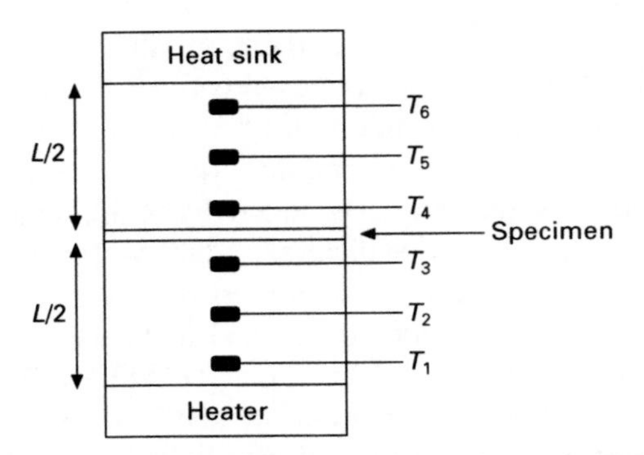

2.15 Measuring setup based on the guarded hot plate method (according to Ref. 17): *L*/2 – a pair of iron contact members, $T_{1\text{-}6}$ – six thermal sensors, a bulb halogen heater immersed in a bottom cylinder and a water cooling system on the top.

energy. The limiting error of the measurement method described previously is known, and can be added to information about the measured value. Thus, in this setup, the thermal conductivity of the adhesive was found to be 1.26 W/m·K, with 0.44 W/m.K standard deviation.[86]

2.6.2 Transient methods of adhesive thermal conductivity measurement

The transient hot-wire technique is a transient dynamic technique based on a linear heat source of infinite length and infinitesimal diameter and on the measurement of the temperature rise at a defined distance from the linear heat source embedded in the test material. As an electric current of a fixed intensity flows through the wire, the thermal conductivity can be derived from the resulting temperature change over a known time interval.[87] Nowadays, four variations of the method are in use:[88] hot wire standard technique, hot wire resistance technique, two-thermocouple technique, and hot wire parallel technique. The theoretical model is the same, and the basic difference among these variations lies in the temperature measurement procedure. This makes it possible to determine the thermal conductivity of a wide range of materials, including cured thermally conductive adhesives.

Modification of the transient hot-wire technique may also concern the shape of the heat source, which can be formed as a strip or a disc. For measuring the thermal properties of materials available as thin slabs, a hot disc can be used. It is assumed that the slabs are thermally insulated in such a way that the heat losses at the boundaries are negligible, compared with the total input of power.[89]

There are also a number of presently existing transient methods of measuring thermal conductivity indirectly – by measurement of a material's thermal diffusivity. The relation between the thermal conductivity λ (W/m·K) and the thermal diffusivity α ($m^2 \cdot s^{-1}$) is given by

$$\lambda = \alpha \cdot \rho \cdot c_p \quad [2.18]$$

where ρ ($kg \cdot m^{-3}$) is the density and c_p ($J \cdot kg^{-1} \cdot K^{-1}$) is the specific heat of the tested composite.

Among the transient methods, the so-called flash method is the most popular in determining thermal diffusivity. The method was described in 1960[90] and was developed over the next few years.[91–93] In this method, a pulse of energy is absorbed on the front face of a specimen and the subsequent temperature change at the rear face is recorded. The front surface of the sample must be uniformly irradiated for a short time compared with the rise time of the back surface temperature. A laser beam is the most popular heat source, but also a xenon lamp or more flexible energy sources can be used (e.g. an electron beam).[94,95] The measured sample usually has the form of a disc with a diameter of a dozen or so millimeters and a few millimeters in thickness. The shape of the time–temperature curve on the back side of the sample is used in analysis.[90,91,96,97] The method's errors result from an assumption that the sample is perfectly insulated from the environment during the test (there is no heat exchange) and the whole energy is absorbed instantaneously (zero pulse width) in a very thin layer of the sample material.

The popular transient method described above does not exhaust all the possibilities of thermal properties measurement. In the photoacoustic technique, a heating source (normally a laser beam) is periodically irradiated on the sample surface. The acoustic response of the gas above the sample is measured and related to the thermal properties of the sample. The method can be used for a single layer on a substrate[98] as well as multilayered materials measurement.[99] The 3ω method employs a metallic strip[100] or wire[101] in intimate contact with the specimen surface. An AC electrical current modulated at a regular frequency ω generates thermal waves in the specimen. Because the electrical resistance of the strip depends on the temperature, the resistance is modulated and it is possible to extract λ from an electrical measurements.[102]

2.6.3 Thermal conductivity of adhesives

Thermal conductivity (W/m·K) of selected adhesives measured by various methods is listed in Table 2.3. The data are taken from papers, not from manufacturers' offers.

As was mentioned previously, not all thermal conductivity measurement methods are characterized by high accuracy and some thermal conductivity values from Table 2.3 may be questionable. Nevertheless, the data cited in the table point out that

Table 2.3 Thermal conductivity (W/m·K) of thermally conductive adhesives with different fillers

Base material	Filler	Filler size	Filler content	Measurement method	Thermal conductivity	Reference
Phenol resin	Ag	3–5 μm/ 3–7 nm	90 wt%	FM (TT) laser	51	28
Epoxy resin	Ag	1–10 μm	88 wt%	FM (TT) laser	20.0	28
Epoxy resin	AlN	7 μm	60 vol%	FM (TT) laser	6.99	15
PVDF	AlN	115 μm	73 vol%	FM (TT) laser	4.53	15
Epoxy resin	AlN	30 μm	62 vol%	HW (TT) line	4.25	16
Epoxy resin	Ag	< 40 μm	42 vol%	GHP (SST)	3.80	52
PVDF	AlN	1.5 μm	73 vol%	FM (TT) laser	3.72	15
Epoxy resin	Al_2O_3 +CNT	45.3+9.0 μm/CNT	60 vol%/ 3 vol%	GHP (SST)	2.6	57
Epoxy resin	Al_2O_3	45.3+9.0 μm/36 nm	60 vol%/ 3 vol%	GHP (SST)	2.1	57
Epoxy resin	Al_2O_3	45.3 μm + 9.0 μm	60 vol%	GHP (SST)	1.7	57
Epoxy resin	Diamond	1 μm	60 wt%	FM (TT) laser	1.0	103
Epoxy resin	BN	1 μm	30 wt%	FM (TT) laser	1.0	103
Epoxy resin	CNT		3 wt%	FM (TT) xenon	0.7	45
Epoxy resin	GN		20 wt%	FM (TT) xenon	0.7	45
Epoxy resin	CNT + CB	ø 20–40 nm +22–60 nm	2 wt% + 10 wt%	FM (TT) laser	0.60	42
Epoxy resin	Al_2O_3	0.3 μm	60 wt%	FM (TT) laser	0.60	103
Epoxy resin	Al_2O_3	27–43 nm	20 vol%	HW (TT)	0.56	104
PVDF	CNT	ø 0.7–1.2 nm	49 vol%	FM (TT) laser	0.54	105
Epoxy resin	CNT	ø 20–40 nm	2 wt%	FM (TT) laser	0.50	42
Epoxy resin	Al_2O_3	<25 μm	50 wt%	GHP (SST)	0.40	14
Epoxy resin	Al_2O_3	25 nm	50 wt%	GHP (SST)	0.36	14
Epoxy resin	CNT	ø 10–30 nm	6.8 vol%	HW (TT)	0.36	104
Epoxy resin	CNT		5.0 wt%	FM (TT) xenon	0.34	48
Epoxy resin	Al_2O_3	1 μm	10 vol%	HW (TT)	0.32	104
Epoxy resin	SiC	2 μm	40 wt%	FM (TT) xenon	0.30	45
Epoxy resin	CNT	ø ~ 2.8 nm	0.5 wt%	HW (TT) disc	0.25	106
Epoxy resin	CNT	ø 2.1 nm	0.5 wt%	GHP (SST)	0.25	49

Materials: PVDF, poly(vinylidene fluoride); BN, boron nitride; CNT, carbon nanotube; GN, graphite nanoplates; CB, carbon black. Measurement methods: SST, steady-state techniques; TT, transient techniques; GHP (SST), guarded hot plate; HW (TT), hot wire; FM (TT), flash.

- higher λ of adhesive is obtained when micro-sized, not nano-sized, particles of filler are used,
- the bulk thermal conductivity of the filler, when particles are of micrometer size, does not influence the λ of the adhesive significantly,
- the higher the filler content, the better the thermal conductivity of adhesive,
- carbon nanotubes (very high λ) in the filler composite do not improve

the adhesive significantly in thermal conductivity because of their low content.

Theoretically, according to Equations 2.6 and 2.7, improvement of thermal conductivity of adhesives saturated by filler with low thermal resistance Θ_λ can be achieved by:

(i) lowering of the numbers and decrease of thermal contact resistance between particles making a chain for heat transport in the direction of temperature gradient (Θ_{TC});
(ii) increasing the number of parallel contact paths in the direction of the temperature gradient.

The first step of conductivity improvement can be done by filler particle phase or shape changing during the base material polymerization (see Fig. 2.12) or additional thermal process. There are investigations of new formulations which incorporate novel fluxing polymeric resins in combination with a blend of metal alloy filler particles. During the curing process, the metal filler particles sinter together and are capable of forming 'solder-like' metallurgical connections to a variety of metal surfaces, resulting in a lowering of thermal contact resistance values. These materials have both high electrical and thermal conductivity – closer to solder materials than most polymer-based materials but with the processing advantages of the best polymeric adhesives.[107]

The idea of additional thermal processing has come from experiences with electrically conductive adhesives. An adhesive based on epoxy resin filled with a mix of silver flakes (Fig. 2.8a) and powder (Fig. 2.8b) was tested. The content of filler in the adhesive was 65.4 wt%. After 'standard' curing (30 min at 180 °C), the adhesive was additionally annealed for 2 hours at 180 °C. After this process, scanning electron microscopy inspection revealed an increase in size of the particles and a decrease of electrical resistance of about two orders of magnitude.[25] Probably the observed effect was caused by recrystallization of the silver. In general, recrystallization begins when the temperature exceeds the so-called recrystallization threshold T_r, which is expressed by the following equation

$$T_r \approx 0.4 \cdot T_m \qquad [2.19]$$

where T_m is the melting point (K). The applied additional post-curing temperature was about 37 % of the silver melting point temperature and a process of changing the shape of particles may occur, as the value of T_r strongly depends on material purity and for extremely pure metals, recrystallization can be observed at a much lower temperature. A similar thermal effect was observed in the case of adhesives with nanosilver annealed at 180 °C.[108] The particles were fused through their surface and many dumbbell-type particles could be found. The morphology was similar to the typical morphology of an initial stage in the sintering process of ceramic, metal and polymer powders.

Both the electrical and thermal conductivity increase has been stated for a typical isotropic conductive adhesive composed of an epoxy base material and silver filler (spherical and flake-shaped particles) content of 85 wt%.[109] As recommended by the manufacturer, a suitable curing condition was 150 °C for 0.5 hours. After this process, the adhesive was additionally annealed in 200 °C during 30 hours. The sample under investigation exhibited anisotropy in the thermal conductivity, due to an oriented dispersion of flake-shaped filler particles. In a direction parallel to the squeegee direction, the thermal conductivity, increased by about 85 % due to the additional thermal process.

When carbon nanotubes are dispersed in the polymer, decreasing Θ_{TC} can be achieved by functionalization of their surfaces. Experimental results have suggested[45] that such treatment can significantly reduce interfacial thermal resistance between filler and epoxy resin by forming stronger chemical bonding across the interface. Surface treatment improved the interaction between filler and polymer matrix, which can significantly reduce phonon scattering, and also increase thermal conductivity.

Increasing the number of parallel contact paths in the direction of the temperature gradient can be achieved by applying hybrid fillers. The idea is presented in Fig. 2.9, in which mixed fillers (micro-sized silver flakes and carbon nanotubes) are presented. However, the low content of nanotubes (about of 0.5 wt%) weakly influences the total thermal conductivity of the composite, and efforts of research labs are aimed at achieving a significant increase of the CNT content.[31] A similar role may be played by nanowires, doped by the composite containing micro-sized silver particles[29,30] and a suitable ratio of filler with various particle sizes.[57]

It seems to be necessary to develop a systematic method of producing various multiple junctions of 'long' nano fillers to establish a 3D network instead of a set of singular structures. There is information about experiments aimed at establishing a junction between crossed nanotubes, a first attempt at producing such web-like CNTs having been achieved with electron beam, ion beam, laser or other techniques of nano-welding.[110] It is reported that under a high-voltage of 1.25 MeV at specimen temperatures of 800 °C from the random criss-crossing distribution of individual nanotubes and nanotube bundles on the specimen grid, several contact points could be identified where tubes were crossing and touching each other. After a few minutes of irradiating and annealing, different shapes of junctions were established.[111] It was demonstrated that also ion irradiation should result in the welding of crossed nanotubes and junctions were obtained.[112]

2.7 Conclusions

Thermally conductive adhesives formulated as composites consist of a polymer base material matrix and a thermally conducting filler. All commercially available polymeric materials, epoxy or other type, thermoset or thermoplastic, have very low thermal conductivity, in the range below 0.3 W/m·K. In practice, only the filler material is responsible for heat transport. Particles of this material in the form of micro- or nanometer-sized balls, flakes, wires, fibers etc., are dispersed randomly in the matrix. Conduction is provided by conductive additives, since high conductivity requires high filler content, considerably above the percolation threshold. It is believed that at this concentration, all conductive particles contact each other and form a 3D network.

For a single particle of filler, its bulk conductivity is determined by the material used. But independently of the filler material with high or even very high thermal conductivity (pure silver – about 420 W/m·K, diamond – 2000 W/m·K, or individual multi-wall carbon nanotube – over 3000 W/m·K) the thermal conductivity of formulated adhesives usually does not exceed several W/m·K. This enormous difference between thermal conductivities is due to constraint by the existing thermal contact resistance between filler particles. It is the main limit for heat transport inside composite formulations. When the adhesive is applied between two surfaces with different temperatures, additionally, the bond thermal resistance of contacts between adhesive layer and joined elements may influence the thermal resistance of the whole joint significantly – even at the level of a dozen or so percent.

The main role of the polymer base materials of the adhesive is to form mechanical bonds at interconnections. This is done after the polymerization process when an adhesive has the form of a solid body. During this process, the polymer matrix may strongly influence the thermal conductivity of the adhesive by its shrinkage. Polymer shrinkage makes real contact, and then stress, between filler particles, which causes elastic or plastic strain of the contact members. As a result, the contact area between particles increases and thermal contact resistance between them decreases. Unfortunately, the contact pressure occurring between filler particles may relax with time and thermal conductivity of the adhesive may become unstable.

There are many methods of adhesive thermal conductivity measurement, with different level of accuracy. But independently on this method and measured values reported in the literature, general observation can conclude that:

- higher conductivity is obtained when micro-sized, not nano-sized, particles of filler are used,
- the bulk thermal conductivity of the filler when particles are of micrometer size does not influence the thermal conductivity significantly,

- the higher the filler content, the better the thermal conductivity of the adhesive.

Theoretically, improvement of thermal conductivity of adhesives saturated by filler with sufficiently high conductivity can be achieved by:

- lowering the numbers of and decreasing the thermal contact resistance between particles making chains for heat transport in the direction of the temperature gradient,
- increasing the number of parallel contact paths in the direction of the temperature gradient.

The lessening of the numbers of contact resistances can be done by using higher-size filler particles and by shortening the distance between joining surfaces. The thermal contact resistance can be diminished by filler particle phase or shape changing during base material polymerization or post-curing processes. Increasing the number of parallel contact paths in the direction of the temperature gradient can be achieved by applying hybrid filler.

2.8 References

1. Bar-Cohen A., Watwe A., Seetharamu K.N. 'Fundamentals of Thermal Management', in *Fundamentals of Microsystem Packaging* (ed. R.R. Tummala), New York: McGraw-Hill, 2001, pp. 212–263.
2. Chen G., Phonon heat conduction in nanostructures. *International Journal of Thermal Science* (2000) **39**: 471–480.
3. Damasceni A., Dei L., Guasti F. Thermal behaviour of silver-filled epoxy adhesives: technological implications in microelectronics. *Journal of Thermal Analysis and Calorimetry* (2001) **66**: 223.
4. Prasher R. Thermal interface materials: historical perspective, status, and future directions. *Proceedings of the IEEE* (2006) **94**(8): 1571.
5. Nan C.-W., Birringer R., Clarke D.R., Gleiter H. Effective thermal conductivity of particulate composites with interfacial thermal resistance. *Journal of Applied Physics* (1997) **81**(10): 6692.
6. Koledintseva M.Y., DuBroff R.E., Schwartz R.W. A Maxwell Garnett model for dielectric mixtures containing conducting particles at optical frequencies. *Progress in Electromagnetics Research* (2006) **63**: 223–242.
7. Mallet P., Guérin C.A., Sentenac A. Maxwell-Garnett mixing rule in the presence of multiple scattering: Derivation and accuracy. *Physical Review* (2005) **B72**: 14205.
8. Levy O., Stroud D. Maxwell Garnett theory for mixtures of anisotropic inclusions: Application to conducting polymers. *Physical Review* (1997) **B56**(13): 8035.
9. Ju S., Li Z.Y. Theory of thermal conductance in carbon nanotube composites. *Physics Letters A* (2006) **353**: 194–197.
10. Nan C.-W., Shi Z., Lin Y. A simple model for thermal conductivity of carbon nanotube-based composites. *Chemical Physics Letters* (2003) **375**: 666–669.
11. Kanuparthi S., Zhang X., Subbarayan G., Sammakia B.G., Siegmund T., Gowda A., Tonapi S. Random network percolation models for particulate thermal interface

materials. *10th Thermal and Thermomechanical Phenomena in Electronics Systems Conference*, San Diego 2006, pp. 1192–1198.
12. Rong M., Zhang M., Liu H., Zeng H. Synthesis of silver nanoparticles and their self-organization behavior in epoxy resin. *Polymer* (1999) **40**(22): 6169.
13. Deng F., Zheng Q. Interaction models for effective thermal and electric conductivities of carbon nanotube composites. *Acta Mechanica Solida Sinica*, (2009) **22**(1).
14. Fan L., Su B., Qu J., Wong C.P. Effects of Nano-sized Particles on Electrical and Thermal Conductivities of Polymer Composites, *9th Int'l Symposium on Advanced Packaging Materials; Proc., Prop. and Interfaces*, Atlanta, Georgia, 2004, p. 193.
15. Xu Y., Chung D.D.L., Mroz C. Thermally conducting aluminium nitride polymer-matrix composites. *Composites* (2001) **A32**: 1749.
16. Bujard P., Ansermet J.P. Thermally conductive aluminium nitride-filled epoxy resin, *5th Semiconductor Thermal and Temperature Measurement Symposium* (1989) 126.
17. Falat T., Felba J., Wymyslowski A. Improved method for thermal conductivity measurement of polymer based materials for electronic packaging. *28th International Conference of International Microelectronics and Packaging Society – Poland Chapter*, Wroclaw (2004) 219.
18. Holm R. *Electric Contacts – Theory and Application*. Springer, Berlin, 1967.
19. Wymyslowski A., Friedel K., Felba J., Falat T. An experimental-numerical approach to thermal contact resistance. *9th International Workshop on Thermal Investigations of ICs and Systems*. Aix-en-Provence (2003) 161.
20. Friedel K., Wymyslowski A. An approach to numerical simulation of thermal contact problems in modern electronic packages. *4th International Conference on Thermal and Mechanical Simulation and Experiments in Micro-Electronics*, Aix-en-Provence (2003) 183.
21. Salgon J.J., Robbe-Valloire F., Blouet J., Bransier J. A mechanical and geometrical approach to thermal contact resistance. *International Journal of Heat and Mass Transfer*, 1121–1129.
22. Wymyslowski A., Falat T., Friedel K., Felba J. Numerical simulation and experimental verification of the thermal contact properties of the polymer bonds. *5th International Conference on Thermal and Mechanical Simulation and Experiments in Microelectronics and Microsystems*, Brussels (2004) 177.
23. Hasselman D.P.H., Donaldson K.Y., Barlow F.D., Elshabini A.A., Schiroky G.H., Yaskoff J.P. Dietz R.L. Interfacial thermal resistance and temperature dependence of three adhesives for electronic packaging. *IEEE Transactions on Components and Packaging Technologies*, (2000) **23**(4): 633.
24. Campbell R.C., Smith S.E., Dietz R.L. Measurements of adhesive bondline effective thermal conductivity and thermal resistance using the laser flash method. *15th IEEE Semiconductor Thermal Measurement and Management Symposium* (1999) p. 83.
25. Felba J., Friedel K., Guenther B., Mościcki A., Schäfer H. The influence of filler particle shapes on adhesive joints in microwave applications. *2nd International* IEEE *Conference on Polymers and Adhesives in Microelectronics and Photonics*, Zalaegerszeg (2002) p.1.
26. www.amepox-mc.com.
27. Felba J., Schaefer H. 'Materials and Technology for Conductive Microstructures', in *Nanopackaging: Nanotechnologies and Electronics Packaging*, (ed. J. Morris), Berlin. Springer, 2008.

28. Ukita Y., Tateyama K., Segawa M., Tojo Y., Gotoh H., Oosako K. Lead-free mount adhesive using silver nanoparticles applied to power discrete package. *International Symposium on Microelectronics IMAPS*, Long Beach, 2004, Session WA7.
29. Chen C., Wang L., Li R., Jiang G., Yu H., Chen T. Effect of silver nanowires on electrical conductance of system composed of silver particles. *Journal of Materials Science* (2007) **42**(9): 3172.
30. Wu H.P., Liu J.F., Wu X.J., Ge M.Y., Wang Y.W., Zhang G.Q., Jiang J.Z. High conductivity of isotropic conductive adhesives filled with silver nanowires. *International Journal of Adhesion and Adhesives* (2006) **26**: 617.
31. EUREKA/EURIPIDES project 'Carbon Nanotubes/epoxy composites' (2007–2011), acronym CANOPY. www.euripides-eureka.eu.
32. Iijima S. Carbon nanotubes: Past, present, and future. *Physica* (2002) B: 1.
33. Terrones M. Carbon nanotubes: Synthesis and properties, electronic devices and other emerging applications. *International Materials Reviews* (2004) **49**(6): 325–377.
34. Popov V.N. Carbon nanotubes: Properties and application. *Materials Science and Engineering* (2004) **R43**: 61–102.
35. Kim P., Shi L., Majumdar A., McEuen P.L. Mesoscopic thermal transport and energy dissipation in carbon nanotubes. *Physica* (2002) **B323**: 67.
36. Pecht M., Agarwal R., McCluskey P., Dishongh T., Javadpour S., Mahajan R. *Electronic Packaging Materials and Their Properties*, Washington: CRC Press LLC, 1999.
37. Fujii M., Zhang X., Xie H., Ago H., Takahashi K., Ikuta T,. Abe H., Shimizu T. Measuring the thermal conductivity of a single carbon nanotube. *Physical Review Letters* (2005) **PRL95**: 065502.
38. Falat T., Platek P., Felba J. Molecular dynamics study of the chiral vector influence on thermal conductivity of carbon nanotubes. *11th Electronics Packaging Technology Conference*, Singapore, 2009.
39. Bal S., Samal S.S. Carbon nanotube reinforced polymer composites – A state of the art. *Bulletin of Materials Science* (2007) **30**(4): 379–386.
40. Peigney A., Laurent Ch., Flahaut E., Bacsa R.R., Rousset A. Specific surface area of carbon nanotubes and bundles of carbon nanotubes. *Carbon* (2001) **39**: 507.
41. Lee T.-M., Chiou K.-C., Tseng F.-P., Huang C.-C. High thermal efficiency carbon nanotube–resin matrix for thermal interface materials. *55th Electronic Component and Technology Conference*, Lake Buena Vista, 2005: 55.
42. Zhang K., Xiao G.-W., Wong C.K.Y., Gu H.-W., Yuen M.M.F., Chan P.C.H., Xu B. Study on thermal interface material with carbon nanotubes and carbon black in high-brightness LED packaging with flip-chip technology. *55th Electronic Component & Technology Conference*, Lake Buena Vista, 2005: 60.
43. Titus E., Ali N., Cabral G., Gracio J., Ramesh Babu P., Jackson M.J. Chemically functionalized carbon nanotubes and their characterization using thermogravimetric analysis, fourier transform infrared, and raman spectroscopy. *Journal of Materials Engineering and Performance* (2006) **15**(2): 182.
44. Heimann M., Wirts-Ruetters M., Boehme B., Wolter K.-J. Investigations of carbon nanotubes epoxy composites for electronics packaging. *58th Electronic Components and Technology Conference*, Orlando, 2008, p. 1731–1736.
45. Qizhen Liang, Wei Wang, Kyoung-Sik Moon, Wong C.P. Thermal conductivity of epoxy/surface functionalized carbon nano materials: *59th Electronic Components and Technology Conference*, San Diego, 2009, p. 460–464.

46. Schadler L.S., Giannaris S.C., Ajayan P.M. Load transfer in carbon nanotube epoxy composites. *Applied Physics Letters* (1998) **73**: 3842.
47. Hong W.-T., Tai N.-H. Investigations on the thermal conductivity of composites reinforced with carbon nanotubes. *Diamond and Related Materials* (2008) **17**: 1577.
48. Thostenson E.T., Chou T.-W. Processing–structure–multi-functional property relationship in carbon nanotube/epoxy composites. *Carbon* (2006) **44**: 3022–3029.
49. Gojny F.H., Wichmann M.H.G. Fiedler B., Kinloch I.A., Bauhofer W., Windle A.H., Schulte K. Evaluation and identification of electrical and thermal conduction mechanisms in carbon nanotube/epoxy composites. *Polymer* (2006) **47**: 2036.
50. Song Y.S., Youn J.R. Evaluation of effective thermal conductivity for carbon nanotube/polymer composites using control volume finite element method. *Carbon* (2006) **44**: 710–717.
51. Bagchi A., Nomura S. On the effective thermal conductivity of carbon nanotube reinforced polymer composites. *Composites Science and Technology* (2006) **66**: 1703.
52. Bolger J.C. Prediction and measurement of thermal conductivity of diamond filled adhesives. *42nd Electronic Components and Technology Conference* (1992) 219.
53. Graebner J.E., Jin. S. Chemical vapor deposited diamond for thermal management. *JOM* (1998) **50**(6): 52.
54. Brierley C.J. Thermal management with diamond. *IEE Colloquium on Diamond in Electronics and Optics*, 1993.
55. Dettmer E.S., Romenesko B.M., Charles H.K., Carkhuff B.G., Merrill D.J. Steady-state thermal conductivity measurements of A1N and SiC substrate materials. *IEEE Transactions on Components, Hybrids and Manufacturing Technology* (1989) **12**(4): 543.
56. Fan L., Su B., Qu J., Wong C.P. Electrical and thermal conductivity of polymer components containing nano-sized particles. *Electronic Components and Technology Conference*, Las Vegas, 2004, p. 148.
57. Sanada K., Tada Y., Shindo Y. Thermal conductivity of polymer composites with close-packed structure of nano and micro fillers. *Composites: Part A* (2009) **40**: 724–730.
58. Lee G.-W., Park M., Kim J., Lee J.I., Yoon H.G. Enhanced thermal conductivity of polymer composites filled with hybrid filler. *Composites: Part A* (2006) **37**: 727–734.
59. Gilleo K. 'Introduction to Conductive Adhesive Joining Technology', in *Conductive Adhesives for Electronic Packaging* (edited by Johan Liu), Electrochemical Publications Ltd, 1999.
60. Li L., Morris J.E. 'Curing of Isotropic Electrically Conductive Adhesives', in *Conductive Adhesives for Electronic Packaging* (edited by Johan Liu), Electrochemical Publications Ltd, 1999.
61. Fukushima K., Takahashi H., Takezawa Y., Hattori M. Itoh M., Yonekura M. High thermal conductive epoxy resins with controlled high-order structure. *Annual Report Conference on Electrical Insulation and Dielectric Phenomena*, Boulder, Colorado, 2004, p. 340.
62. Su B., *Electrical, Thermomechanical and Reliability Modeling of Electrically Conductive Adhesives*, Doctoral Thesis, Georgia Institute of Technology, 2006.

63. Fałat T., Felba J., Wymysłowski A., Jansen K.M.B., Nakka J.S. Viscoelastic characterization of polymer matrix of thermally conductive adhesives. *1st Electronics Systemintegration Technology Conference*, 2006, Dresden, pp. 773–781.
64. Morris J.E. 'Conduction Mechanism and Microstructure Development in Isotropic, Electrically Conductive Adhesives', in *Conductive Adhesives for Electronic Packaging* (edited by Johan Liu), Electrochemical Publications Ltd, 1999.
65. Mościcki A., Felba J., Sobierajski T., Kudzia J. Snap curing electrically conductive formulation for solder replacement applications. *Journal of Electronic Packaging* (ASME) (2005) **127**: 91.
66. Falat T. *Heat Transfer Analysis in Composite Materials Filled with Micro and Nano-sized Particles*, Doctoral Thesis (in Polish), Faculty of Microsystem Electronics and Photonics, Wroclaw University of Technology, Wroclaw, 2007.
67. Felba J., Falat T. Thermally conductive adhesives for microelectronics – barriers of heat transport. *6th International IEEE Conference on Polymers and Adhesives in Microelectronics and Photonics, Polytronic, 2007*, Odaiba-Tokyo, Japan, p. 228.
68. Falat T., Wymysłowski A. Kolbe J., Jansen K.M.B., Ernst L. Numerical approach to characterization of thermally conductive adhesives. *Proceedings of EuroSimE 2006 Conference*, Como, 2006, p. 290.
69. Felba J., Fałat T., Wymysłowski A. Influence of thermo-mechanical properties of polymer matrices on the thermal conductivity of adhesives for microelectronic packaging. *Materials Science–Poland* (2007) **25**(1): 45–55.
70. Jansen K.M.B., Wang L., Yang D.G., van'T Hof C., Ernst L.J., Bressers H.J.L., Zhang G.Q. Constitutive modeling of moulding compounds. *Proceedings of IEEE 2004 Electronic Components and Technology Conference*, Las Vegas, Nevada, 2004, p. 890.
71. Jansen K.M.B., Wang L., van'T Hof C., Ernst L.J., Bressers H.J.L., Zhang G.Q. Cure, temperature and time dependent constitutive modeling of moulding compounds. *Proceedings of the 5th International Conference on Thermal and Mechanical Simulation and Experiments in Micro-electronics and Micro-systems, EuroSimE*, Brussels, 2004, p. 581.
72. Milosheva B.V., Jansen K.M.B., Janssen J.H.J., Bressers H.J.L., Ernst L.J. Viscoelastic characterization of fast curing moulding compounds. *6th Int. Conf. on Thermal, Mechanical and Multiphysics Simulation and Experiments in Micro-electronics and Micro-systems, EuroSimE*, Berlin, 2005, p. 462.
73. Fałat T., Wymysłowski K., Kolbe J. Numerical approach to characterization of thermally conductive adhesives. *Microelectronic Reliability*(2007) **47**: 342–346.
74. Li H., Jacob K.I., Wong C.P. An improvement of thermal conductivity of underfill materials for flip-chip packages. *IEEE Transactions on Advanced Packaging* (2003) **26**(1): 25.
75. Macek S., Rocak D., Sebo P., Stefanik P. The use of polymeric adhesives in bonding power hybrid circuits to heat sinks. *23rd International Spring Seminar on Electronics Technology ISSE*, Balatonfured, 2000, p. 185.
76. Dietz R., Robinson P., Bartholomew M., Firmstone M. High power application with advanced high k thermoplastic adhesives. *11th European Microelectronics Conference*, Venice, 1997, p. 486.
77. Zarr R.R. A history of testing heat insulators at the National Institute for Standards and Technology. *ASHRAE Transactions* (2001) **107**(2).
78. Yi He, Rapid thermal conductivity measurement with a hot disk sensor. Part 1. Theoretical considerations. *Thermochimica Acta* (2005) **436**: 122–129.

79. Tye R.P. Thermal conductivity. *Nature* (1964) **205**: 636.
80. Smith D.R., Hust J.G., Van Poolen L.J. *A Guarded-hot-plate Apparatus for Measuring Effective Thermal Conductivity of Insulations Between 80 K and 360 K*. US National Bureau of Standards, 1982, NBSIR 81-1657.
81. Stacey C. NPL vacuum guarded hot-plate for measuring thermal conductivity and total hemispherical emittance of insulation materials. *Insulation Materials: Testing and Applications* (2002) **4**, ASTM STP 1426.
82. Teertstra P. Thermal conductivity and contact resistance measurements for adhesives. *Proceedings of IPACK2007*, ASME InterPACK '07, Vancouver, 2007.
83. Carlberg B., Wang T., Fu Y., Liu J., Shangguan D. Nanostructured polymer–metal composite for thermal interface material applications. *58th Electronic Components and Technology Conference*, Orlando, 2008, pp. 191–197.
84. Zhang K., Yuen M.M.F., Wang N., Miao J.Y., Xiao D.G.W., Fan H.B. Thermal interface material with aligned CNT and its application in HB-LED packaging. *56th Electronic Components and Technology Conference*, San Diego, 2006, pp. 177–182.
85. Munari A., Xu J., Dalton E., Mathewson A., Razeeb K.M. Metal nanowire–polymer nanocomposite as thermal interface material. *59th Electronic Components and Technology Conference*, San Diego, 2009, pp. 445–452.
86. Platek B., Falat T., Felba J. An accurate method for thermal conductivity measurement of thermally conductive adhesives. *34th International Spring Seminar on Electronics Technology*. Tatranská Lomnica, 2011.
87. Franco A. An apparatus for the routine measurement of thermal conductivity of materials for building application based on a transient hot-wire method. *Applied Thermal Engineering* (2007) **27**: 2495–2504.
88. dos Santos, W.N. Advances on the hot wire technique. *Journal of the European Ceramic Society* (2008) **28**: 15.
89. Gustavsson M., Karawacki E., Gustafsson S.E. Thermal conductivity, thermal diffusivity, and specific heat of thin samples from transient measurements with hot disk sensors. Review Scientific Instrumentation (1994) **85**(12): 3856–3859.
90. Parker W.J., Jenkins R.J., Butler C.P., Abbott G.L. Flash method of determining thermal diffusivity, heat capacity, and thermal conductivity. *Journal of Applied Physics* (1961) **32**: 1679.
91. Cowan R.D. Pulse method of measuring thermal diffusivity at high temperatures. *Journal of Applied Physics* (1963) **34**: 926.
92. Taylor R.E., Cape J.A. Finite pulse-time effects in the flash diffusivity technique. *Applied Physics Letters* (1964) **5**: 212.
93. Watt D.A. Theory of thermal diffusivity by Pulse technique. *British Journal of Applied Physics* (1966) **17**: 231.
94. Walter A.J., Dell R.M., Burgess P.C. The measurement of thermal diffusivities using a pulsed electron beam. *Rev. Int. Hautes Tempér. Réfract.* (1970) **7**: 271.
95. Falat T., Felba J. Electron beam as a heat source in thermal diffusivity measurement of thermally conductive adhesives. *Electrotechnica and Electronica* (2006) **5–6**: 189.
96. Touloukian Y.S., Powell R.W., Ho C.Y., Nicolaou M.C. 'Thermal Diffusivity', in *Thermophysical Properties of Matter*, (1973) p. 10: New York–Washington: IFI/PLENUM.
97. Clark III L.M., Taylor R.E. Radiation loss in the flash method for thermal diffusivity. *Journal of Applied Physics* (1975) **46**: 714.

98. Rosencwaig A., Gersho A. Theory of the photoacoustic effect with solids. *Journal of Applied Physics* (1976) **47**: 64.
99. Hu H., Wang X., Xu X. Generalized theory of the photoacoustic effect in a multilayer material. *Journal of Applied Physics* (1999) **86**: 3953.
100. Kim J.H., Feldman A., Novotny D. Application of the three omega thermal conductivity measurement method to a film on a substrate of finite thickness. *Journal of Applied Physics* (1999) **86**(7): 3959.
101. Dames C., Chen G. 1ω, 2ω, and 3ω methods for measurements of thermal properties. *Review of Scientific Instruments* (2005) **76**: 124902.
102. Cahill D.G., Pohl R.O. Thermal conductivity of amorphous solids above the plateau. *Physical Review* (1987) **B35**(8): 4067–4073.
103. Woong Sun Lee, Jin Yu. Comparative study of thermally conductive fillers in underfill. *Diamond and Related Materials*, (2005) **14**: 1647–1653
104. Ekstrand L., Kristiansen H., Liu J. Characterization of thermally conductive epoxy nano composites, *28th International Spring Seminar on Electronics Technology*, 2005, pp. 19–23
105. Yunsheng Xu, Gunawidjaja Ray, Beckry Abdel-Magid. Thermal behavior of single-walled carbon nanotube polymer–matrix composites. *Composites: Part A* (2006) **37**: 114–121.
106. Gojny F.H., Wichmann M.H.G., Fiedler B., Kinloch I.A., Bauhofer W., Windle A.H., Schulte K. Evaluation and identification of electrical and thermal conduction mechanisms in carbon nanotube/epoxy composites. *Polymer* (2006) **47**: 2036–2045.
107. Grieve A., Capote M.A., Soriano A. New adhesive materials with improved thermal conductivity. *56th Electronic Components and Technology Conference*, San Diego, 2006, pp. 946–951.
108. Yi Li, Myung Jin Yim, Kyung Sik Moon, Wong C.P. Nano-scale conductive films with low temperature sintering for high performance fine pitch interconnect, *57th Electronic Components and Technology Conference*, Reno, 2007, pp. 1350–1355.
109. Inoue M., Muta H., Maekawa T., Yamanaka S., Suganuma K. Temperature dependence of electrical and thermal conductivities of an epoxy-based isotropic conductive adhesive. *Journal of Electronic Materials* (2008) **37**(4): 462–468.
110. Sahin S., Yavuz M., Zhou Y.N. 'Introduction to Nanojoining', in *Microjoining and Nanojoining* (ed. Y. N. Zhou), Woodhead Publishing Limited, Cambridge, UK, 2008.
111. Charlier J.-C., Terrones M., Banhart F., Grobert N., Terrones H., Ajayan P.M. Experimental observation and quantum modeling of electron irradiation on single-wall carbon nanotubes. *IEEE Transactions on Nanotechnology* (2003) **2**(4): 349.
112. Wang Z., Yu L., Zhang W., Ding Y., Li Y., Han J., Zhu Z., Xu H., He G., Chen Y., Hu G. Amorphous molecular junctions produced by ion irradiation on carbon nanotubes. *Physics Letters* (2004) **A324**: 321.

3
Anisotropic conductive adhesives in electronics

J. W. C. DE VRIES and J. F. J. M. CAERS,
Philips Applied Technologies, The Netherlands

Abstract: This chapter presents an industrial approach to anisotropic conductive adhesive interconnection technology. The focus is on the long-term performance for this type of electrical interconnection as an alternative to soldered joints of very fine pitch (less than 100 μm). Beginning with a brief description of what an adhesive bond actually is, as compared with a soldered joint, the most important material properties and the critical stress factors are considered. They are illustrated by discussing some case studies: heat seal connectors and ultra-thin ball grid array packages.

Key words: electrical interconnections, flexible foil substrates, long-term behavior, degradation mechanism, evaluation methods, reliability.

3.1 Introduction

Adhesives are often considered as alternatives for soldered interconnections; and there are many different application areas and likewise a multitude of embodiments. Here we will focus on the electrical connection of flip chips on flexible substrates, and heat seal interconnections, as they also include flexible printed circuit boards.

The main drives for investigating adhesives as alternative first-level interconnections to soldering for flip chip packages on flexible substrates are threefold: the process flow is relatively simple, the technology is lead-free, and when compared to soldering there is the possibility to apply the interconnection technology to components with a very fine pitch – in this work down to 40 μm or even less. The questions that one has to answer cover the following topics: (i) compatibility of the adhesive interconnection process with reflow soldering (this is certainly one of the most harsh assembly process steps), (ii) the long-term reliability aspects of these novel interconnections (which immediately opens the question to the relevant failure mechanisms and the appropriate test and analytical methods), (iii) what, if any, is the limiting pitch or spacing for using anisotropic adhesives, and (iv) when should one begin to apply non-conductive adhesive? Special points of attention are then the bonding process window, in terms of temperature and pressure, and the proper combination of materials.

As mentioned, the focus will be upon flip chip on flex and heat seal connectors; other possible application fields are flip chip on board and chip on glass. We will address these only in so far as material is available to illustrate the main topic of this chapter. Further, the authors would like to emphasize that the contents of this chapter – mainly from their own expertise, but illustrated with information from other authors as well – are the result of efforts to develop industrial processes. This implies that, for instance, the adhesive bonds are evaluated as to their endurance behavior, or that the degree of curing is not very high and post-curing can occur during subsequent thermal process steps.

3.1.1 Organization of the chapter

In order to better explain what anisotropic adhesive (ACA, ACF, ACP, see list of abbreviations in the Appendix at the end of this chapter) interconnections are and how these perform, their nature is described first, directly followed by a summary account of the literature. The third section contains basic information about materials and processing. In line with our approach to focus on the industrial aspects of ACF-bonds, such as their reliability, the critical stress loads will be discussed and illustrated with examples in the fourth section. As a result of the performance under various types of stress, the particular sensitivity of adhesive interconnections to moisture and temperature justifies a dedicated section on evaluation methods. Much of what has been treated up to this point culminates in the case studies of the sixth section: heat seal connectors and an ultra-thin ball grid array package. Some concluding remarks are formulated in the final section.

3.2 Nature of adhesive bond

Dealing with items such as electrically conductive adhesive interconnections, including their short- and long-term performance, necessitates having knowledge about the nature of electrical adhesive bonds. This may be best explained by contrasting them to soldered electrical interconnections. In the latter case, an alloy is formed between the solder material that melts in the soldering process, and the metallization of the two parts that have to be connected. The solder is thus in between the two parts and is a shackle of the conductive path. In Fig. 3.1 this is displayed in a simplified way. Although the matrix of the adhesive, and in case of conductive adhesives also the filler particles, will certainly be between the two parts that have to be connected, the bulk of the adhesive is outside and around the bond pads, and for good reason, because the adherence to the adjacent parts will pull these together, in particular when, during the curing of the adhesive, it shrinks. This is shown in Fig. 3.1. Thus the adhesive joint exerts a force onto the two parts and the higher this force, the lower the contact resistance. See Fig. 3.2 for an example.

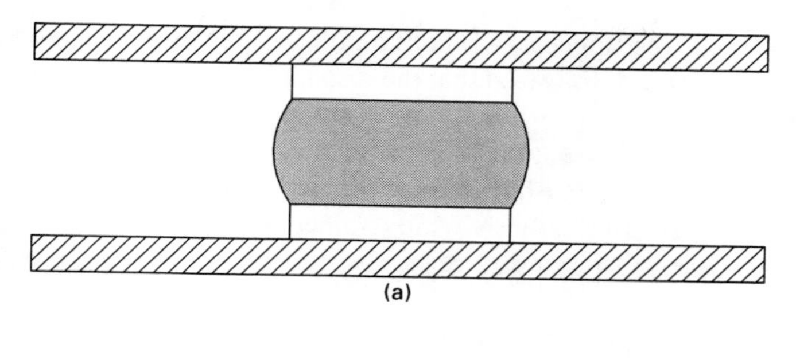

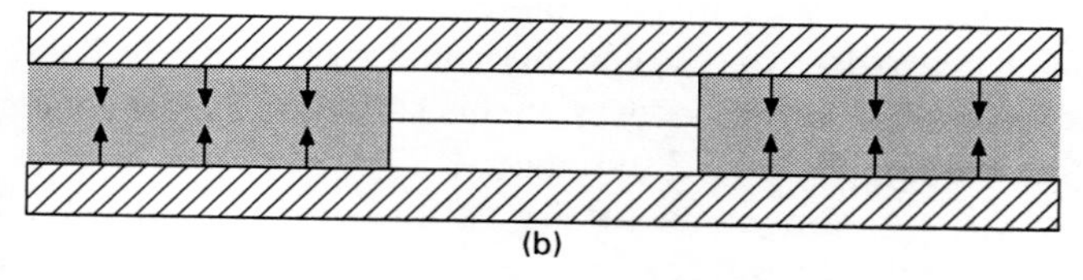

3.1 Schematic of a soldered (a) and an adhesive (b) interconnection. The parts are indicated by hatching, bond pads are white, solder and adhesive are dark grey. The arrows represent the clamping force.

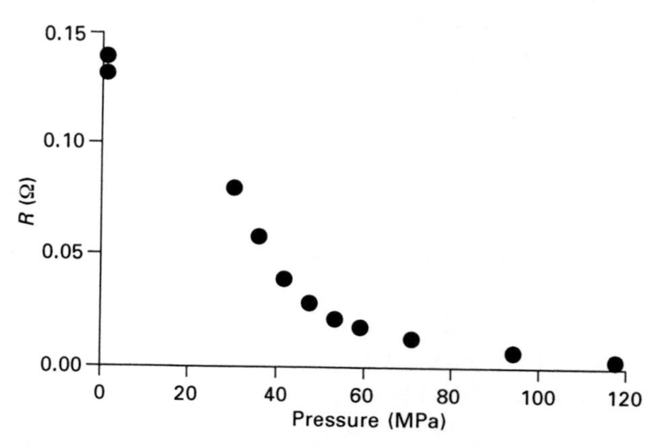

3.2 In situ measured resistance, *R*, of daisy-chained adhesive interconnection as a function of applied pressure.

Wu *et al.* (1999) formulated a physical model for the contact resistance of conductive adhesives. They incorporated the deformation of the conducting filler particles as a function of the applied force, to arrive eventually at:

$$R \approx c \ln(p_0/p) \qquad [3.1]$$

Here, p is a measure of the applied pressure and c, p_0 are constants. The functional dependence is indeed very similar to the experimental data shown in Fig. 3.2. Alternatively, the deformation of an ensemble of particles that have a normal (Gaussian) particle size distribution has been used to derive an expression for the resistance as a function of the applied force (Divigalpitiya and Hogerton, 2003). It followed that the resistance is inversely proportional to the force as $R \sim F^{-n}$, with $n \approx 1$. These authors concluded that the width of the particle size distribution is a better measure for the quality of the interconnection than the particle density on the bond pad. Similar work was done on non-conductive adhesive interconnections (Caers *et al.*, 2003c). Also, in this case, the resistance depended inversely upon the pressure: $R \sim P^{-0.92}$.

This notion is helpful to better understand the bonding process and the subsequent behavior of the adhesive interconnection. Any actor that causes the clamping force to decay is a potential degradation factor. In Table 3.1 the results of a literature scan that focuses on this particular aspect are compiled; the following paragraphs give a summary.

There is fair agreement as to the failure mechanism of adhesive interconnections, either on rigid or flexible carriers. In general, it is agreed that high temperature – humidity conditions are far more detrimental to reliability than are constant or cyclic temperature test conditions. More specifically, cyclic exposure to humid circumstances invokes a kind of fatigue that causes much faster degradation than constant humidity exposure. Basically, degradation of the electrical interconnect is attributed to a decrease in the contact area, or to particle dimension, where larger particles in the adhesive set back the reliability more than do smaller particles. Thermally induced degradation of the polymer matrix is also mentioned. This, as such, or in combination with the presence of air bubbles, is supposed to affect the adhesion strength. The said air bubbles are assumed to lead to corrosion of the conductors, or they cause local variations in thermal expansion mismatches. A three-layered foil might lead to variations in contact pressure, by virtue of the adhesive layer between the metallization and the polymer substrate, with bad consequences for reliability. This is also the case with board warpage. Relaxation of built-in elastic stress causes the contact resistance to increase. Finally, fully cured adhesives are reported to be more reliable than partly cured samples.

In a few cases, the resistance as measured on-line has been compared with that obtained from the conventional off-line method. However, the conclusions do not always agree with one another, as may be seen from a comparison of a few such investigations (Liu and Lai, 1999, 2002; Caers *et al.*, 2003a, b; De Vries, 2004). Remarkably few publications give failure distributions to check the presence of single or multiple failure mechanisms, or to estimate lifetimes.

It is important to note that comparatively little has been published on the reflow soldering behavior of flip chip on foil substrates and flip chip on FR4 assemblies. Flip chip bonded with anisotropic conductive adhesive on FR4 have been subjected to a reflow step followed by a thermal cycling test (Seppälä *et al.*, 2001). These authors observed a decrease of the contact

Table 3.1 Literature overview with key issues. ICA/P: isotropic conductive adhesive/paste, ACF/AP: anisotropic conductive film/adhesive (paste), NCF/A: non-conductive film/adhesive (paste), ENIG: electroless Ni immersion Au. Bonding pressures are indicated relative to the total bump area. PA: polyamide, PI: polyimide, c: cycles

Reference	Adhesive	Substrate	Metallization	Bump	Tests	Remarks
Li *et al.* (1995)	ICA-Ag	PCB	Cu + pass.	Pretinned Passivated Cu	85 °C/85%RH/1000 h	
Lai and Liu (1996)	ACA-Ni	FR4, Flex	Flash Au	Ni/Au Bumpless	85 °C/85%RH/1000 h 125 °C/1000 h –40 °C/+125 °C/1000 c –10 °C/+65 °C/80–95%RH	
Liu (1996)	ACA	Any	Any	Any	85 °C/85%RH/1000 h	Model
Olliff *et al.* (1997)	ICA-Ag (?)	None	Cu	Cu	Shear	Thermoplastic
Liu (1998)	ACA, ICA	Any	Any	Any	Any	Overview
Lam *et al.* (1999)	ACA, ICA	Rigid	—	—	—	Model
Rusanen and Lenkkeri (1999)	ICA	FR4 Al_2O_3	Thick film Au Cu/Ni/Au	Electroplated Au	–40 °C/125 °C/2000–2600 c	Creep model
Miessner *et al.* (1999)	ACA	FR4	Cu/Ni/Au	Ni/Au Au-stud Electroplated Au	85 °C/85%RH 125 °C –55 °C/125 °C moisture and reflow	200–300 μm pitch 50–150 g/bump 88–136 bumps
Liu and Lai (1999)	ACF-Ni ACF-polym. NiAu NCF	FR4	Cu/Ni/Au	ENIG	–40 °C/+125 °C/3000 c	100 μm pitch 7–50 MPa
Jagt (1999)						Overview
Jagt *et al.* (2000)	ACA-polym. Au, Au, Ni ACF-polym. Au, Ni	Flex	—	Ni/Au Au	–55 °C/+125 °C/2200 c 85 °C/85%RH/1680 h 120 °C/85%RH/580 h	300–80 μm pitch On/off-line Modeling
Janssen and Kums (2000)	ACA-Ni	Flex	Cu/Ni/Au	Ni/Au	–55 °C/+125 °C/525 c 85 °C/85%RH/1000 h	200–300 μm pitch 50–100 MPa

Continued

Table 3.1 Continued

Reference	Adhesive	Substrate	Metallization	Bump	Tests	Remarks
Jokinen *et al.* (2000)	ACF-Ni ACF-polym. NiAu	PEN-flex	Cu/Ni/Au	Au	–30 °C/85 °C/200 c + –40 °C/125 °C/200 c	250 μm pitch 10–45 MPa
Nysaether *et al.* (2000)	ICA	FR4	Cu	ENIG Au	–55 °C/125 °C/4000 c	450 μm pitch
Yim and Paik (2001)	ACA	FR4	Au	Au-stud	85 °C/85%RH/1000 h 85 °C/1000 h –60 °C/150 °C/700 c	Silica filler addition
Palm *et al.* (2001a)	ACF-polym. NiAu ACF-Ni, ACA-Ni	Flex	Cu/Au	Au	85 °C/85%RH/2000 h –40 °C/+125 °C/1880 c	80 μm pitch 30–60 MPa (?)
Palm *et al.* (2001b)	ACF-polym. NiAu ACA-Ag	PI-flex PEN-flex EG-flex	Cu/Ni/Au	Au	–40 °C/+125 °C/1000 c	80 μm pitch 30–60 MPa (?)
Seppälä *et al.* (2001)	ACF-polym. Au ACF-polym. Au ins.	FR4	Cu/Ni/Au	ENIG	(reflow+) –30 °C/+85 °C/500 c –40 °C/+125 °C/500 c	200–400 μm pitch 92–158 MPa
Ling *et al.* (2001)	ACA, NCP, ICP, solder	Rigid	—	Au-stud Au/Ni-plated die Solder bumped die	–55 °C/+125 °C/1000 c	200 μm pitch
Chang *et al.* (2001)	ACF-resin Au	PI-flex	Cu/Au	Au	–20 °C/+70 °C-500 c 60 °C-95%RH/500 h	Process study 80 μm pitch 60–140 MPa
Jokinen and Ristolainen (2001)	ACF-Ni	PEN-flex	Cu/Ni/Au	Au	–40 °C/+125 °C/1000 c 85 °C/85%RH/504 h bending	250 μm pitch 50 μm thick IC 60–100 MPa
Määttänen *et al.* (2002)	ACF-resin NiAu	PI-flex	Cu/Ni/Au	Au	–40 °C/+125 °C/1000 c 85 °C/85%RH/1000 h	54 μm pitch 40 MPa(?)

Van Wuytswinkel *et al.* (2002)	Various	Rigid	Various	—	−40 °C/+150 °C/750 85 °C/85%RH/1000 h 150 °C/1000 h	Corrosion inhibitors
Lawrence Wu and Chau (2002)	ACF-polym. Ni	Glass	ITO	Au	85 °C/90%RH/1000 h 85 °C, 100 °C, 125 °C/500 h	100 μm pitch 1–100 MPa
Liu and Lai (2002)	ACF-Ni ACF-plastic NiAu NCF	FR4	Cu/Ni/Au	ENIG	−40 °C/+125 °C/3000 c	100 μm pitch 7–50 MPa
Yin *et al.* (2003)	ACF-polym. NiAu	Flex	Cu/Ni/Au	Ni/Au Au	(Reflow profiles+) 85 °C/85%RH/500 h	Process study 80 μm pitch(?) 87 MPa
Caers *et al.* (2003a)	ACA	Flex	Cu/Ni/Au	Au	85 °C/85%RH/1000 h	300 μm pitch On/off-line model
Caers *et al.* (2003b)	ACF NCF	Flex	—	—	—	Overview Failure mechanism Material properties model
Caers *et al.* (2003c)	NCA	Flex	Cu/Ni/Au	Au	85 °C/85%RH/1800 h	300 μm pitch On/off-line Materials properties model
Palm *et al.* (2003)	ACF-resin NiAu	Espanex Upilex EG G10	Cu/Ni/Au	Au	85 °C/85%RH/1000 h −40 °C/+125 °C/1000 c	80 μm, 54 μm pitch
De Vries and Janssen (2003)	ACF-polym. Au	Espanex DuPont AP	Cu/Ni/Au	Ni/Au	MSL/reflow 85 °C/85%RH/3000 h	200–300 μm pitch 100–200 MPa Failure mechanism
De Vries (2004)	ACF-polym. Au	DuPont AP	Cu/Ni/Au	Ni/Au Au	85 °C/85%RH/3600 h −40 °C/+125 °C/3600 c	150–300 μm pitch On/off-line Failure mechanism

Continued

Table 3.1 Continued

Reference	Adhesive	Substrate	Metallization	Bump	Tests	Remarks
Kim *et al.* (2004)	ACF-polystyrene NiAu	Flex on glass			80 °C/85%RH/1000 h –15 °C/+100 °C/1000 c	80 μm pitch, various curing conditions
Wu *et al.* (2004)	ACF-resin NiAu ACF-resin Au	PA-flex glass	Cu/Au	Au-plated	Impact test	~80 μm pitch? 30–110 MPa 150–230 °C
Cao *et al.* (2005)	ACA, various fillers	Flex	Cu/Ni/Au	ENIG	120 °C/85%RH/272 h	100 μm pitch ACA formulation
Frisk and Ristolainen (2005)	ACF-polym. Au hard Ni	LCP-flex	Cu/Ni/Au	Au	85 °C/85%RH/1000 h –40 °C/+85 °C/1000 c	250 μm pitch On-line monitoring
De Vries *et al.* (2005a)	ACF-polym. Au NCF	Espanex	Cu/Ni/Au	Au	MSL/reflow –55 °C/+125 °C/3600 c 85 °C/85%RH/1600 h 110 °C/85%RH/1440 h	100 μm pitch 200–400 MPa Failure mechanism
Chan *et al.* (2006)	NCP, NCF	PI-flex	Cu/Ni/Au	Au	–55 °C/+125 °C/400 c	~90 μm pitch (?) 100–160 MPa
Chiang *et al.* (2006)	NCA	PI-flex	Cu/Ni/Au	Au	121 °C/100%RH/192 h 150 °C/500 h	60 μm pitch Processing conditions
Ikeda *et al.* (2006)	ACF-Ni		Cu/Ni/Au	Au	MSL/reflow	200 μm pitch Delamination behavior Moisture absorption
Kim and Jung (2007a)	ACF-polym. NiAu	PI-flex	Cu/Ni/Au	Au	–40 °C/+125 °C/500 c with bond gap measurement	~140 μm pitch (?) 40–80 MPa

Kim and Jung (2007b)	ACF-polym. NiAu	PI-flex	Cu/Ni/Au	Au	No-preconditioning/ reflow 220 °C	~140 μm pitch (?) 40–80 MPa
Majeed *et al.* (2007)	ACA-NiAu ACF-soft Au	PI-flex	Cu/Ni/Au	Au-stud	85 °C/85%RH/700 h –55 °C/+125 °C/3000 c	200 μm pitch On/off-line
Kim *et al.* (2008a)	ACF-polym. NiAu NCF	Flex	Cu/Ni/Au	Au	No-preconditioning/ reflow	200 μm pitch 60–80 MPa
Kim *et al.* (2008b)	ACF-polym. NiAu	Flex	Cu/Ni/Au	Au	–55 °C/+125 °C/500 c	~140 μm pitch 50–90 MPa FE-analysis
Lu and Chen (2008)	ACA	PI-flex	Cu/Ni/Au	Au	85 °C/85%RH/1000 h bend test	80 μm pitch Thin die (50 μm)
Frisk and Cumini (2009)	ACA-polym. Au	FR4 LCP	Cu/Ni/Au	Au	–40 °C/+125 °C/20000 c	250 μm pitch

resistance directly after reflow which was tentatively ascribed to post curing. During the thermal cycling test, more failures were found in assemblies that had been subjected to the reflow step. In a study at the department of electronic engineering of the City University of Hong Kong, ICs with a pitch of 80 μm were assembled (Yin *et al.*, 2003). After reflow testing at various peak temperatures, contact resistances increased by some 30% to 100%, depending on the height of the bump. During subsequent humidity testing, resistances increased further, which was attributed to the decreasing contact areas (between the conducting particles and the bond-pads) by swelling of the adhesive as it took up water. The number of open joints was 40% after reflowing at 260 °C, 5% at 230 °C, and zero at 210 °C. Subsequent accelerate damp heat testing (85 °C/85%RH) supposedly led to swelling of the adhesive and thus caused the contact resistance to increase. However, in the reflowed assemblies this effect was much smaller or even almost absent. Probably, according to these authors, reflowing leads to full curing of the adhesive, which makes them resistant to absorption of water. In addition the work deals with an extensive process study concerning the bonding conditions. Ikeda *et al.* (2006) studied the delamination behavior in 200 μm-pitch assemblies under moisture–reflow conditions. To this end they introduced a so-called 'delamination toughness factor', which is the total stress intensity factor of an interface crack between two materials in the presence of a beginning delamination. Kim *et al.* (2007b, 2008a) discussed the reflow behavior but without preconditioning. According to their results, the contact resistance of non-conductive adhesive interconnections remained constant after reflow while with anisotropic conductive adhesives the resistance increased. But after repeated reflow treatments, the non-conductive versions did not exhibit Ohmic behavior anymore, while the ACF-samples still did. In fact not one of these publications has actually included the JEDEC-MSLA test. Still, as the assemblies will very probably have taken up moisture from the air, the reflow step effectively may be compared to a level 6 test (see Table 3.5 later in this chapter).

As for the pitch, the groups of Elcoteq and the University of Tampere reported on pitches of 80 μm and 54 μm (Palm *et al.*, 2001a, b; Määttänen *et al.*, 2002). The bonding pressure was in the range of 30–60 MPa, which is lower than our settings, which were in the range of several hundreds of MPas (see Section 3.3.2). The 80 μm-pitch samples showed some failures after 740–1760 hours in 85 °C/85%RH-testing, depending on the type of adhesive. In the thermal cycling test at –40 °C/+125 °C after 1000 cycles, one adhesive showed failures, while another type performed well for 1880 cycles when the test was terminated. The different reliability results of these adhesives were tentatively ascribed to the number of particles trapped between the bumps and the tracks. The influence of various types of foil was investigated. From thermal cycling tests, the layered structure of the foil affected the reliability. In a triple layer foil (substrate–adhesive–copper), the bond pads occasionally sunk into the adhesive of the foil. Further conclusions were that a larger contact area is beneficial to reliability and full metal particles give lower contact resistances than metal coated polymer particles. The smaller pitch of 54 μm showed that the degree of curing is important for reliability. In humidity testing, the swelling of the adhesive diminished the contact area, which might explain the better reliability of samples with a larger contact area. A few other publications deal with reflow testing on fine pitch structures. 80 μm-pitch assemblies have been described, and one extensively discussed the effects of bonding pressure, temperature and time, on the contact resistance. Another studied curing temperature, with stress test results only after completion of the tests. Still also here the bonding pressures were moderate.

More recently, the mechanical robustness of adhesive interconnections has received attention, with impact testing (Wu *et al.*, 2004) and static bending (Lu and Chen, 2008) being dealt with. The impact test results showed that interface cracks account for failure in chip-on-glass assemblies, whereas cohesive failure occurs in chip-on-foil samples. From the bend tests it was concluded that if bonding is done at a temperature of 160 °C, the adhesive film delaminates from the bump of the die, while after bonding at 190 °C the die breaks.

As for modeling of the driving mechanisms, Liu (1996) derived a rather simple model to account for both oxidation of the metallic interfaces and reduction of the contact area, for instance by crack growth. As mentioned previously, the former plays a minor role compared to the latter. Reduction of the contact area is supposed to lead to a rapid increase of the resistance after a certain period of time:

$$R' \approx \frac{1}{1 - t/t_0} \qquad [3.2]$$

In contrast, oxidation causes a gradual but moderate increase:

$$R' \approx 1 + a\sqrt{t/t_0} \qquad [3.3]$$

In these two equations, R' is the normalized resistance, t is the elapsed time and t_0 the total test time. From a kinetic chemical model, Lam *et al.* (1999) have determined that, on rigid substrates, the loss of joint strength is a combined effect of moisture and stress. The degradation proceeds from low water concentration and high stress level to a large water concentration at a low stress level. Rusanen and Lenkkeri (1999) have addressed the creep in adhesives. By expressing the creep strain as a linear function of time, the authors related the number of cycles to failure to the applied strain through the Manson-Coffin equation. From finite element simulations, Kim *et al.* (2008b) confirmed that tensile stress in a direction perpendicular to the bonding plane (peeling stress) accounts for failures in adhesively bonded packages.

3.3 Materials and processing

3.3.1 Materials

Within the limitation outlined in the introduction, relevant materials are the flexible substrate, the dies with bumps, and the adhesives. Flexible foil for the electronic applications that are the subject of this chapter can be divided in three-layer and two-layer types. The three-layer foils have a polymeric base material with metallic traces attached by an adhesive layer. Such an adhesive interlayer may form a risk in the bonding process of very narrow structures because the bond pads may slide away (see Section 3.2). The two-layer foils, in contrast, have the metallic conductors attached adhesive-less to the base material. For the present treatment of adhesive interconnections applied to very fine pitches of 40 μm, three-layer foils are not very suitable. The traces are usually Cu with a Ni/Au-finish, but Al or Cu can be used as low-cost alternative metallizations.

For the investigations described in this work, the dies and the flexible foils were designed to incorporate daisy chains to monitor the connectivity of the interconnections. But the die-to-foil assembly also contained a number of single bump contacts of which resistance could be measured using a four-wire method (see Fig. 3.3 for an example of such a structure). The used flexible foils with a pitch of 100 μm had a polyimide layer of 50 μm

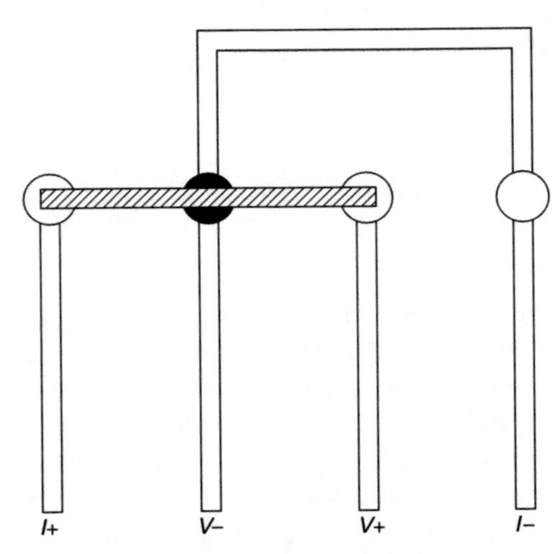

3.3 Typical four-wire structure to measure the resistance of the black interconnect. Circles represent interconnections, white bars are tracks on the substrate, hatched bar is track on the die.

thick, a solder resist layer of 25 μm thick, and in total a 15 μm thick Cu/Ni/Au metallization. For the 60 μm and 40 μm pitch type, the polyimide was 40 μm thick, solder resist was 12 μm, and the metallization consisted of 12 μm Cu, 2 μm Ni, and flash Au of 0.06 μm. In Figs 3.4 and 3.5, the

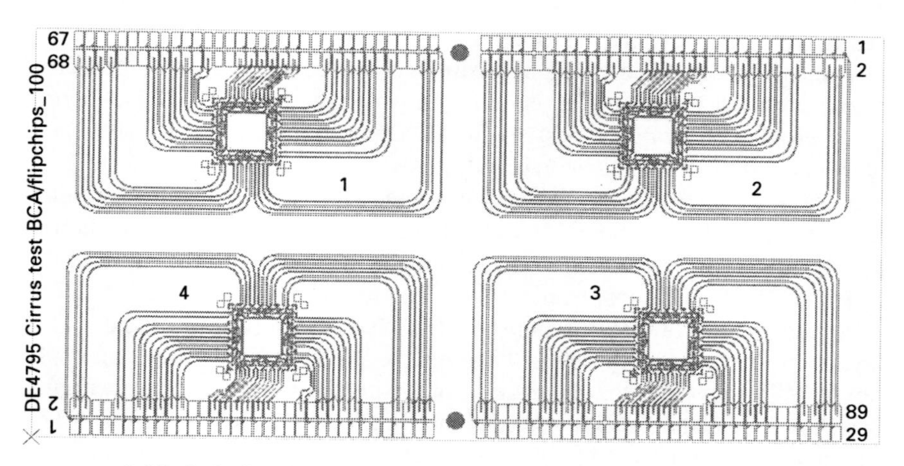

3.4 Foil design for flip chip with 100 μm pitch. Four flip chips can be placed, various daisy chains and single contact resistances can be measured.

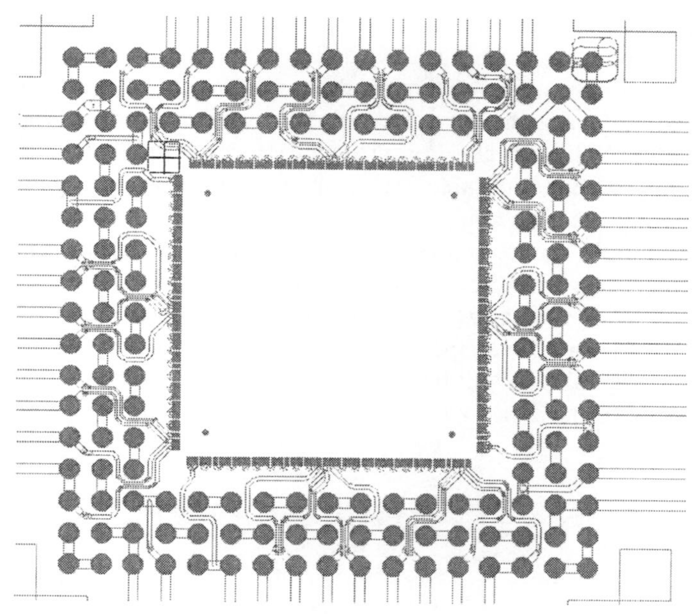

3.5 Detail of Fig. 3.4 showing footprint of flip chip. Several traces are visible for daisy chain and single contact measurements. The design has three rows for second-level solder bumps.

layout of such foils is shown. Similarly, for heat seal connectors, a four-wire contact resistance measurement can be done using a dedicated test structure as given in Fig. 3.6. To overcome problems with contacting the heat seal connector, the two foil edges are bonded to the printed circuit board too. This way, a triple-bond structure is created; the contact resistance of the central interconnects can be monitored.

Adhesives can be divided according to the content of filler particles. Those with a high content of conducting particles, typically 70% or more, are isotropic conductive adhesives (ICA). The filler material is usually made of silver flakes. For much lower filling factors, in the range of 5–15% or even less, the conducting particles become effectively isolated from each other, thus blocking conduction in the lateral direction, and only conduction in the direction perpendicular to the metallic surfaces of the chip and the foil can occur. These are anisotropic conductive adhesives (ACA). The filler can consist of hard particles such as Ni, which serve to crush oxide layers on the surfaces that must be connected. Alternatively, one can apply flexible filler material that is made from Au-coated polymer spheres. Figure 3.7 shows an example of such a contact in a 100 μm-pitch assembly. The advantage of metal coated polymer spheres is that they can accommodate the relaxation of the adhesive to some degree. For completeness, non-conductive adhesives (NCA, NCF, NCP, see list of abbreviations in the Appendix at the end of the chapter) have a filler content of nil, but one may regard these essentially as anisotropic conductive adhesives as conduction can occur only in one direction. In Fig. 3.8, the distribution of 96 single-contact resistances is shown for 60 μm-pitch interconnections. For assemblies made with ACF (Au-coated polymer), the values are slightly but clearly lower than for the corresponding NCF-samples. For three adhesives, a selection of the physical properties is compiled in Table 3.2. At the glass transition temperature (T_g) these materials change from a glassy state below the transition temperature to a rubbery state above, and *vice versa*. In fact, the transition temperature is the center of a transition region that can extend over 10–20 degrees. Material properties such as coefficient of thermal expansion (CTE) and storage modulus change dramatically from one to the other state. In the next section on processing, it will be seen that the transition temperature depends on the curing of the adhesive and aging.

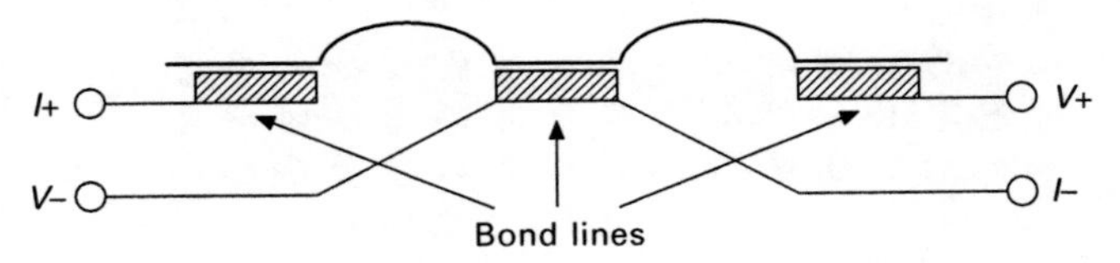

3.6 Test structure for four-wire resistance measurement for heat seal connectors.

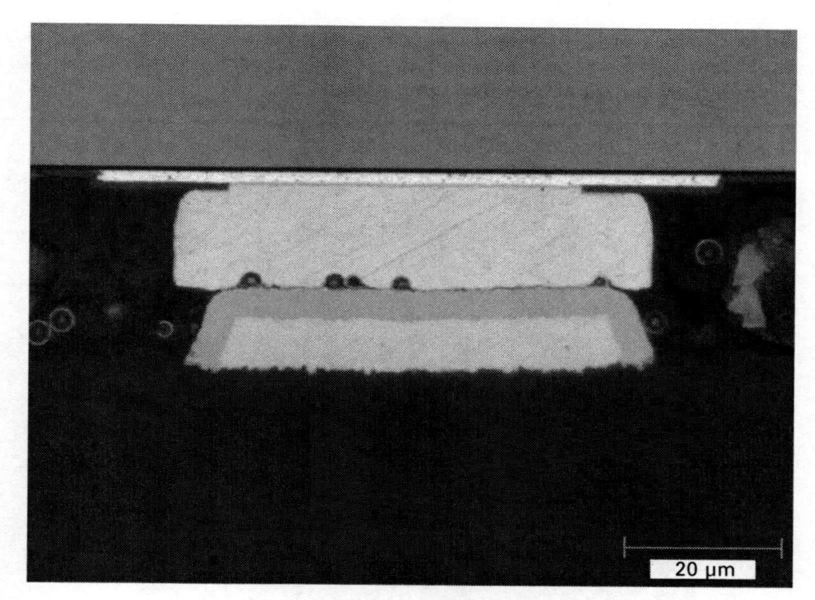

3.7 Anisotropic adhesive contact of die (grey mass on top) with Au-bump on flexible foil with Cu/Ni/Au-bond pad. Several Au-coated polymer filler particles are visible inside and outside the contact.

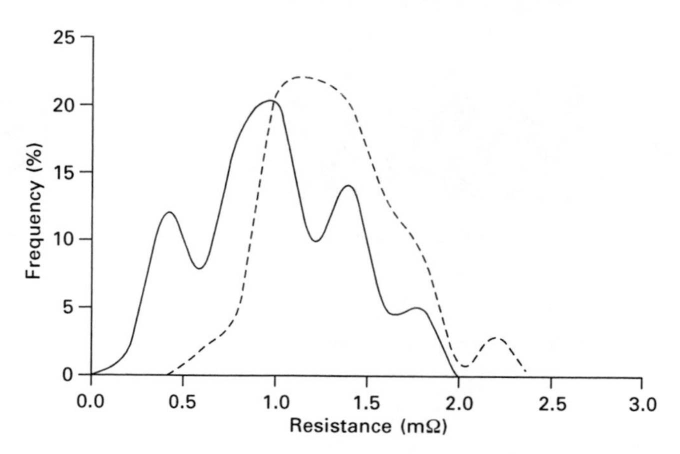

3.8 Distribution of 96 single-contact resistance values of 60 μm die on flexible foil. Solid line: ACF. Dashed line: NCF.

Dies with various types of contact bumps exist. In case of relatively hard Ni/Au-bumps (see Fig. 3.9), the bumps will keep their shape and the contact with the bond pad on the flexible foil is not as intimate as when using much softer galvanic Au-bumps or Au-stud bumps (see Figures 3.10 and 3.11). The Au-bump depicted in Fig. 3.10 even follows the shape of the bond pad.

Table 3.2 Physical properties of conductive adhesives (suppliers' data). Thermal expansion coefficient (α) below and above glass transition temperature (T_g), and Young's modulus (E) at indicated temperatures

Adhesive	α (ppm/K) < T_g	α (ppm/K) > T_g	T_g (°C)	E (GPa)
S-1	63	130	125	1.2 (40 °C) 0.02 (150 °C)
H-3	55	550	145	1.4 (40 °C)
T-4	65		132	3.9

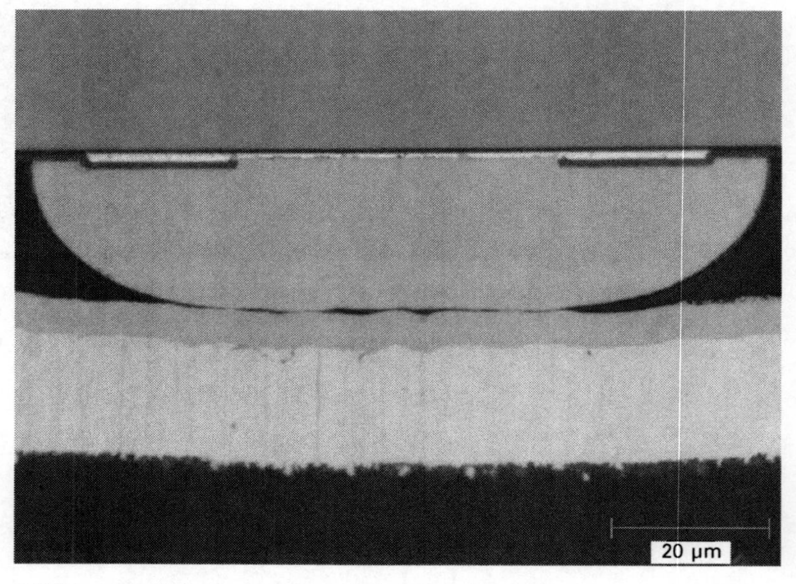

3.9 Die with NiAu-bumps bonded on flexible foil.

In most cases described in this chapter, the dies have dimensions of 5 × 5 mm^2. In Table 3.3, a number of relevant parameters of these and other test dies are compiled.

3.3.2 Processing

As one can see from Table 3.1, the bonding pressure there is always low in comparison with our own settings, as detailed below. This means that the conductive particles are deformed but they are never pressed into the Au-bumps. The reason for using higher pressure is to achieve a more intimate contact between bump and track, and so open the way to use a non-conducting adhesive instead of an anisotropic adhesive.

A typical processing sequence is as follows. The flexible substrate is

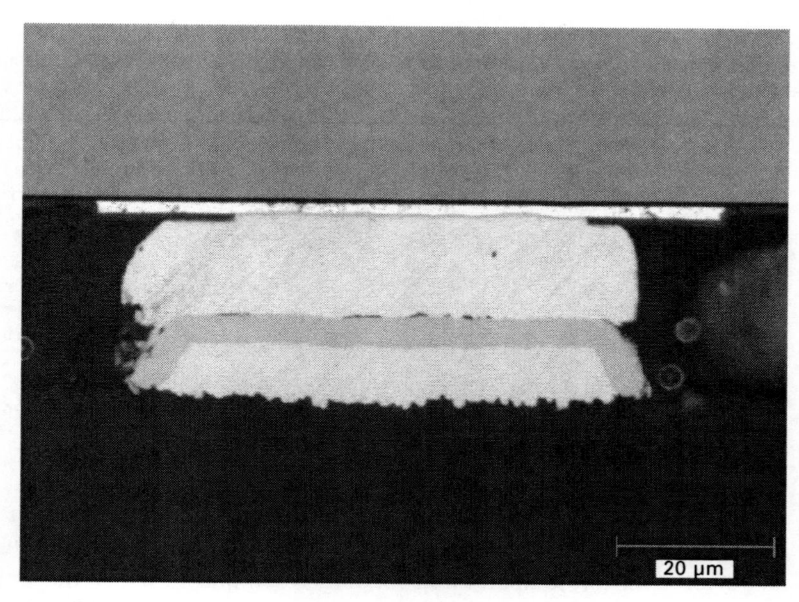

3.10 Die with galvanic Au-bumps bonded on flexible foil.

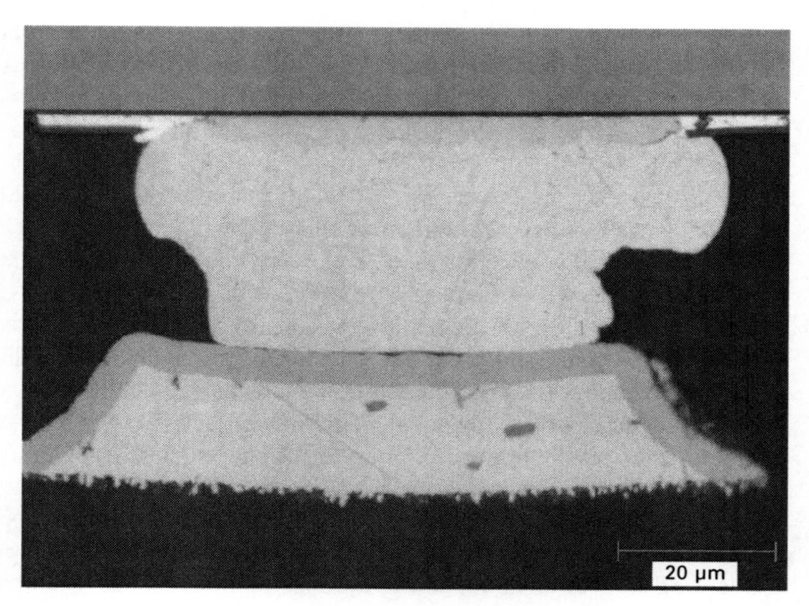

3.11 Die with Au-stud bumps bonded on flexible foil.

placed on the bonding table of a flip chip bonder and stretched over a glass plate by means of a vacuum to ensure its flatness. Then the desired amount of adhesive paste or film is applied to the substrate – in the latter case the

Table 3.3 Geometrical data of dies used in flip chip on foil assemblies. Bump dimensions are given for rectangular shapes (length, width) or octagonal/circular shapes (diameter, Φ)

Size (L×W) (mm)	Pitch (μm)	Bump type	Length (μm)	Width (μm)	Height (μm)	No. of bumps	Total bump area (mm²)
5×5	300	NiAu	$\Phi = 80$		20	56	0.281
5×5	200	NiAu	$\Phi = 80$		20	84	0.402
5×5	200	Au	$\Phi = 80$		20	84	0.402
10×10	150	NiAu	$\Phi = 100$		20	256	2.01
5×5	150	Au	75	60	15	120	0.540
5×5	100	Au	75	60	15	176	0.792
5×5	60	Au	85	35	15	256	0.762
5×5	40	Au	85	20	15	384	0.653

adhesive is already partly cured which makes it suitable for cutting to size and easier handling. The component is aligned and pressed onto the substrate. Together with the temperature, bonding force and time are the important process settings.

If the bonding force is too low, the resulting clamping force (see Section 3.2) is not high enough for a reliable interconnection. Too high a bonding force has the risk of inducing cracks in the Si-die or excessive deformation of the bond pads and the foil. In particular, when thin dies have to be mounted this is a risk, as one can see in Fig. 3.12 for a die that was thinned to 160 μm and then bonded with a 250 N/mm² bump area.

For industrial processes, which are the subject of this chapter, the bonding pressure is usually higher than in many research studies. This was already mentioned in Section 3.2. In the present investigations, bonding was usually done with pressures in the range of 100–400 MPa. To avoid any misunderstanding, the bonding pressure is calculated per contact area between bumps and bond pads, not with regard to the area of the die.

The temperature–time profile determines the curing of the adhesive. For two ACF-materials, the conversion was followed by means of a differential scanning calorimeter. Figures 3.13 and 3.14 show that different adhesives have different curing profiles, but at 180 °C or higher, the differences seem to vanish. The curing condition affects the glass transition temperature. Whereas most suppliers refer to the maximum glass transition temperature after curing at medium temperatures for several minutes, this is not practical in an industrial process where bonding has to proceed in seconds rather than minutes. In that case, the transition temperature can be 10–30 degrees lower. Realistically, in industrial processes the adhesives are seldom fully cured. As will be seen in a later section, during testing at elevated temperature and humidity, post-curing can occur.

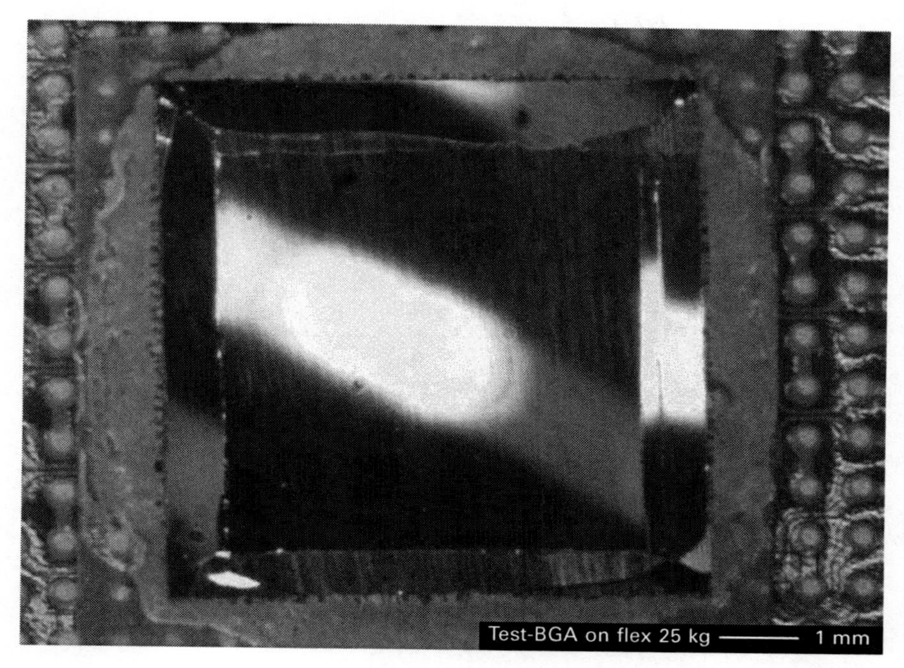

3.12 Thinned die 160 μm (pitch 60 μm), cracked during bonding with 250 N/mm² bump area.

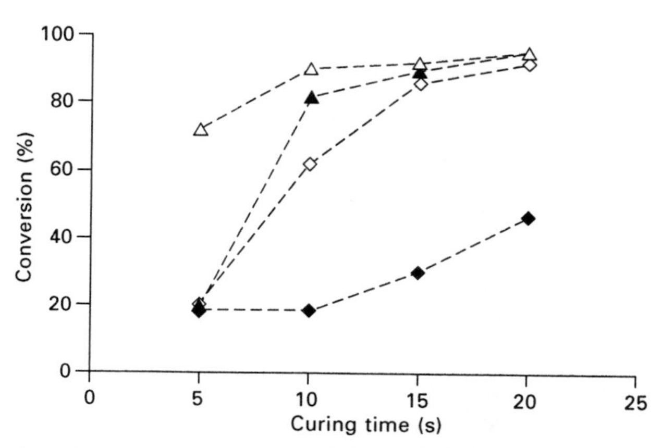

3.13 Curing conversion of ACF type S-1. Temperature: 160 °C (◆), 180 °C (◇), 200 °C (▲), 220 °C (△). The dashed lines are guides for the eye.

3.3.3 Moisture absorption

Although this is a materials issue, the absorption and desorption of water is such an important property of the adhesives and the flexible foils that it will

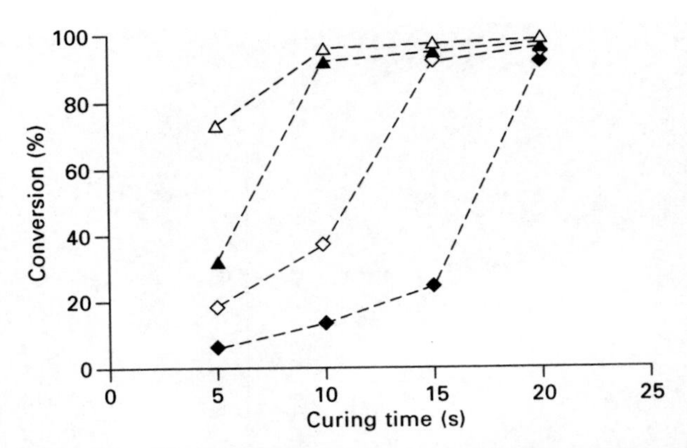

3.14 Curing conversion of ACF type H-3. Temperature: 160 °C (◆), 180 °C (◇), 200 °C (▲), 220 °C (△). The dashed lines are guides for the eye.

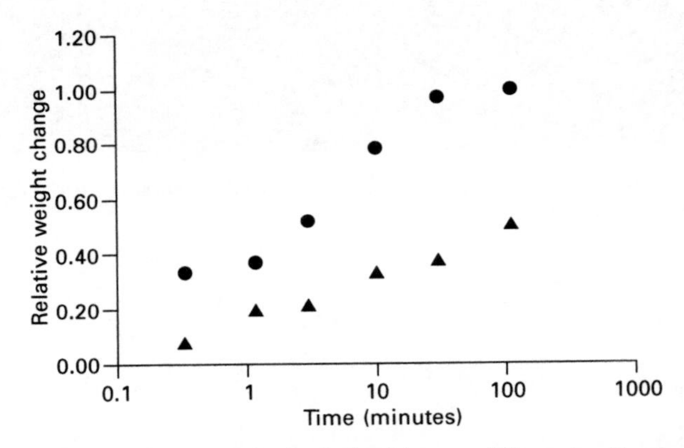

3.15 Moisture uptake in foil with two different adhesives.

be treated separately. Not only the amount of moisture that can be taken up by the materials is a critical issue, rather this must be treated in combination with the desorption and permeability of the foil to moisture (de Vries *et al.*, 2005a). An adhesively bonded chip on foil assembly can take up moisture within a relatively short period of time, as Fig. 3.15 shows – this process depends on the materials properties of the adhesive. Although the amount of moisture the two adhesives can absorb is somewhat larger for type S-1, the diffusion coefficient of type H-3 is so much higher that, in the end, this foil–adhesive combination takes up more moisture. In a reflow process step, the water vaporizes and is forced out of the assembly. Should the diffusion rate of the water through the adhesive be larger than through the foil, then water accumulates at the interface of the adhesive with the foil and the pressure can inflict damage to the interconnections. If this is the other way around, the water may pass safely from the adhesive, through the foil to the outside world. Making a reliable adhesive interconnection depends therefore, among other things, on the proper combination of materials. Desorption curves of water from saturated flexible foils from room temperature to reflow temperatures of 260 °C are compiled in Fig. 3.16. In Table 3.4 the moisture absorption and the diffusion coefficients for selected flexible foils and adhesives are listed. In all cases, the moisture diffusion coefficient of the adhesives is higher than for the flexible foils. The best combination therefore includes a foil with a high diffusion coefficient. Since adhesive-less foils are preferred for their better stability, only two flexible foil types from this list are applicable.

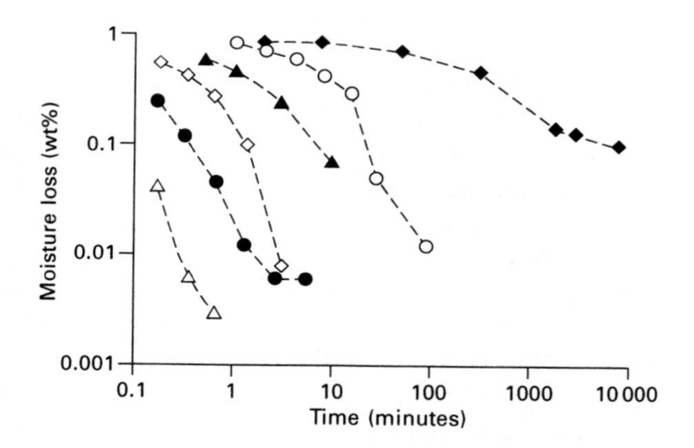

3.16 Desorption of water from flexible foil by drying at various temperatures: 21 °C (◆), 85 °C (○), 110 °C (▲), 160 °C (◇), 210 °C (●), and 260 °C (△). The dashed lines are guides for the eye.

Table 3.4 Moisture absorption (wt%) and diffusion coefficients (*D*) of flexible foils and adhesives

Material	Type	Thickness (μm)	Absorption (wt%)	*D* (10^{-13} m^2/s)
Foil	D-1	50	1.0	0.30
	D-2	50	1.8	0.08
	D-3	50	1.1	0.18
	R-4	50	0.1	0.08
	N-5 adhesive	45	2.0	0.5
	N-6 adhesive-less	50	3.3	0.44
ACF	S-1	30	2.3	0.86
	S-2	30	2.4	0.78
	H-3	40	2.0	2.5

3.4 Critical loading

In this section we will elaborate further on a number of the aforementioned stressors and their effect on the quality and reliability of an adhesive interconnection. In Section 3.2 on the nature of an adhesive bond, an overview of the literature has already been presented. Here, such information will be referenced whenever it is necessary to explain the results.

3.4.1 Bonding

Factors affecting the eventual clamping force inside adhesive interconnections are the bonding force, and the shrinkage of the adhesive during curing and during cooling down to room temperature. The higher the clamping force, the less sensitive the interconnection becomes to changes such as increases in temperature, and swelling of the adhesive because of moisture absorption. This can be seen clearly from Fig. 3.2. Figure 3.17 shows how the contact resistance develops during the bonding process. The contact resistance is monitored *in-situ* during the bonding process; in this example, bonding is done at 150 °C and at 180 °C. The lower part of Fig. 3.17 gives the temperature–force–time profile that was used during the bonding process. The top part of Fig. 3.17 shows how the electrical contact is formed, using the contact resistance as a criterion. If bonding is done at 180 °C, the electrical contact is formed during the bonding process. When the bonding pressure is released, the interconnect cools down and, as a result, the contact resistance decreases because of the T-dependence of the resistance and the shrinkage of the adhesive. If bonding is done at 150 °C, no good electrical contact is formed during the 10 second bond time. When the bonding pressure is released,

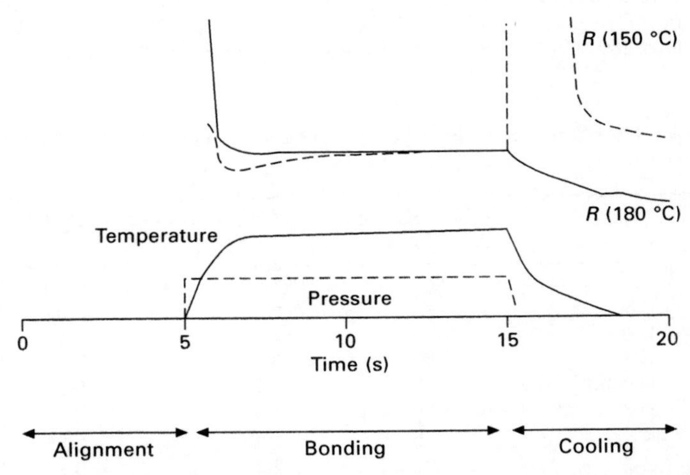

3.17 Evolution of the contact resistance during the bonding process.

the resistance increases. Only during cooling down to room temperature, does the shrinkage of the adhesive create an electrical contact. The eventual contact resistance, however, is higher than for bonding at 180 °C, and one can expect that if this assembly is used at a higher temperature, the electrical contact will be lost. This means that 10 seconds at 150 °C is too short to form a good interconnect for this type of adhesive.

3.4.2 Resistance to reflow soldering

Conductive adhesive interconnections are often used in multi chip modules. This means that afterwards, the modules are soldered to a printed circuit board. Hence, the first process step after bonding that needs attention often is (reflow) soldering of the semiconductor package. Accumulated moisture in the polymeric matrix of the flexible foil and the adhesive will evaporate vigorously and can damage the adhesive bond.

In the electronic assembly industry, it is common practice to classify packages by their robustness in the reflow-soldering process. For this, a standardized method has been defined. Once any type of package has been given such moisture classification or moisture sensitivity level (MSL), process engineers will know how to handle the packages in order not to inflict any damage to the assembly. Usually, one follows the JEDEC-standard (JEDEC, 2008) as shown in Table 3.5. In brief, through this procedure, products are preconditioned by subjecting them to one of the standard or accelerated conditions, and then exposing them three times to the appropriate reflow profile. Prior to preconditioning, and after reflowing, the devices are evaluated for delamination and also one or more relevant parameter is measured; the occurrence of delamination or changes in the measured parameters determine whether or not the moisture sensitivity level is met. If it is met, the test is repeated at the next higher (more severe) level, until eventually the packages fail the test. Then the last successful test level is assigned as the moisture sensitivity level for that particular type of package. For adhesive interconnections, the value of the contact resistance is a suitable criterion since it is sensitive to variations in the clamping force, as explained above.

It must be noted that the adhesives that are discussed in this section are not fully cured, as was already mentioned in the introductory section. Therefore, in contrast to the observations of Yin *et al.* (2003), the adhesives are not completely resistant to absorption of water and thus will take up water with due consequences.

An example for such assessment of the moisture sensitivity level is given in Fig. 3.18 for a foil design with 300 μm and 200 μm pitch dies and NiAu-bumps of 20 μm height bonded with a 150 N/mm^2 bump area, see De Vries and Janssen (2003). For this experiment, the foil labeled D-1 from Table 3.4

Table 3.5 Moisture sensitivity levels according to JEDEC (2008), see text of Section 3.4.2 for an explanation. eV: activation energy of moisture ingress; MET: includes manufacturers' exposure time, default 24 hours. In this work the accelerated equivalent conditions were applied from the '0.40–0.48 eV' column

			Soak requirement				
					Accelerated equivalent		
	Floor life		Standard		0.40–0.48 eV	0.30–0.39 eV	
Level	Time	Conditions	Time (h)	Conditions	Time (h)	Time (h)	Condition
1	Unlimited	≤30 °C/85%RH	168	85 °C/85%RH	NA	NA	NA
2	1 year	≤30 °C/85%RH	168	85 °C/60%RH	NA	NA	NA
2a	4 weeks	≤30 °C/85%RH	696	30 °C/60%RH	120	168	60 °C/60%RH
3	168 h	≤30 °C/85%RH	192 (MET)	30 °C/60%RH	40	52	60 °C/60%RH
4	72 h	≤30 °C/85%RH	96 (MET)	30 °C/60%RH	20	24	60 °C/60%RH
5	48 h	≤30 °C/85%RH	72 (MET)	30 °C/60%RH	15	20	60 °C/60%RH
5a	24 h	≤30 °C/85%RH	48 (MET)	30 °C/60%RH	10	13	60 °C/60%RH
6	Time on label	≤30 °C/85%RH	Time on label	30 °C/60%RH	NA		

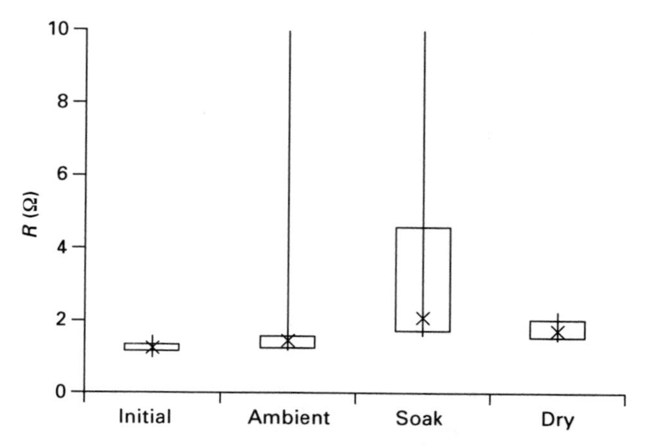

3.18 Daisy chain resistance at indicated conditions (see text). Bonding pressure was 150 N/mm^2 bump area. ACF on foil D-1 (Table 3.4), 300 μm and 200 μm pitch. NiAu-bumps of 20 μm height. Median (×), 2nd and 3rd quartiles (boxes), minimum and maximum values (lines).

was used. Sixty daisy-chained samples were taken and split in three series of twenty. These samples were characterized initially, divided in three groups, and then each set was treated differently: keeping under ambient conditions (20 °C/50%RH/24 hours); soaked (85 °C/85%RH/24 hours); and dried (85 °C/24 hours). Subsequently, the assemblies were fed through a reflow furnace set at a peak temperature of 230 °C. This does not exactly correspond to the JEDEC-levels, but one can clearly see the effect of the moisture content on the resistance. Temperature had a minor effect on the resistance value which had increased also after drying. Storing at ambient also caused one daisy chain to fail, which could have been an early failure. But most impressive is the result after humidity storage. All resistance values increased, and there were three failures.

The second example shows the additional influence of bonding pressure. The same type of foil and adhesive was used as described above. This time the samples did receive the standard preconditioning at MSL2 (see Table 3.5, accelerated equivalent condition). From Fig. 3.19 one sees that increasing the bonding pressure from 50 g/bump to 100 g/bump reduced the initial resistance of the daisy chains. (The dies with 300 μm pitch had 56 bumps, the dies with 200 μm pitch counted 84 bumps.) After preconditioning and reflow, in all cases the resistance increased. But the samples with the higher bonding pressure had still the lowest resistance value. This is in line with Section 3.4.1.

The third example serves to illustrate the aforementioned damage model of moisture being expelled and thus inflicting damage to the interconnection.

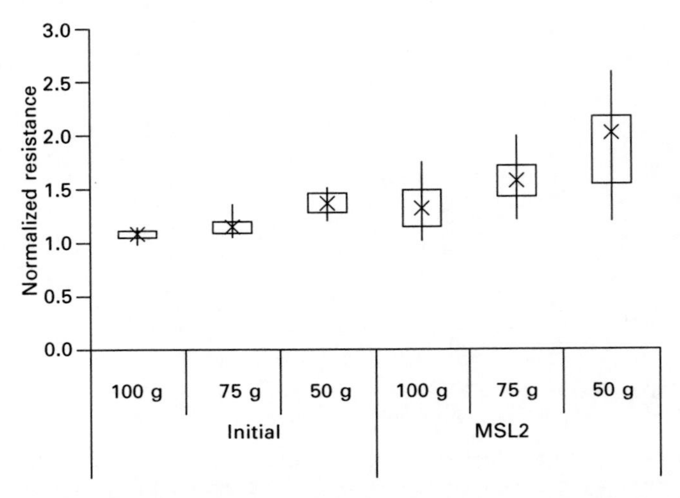

3.19 Normalized resistance at indicated bonding forces (in g/bump) before and after MSL2-test. ACF on foil D-1 (Table 3.4), 300 μm and 200 μm pitch. NiAu-bumps of 10 μm height.

For the purpose of enhancing the flatness of the flexible foil during the bonding process, a thin metallic stiffener was applied to the backside of the foil; see the cross-section in Fig. 3.20. Such assemblies were subjected to the moisture sensitivity test as explained previously – preconditioning (MSL2, see Table 3.5) and three times through a reflow oven. The initial state is comparable to the cross section of Figures 3.7 and 3.10. The adhesive was of type H-3 and the foil was of type N-6 (see Table 3.4). The layout of Fig. 3.4 was used with a pitch of 100 μm. Without the stiffener, the moisture can evaporate through the adhesive and the foil, the resistance after reflow was still good – in Fig. 3.21 a cross-section after the MSL-test is shown. In contrast, with such a stiffener, the accumulated water could not escape fast enough: only the route parallel to the surface of the foil is available, which is two orders of magnitude longer than its thickness. The result was a lift-off of the die with the bump from the bond pad, as one can see in the cross-section shown in Fig. 3.22. From this it is clear that this method of applying a non-permeable stiffener is not feasible, but the results of this experiment support the failure mechanism proposed in Section 3.3.3.

3.4.3 Accelerated testing

Apart from testing the resistance against processing, assessment of the long-term behavior of adhesive interconnections under external stress conditions is relevant. Here, one is led by the nature of an adhesive interconnection to select the most suitable environmental tests. This means that thermal cycling

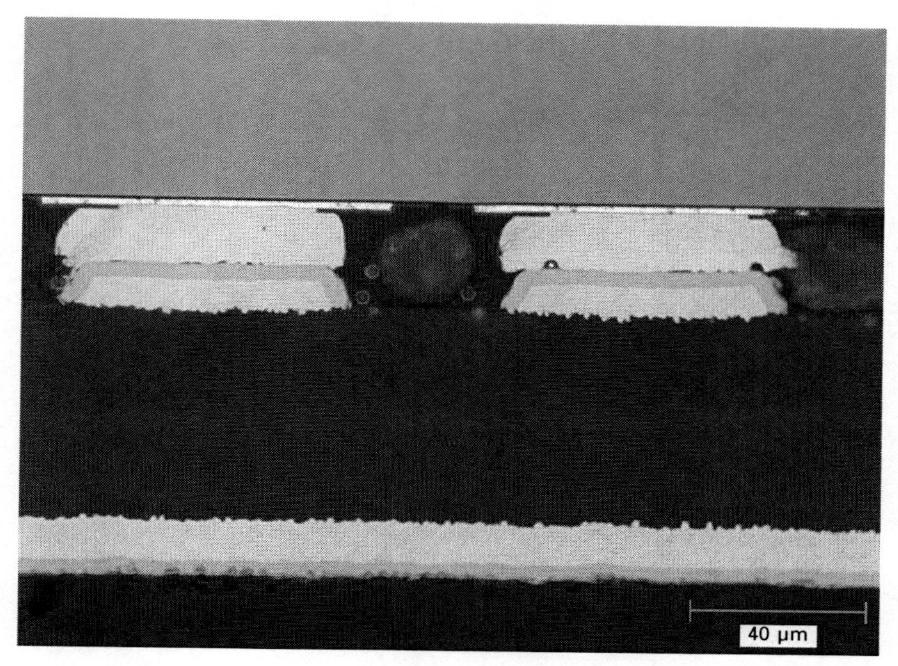

3.20 Assembly before MSL-test with thin metal stiffener opposite to die.

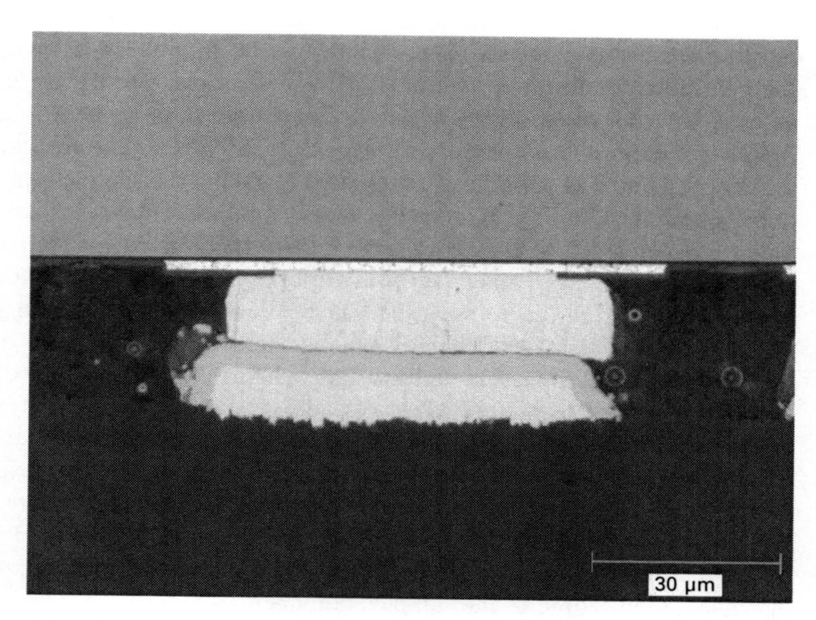

3.21 ACF-interconnection good after MSL2-test; 100 μm pitch.

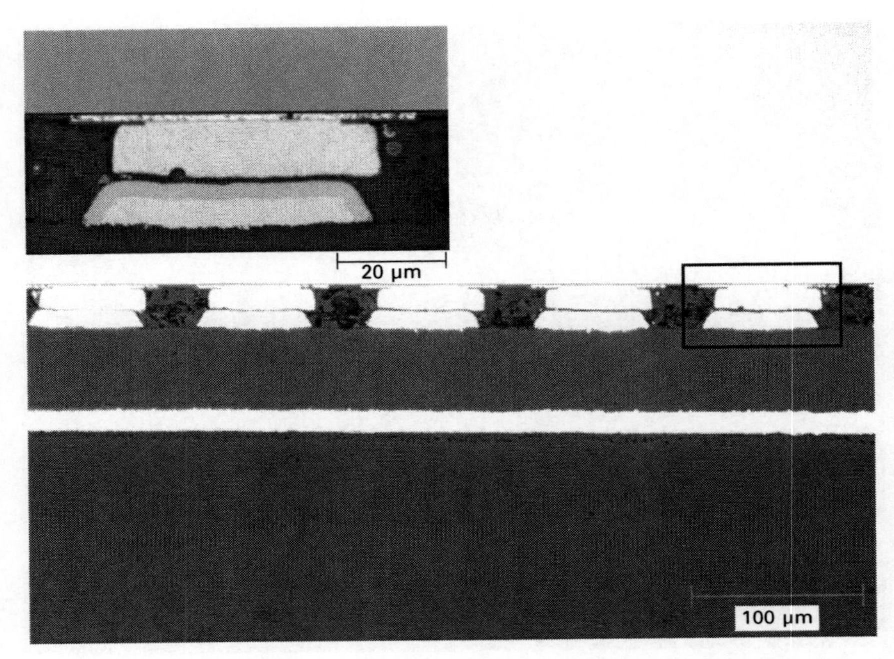

3.22 ACF-interconnection failed after MSL2-test; 100 μm pitch with stiffener. Magnified overview is of portion within the box.

and exposure to a humid environment, either constant or periodic, are the most severe test conditions. In particular, it is of importance to investigate the possibility to design accelerated tests. To this end, similar assemblies as were used for the moisture sensitivity experiments described in Section 3.4.2 were exposed to a damp heat test (85 °C/85%RH). The resistance of the daisy chains was periodically measured by taking the samples out of the climate chamber. In Fig. 3.23 such a series of measurements is shown (de Vries and Janssen, 2003). Each symbol in the graph represents the moment when the samples were taken out of the test for inspection. The differences in the initial values of the resistance have to do with the different length of the daisy chains. The 200–300 μm pitch assemblies consisted of four dies, two for each pitch, and had ten daisy chains on one side of a die and two that extended over three sides of a die. A failure was – arbitrarily – defined as a factor of 1.5 increase of the resistance compared to the initial value. The prolonged exposure to damp heat caused the bond to detach. A gap between the adhesive and the foil of several micrometers occurred and the NiAu-bump was lifted from the bond pad (Fig. 3.24). In Fig. 3.25 the failure distributions are given, which make clear that as the bonding pressure increased, the lifetime of the samples is longer.

To further accelerate the degradation process it is common practice to

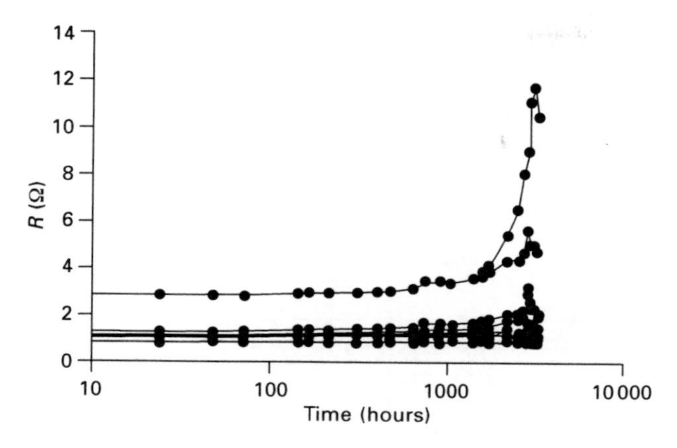

3.23 Daisy chain resistance versus time 85 °C/85%RH; 200–300 μm pitch, 100 g/bump.

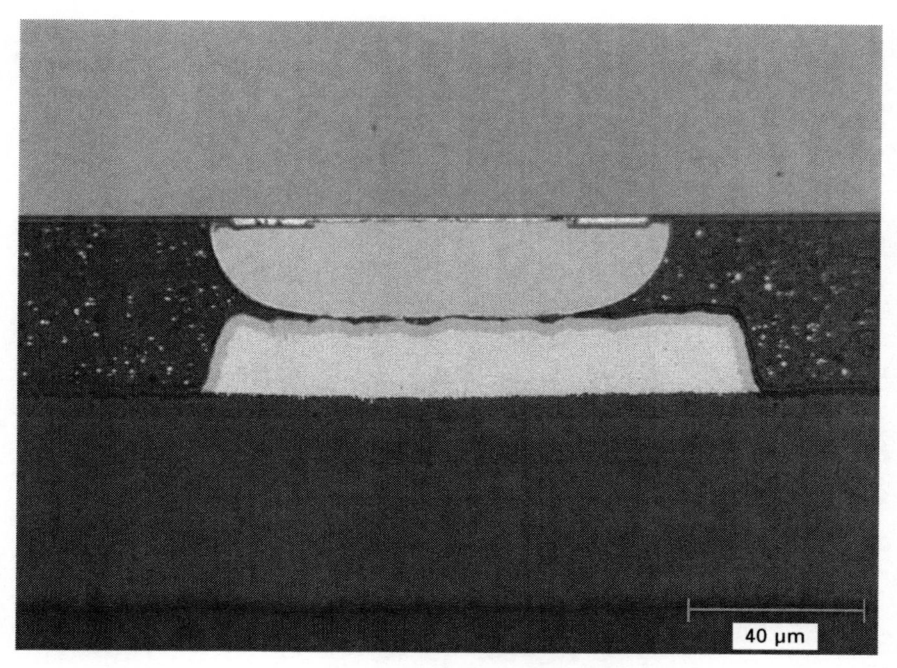

3.24 Failure from damp heat test (85 °C/85%RH). The gap between the adhesive and the foil is visible as a dark band.

increase the stress level, provided that the failure mechanism remains the same. For the damp heat test, this can be achieved only by increasing the temperature. So called 'highly accelerated stress testing' (HAST) is a suitable tool for this. Such equipment allows temperatures of up to 140 °C while the

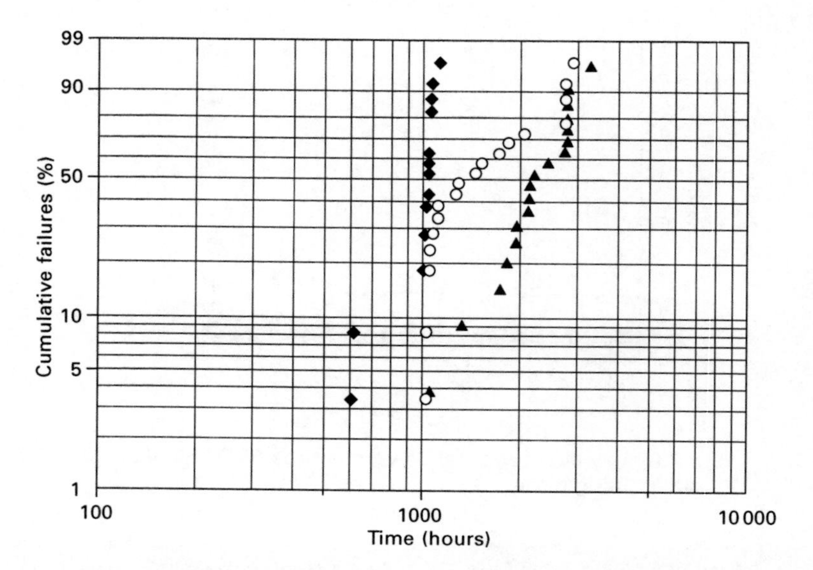

3.25 Weibull failure distributions of 300 μm and 200 μm daisy chains exposed to 85 °C/85%RH. Bonding forces are: (▲) 100 g/mm², (○) 75 g/mm², (◆) 50 g/mm².

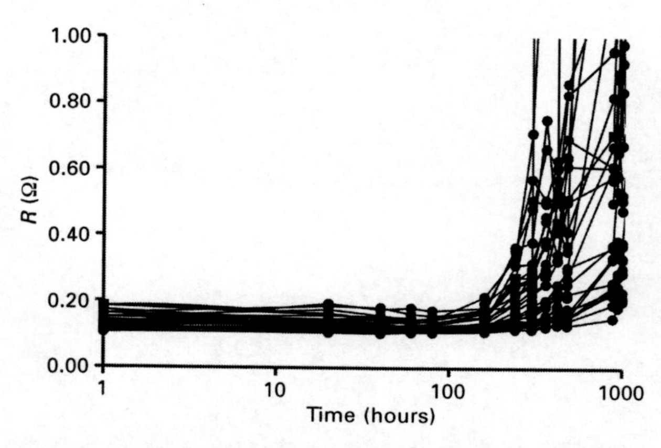

3.26 Single bump resistance versus time 110 °C/85%RH; 100 μm pitch, 250 N/mm².

humidity can be controlled to 96%RH. With respect to the glass transition temperature of the adhesives in the range of 125 °C and higher, the temperature in these highly accelerated test was limited to 110 °C. The resistance of single contacts in a 100 μm-pitch assembly, subjected to 110 °C/85%RH after MSL2 preconditioning and reflow, was recorded periodically and the result is shown in Fig. 3.26 (de Vries *et al.*, 2005a). One can see the same generic behavior as for the larger pitch assembly tested at 85 °C/85%RH shown in

Fig. 3.23. Also in this test, the failure mode is lifting of the contact bump from the bond pad, as one can see in Fig. 3.27. Thus both tests, the standard or accelerated damp heat test of 85 °C/85%RH and the highly accelerated version at 110 °C/85%RH, can be applied to investigating the reliability of conductive adhesive interconnections.

Another illustration to this effect is Fig. 3.28, where the failure distributions of the two damp heat tests on 100 μm assemblies is given for two adhesives. At 85 °C/85%RH, the two adhesives behaved quite similarly, while at 110 °C/85%RH there was some difference. But in both adhesives, the acceleration of the test is clear. The statistical result is: for H-3 a Weibull slope of β = 1 in 110 °C/85%RH and β = 3.8 in 85 °C/85%RH; for S-1 this is β = 0.8 and β = 2.1 respectively.

For reasons of completeness, also temperature cycling results are presented. The same type of assemblies as those of Fig. 3.26 was subjected to a thermal shock test (–55 °C/+125 °C/20′–20′-cycle), after MSL2 and reflow. Two things are worth noting. First, the increase of resistance was far less pronounced than in the humidity test, even though the maximal temperature was higher. Second, one notices the irregular behavior of some contacts which also was found in the humidity tests. There is strong reason to ascribe this to taking

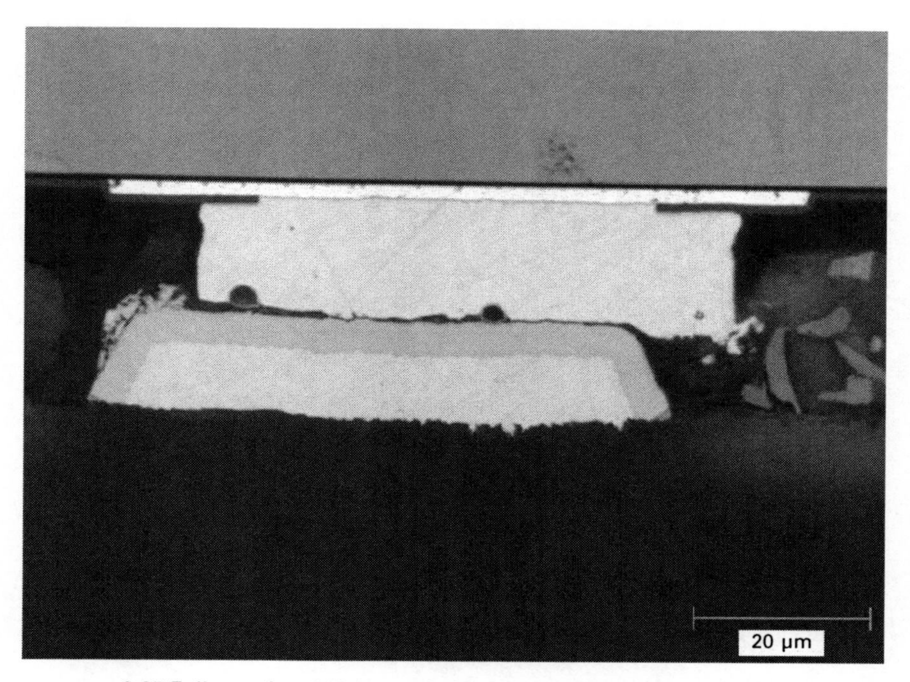

3.27 Failure after 560 hours in highly accelerated stress test (110 °C/85%RH).

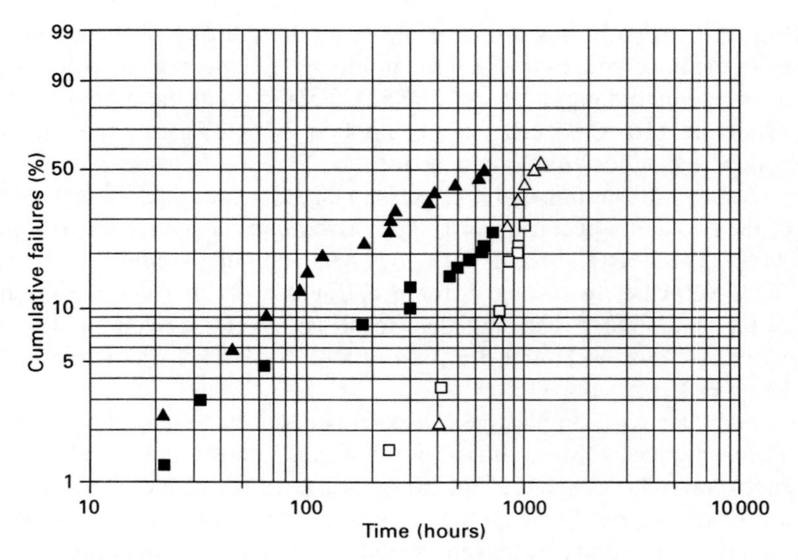

3.28 Weibull failure distributions of 100 μm pitch single-bump contacts after MSL3, 0.16 mm die. Adhesive S FP 110 °C/85%RH (■), 85 °C/85%RH (□). Adhesive H AC 110 °C/85%RH (▲), 85°/85%RH (△).

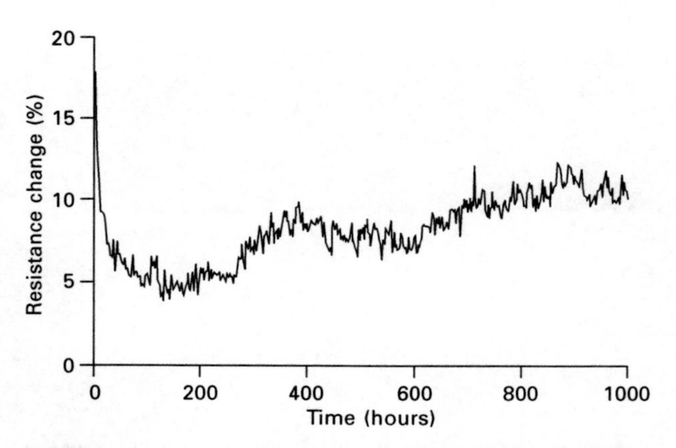

3.29 On-line monitoring of the contact resistance drift in accelerated damp-heat test (85 °C/85%RH).

the samples out of the test chamber for off-line electrical measurements. This will be addressed in Section 3.5 on evaluation methods.

3.4.4 On-line versus off-line monitoring

Figure 3.29 illustrates the power of on-line monitoring of the adhesive interconnects: in this example, showing the resistance drift over time, we

can clearly distinguish three different phases. First, we get a steep increase in contact resistance as a result of temperature increase and of moisture take-up by the adhesive, when going from room temperature conditions to 85 °C/85%RH conditions. In the next phase, the contact resistance shows a decreasing trend, most likely due to a post-cure effect. Finally, the contact resistance starts to increase due to aging at high temperature and humidity conditions. Off-line monitoring would have shown only that the contact resistance did slightly change as a result of the accelerated test.

In Fig. 3.30 we show typical differences as observed between off-line and on-line resistance monitoring. The on-line resistance data show a decreasing trend, suggesting a post-cure. The off-line measurements clearly show an increasing contact resistance. Possible explanations for this difference are damage from handling during off-line monitoring or from the sudden change in temperature and moisture every time when taking the samples out of and putting them back into the test chamber. A similar observation was made during another series of experiments: the samples were continuously monitored on-line, but the test chamber needed to be opened shortly periodically to refill the water reservoir. Figure 3.31 shows the decrease in the measured contact resistance every time the test chamber was opened, which resulted in a decrease in temperature of the samples under test. In some cases, however, a sudden increase in the contact resistance could be observed after opening and closing the chamber, as is shown in Fig. 3.31. Sometimes, this then was the start of a visible further higher degradation rate.

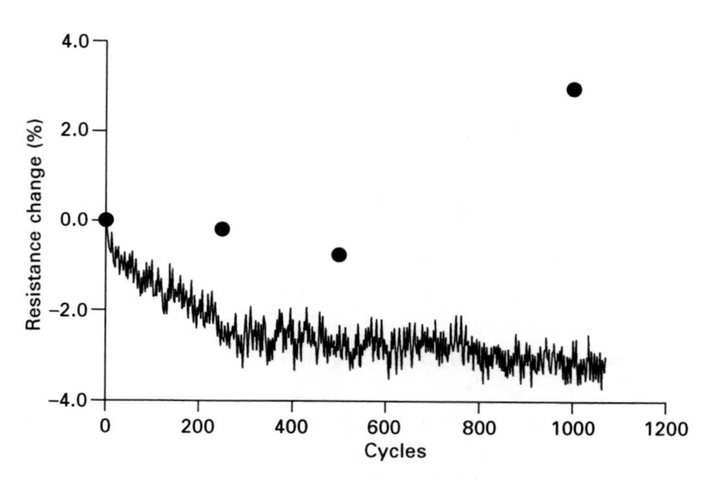

3.30 Resistance drift during thermal shock test −40 °C/+100 °C. On-line (solid line); off-line (●).

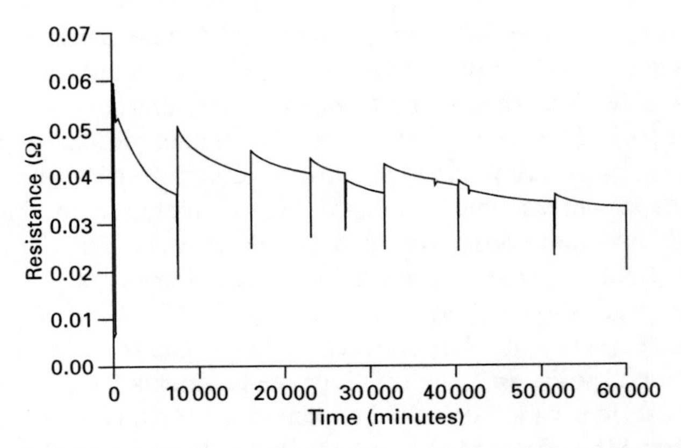

3.31 Effect of periodic opening of the accelerated humidity chamber, resulting in a sudden and permanent increase of the contact resistance.

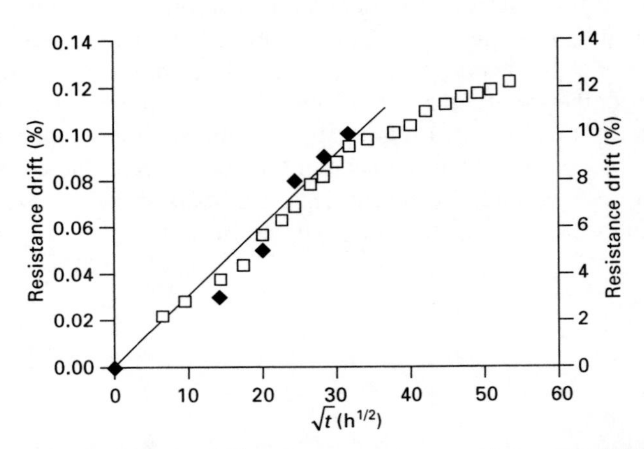

3.32 Resistance drift versus square root of time; 85 °C/85%RH; steady state (left axis, ◆), cyclic (right axis, □).

3.5 Evaluation methods

The various results that were presented in the previous section on critical loading have made clear that adhesive interconnections are, in particular, sensitive to cyclic exposure to moisture and temperature.

The ACA materials that are currently being used are quite stable under steady state humidity conditions. An example for an FCOF interconnect is shown in Fig. 3.32; a test structure as shown in Fig. 3.3 was used to monitor the contact resistance drift on-line. For this case, the average change in contact resistance over 1000 hours was only ≈ 1000 ppm. Therefore, this test method is not suited for optimizing the bonding process or selection of new types of adhesive materials as it is not discriminating and takes too long. As discussed earlier, the FCOB interconnects are more sensitive to cyclic temperature and humidity loading. Figure 3.33 shows a typical example of the contact resistance drift for a FCOF with ACA, brought to 85 °C and changing the relative humidity level cyclically from 30% to 85%, keeping the temperature constant. Both steady state accelerated humidity test and cyclic humidity test (85 °C and relative humidity ranging between 30% and 85%) follow approximately square root time dependence (see Fig. 3.32). Comparing the graphs for steady state and cyclic humidity one learns that the degradation rate for the cyclic humidity test is roughly two orders of magnitude higher than that for the steady state situation.

The relatively high response of the contact resistance of adhesive interconnects to a changing environment, offers the opportunity to use the response of the contact resistance as a possible criterion for the robustness of the interconnect. The resistance–pressure relation in Fig. 3.2 suggests that the initial resistance is a good indication of the quality and reliability of the bump-to-pad contact. However, the initial resistance may be obscured by process variables such as misalignment, etc., when a satisfactory high-pressure contact may display high resistance. This can be avoided if the change of resistance (response) corresponding to a small change in the contact pressure is used as a criterion instead. For a minor change in the contact pressure, there will be only a small response from a good quality contact pair but a large response from a poor contact pair. In reality, the small change in the contact pressure can be conveniently introduced through humidity cycling, making use of the hygroscopic swelling characteristic of the adhesive. The response can be determined from a simple test, lasting less than a day. In our experiments, the response was calculated from the cyclic humidity test, as the difference between the average contact resistance at the start of the

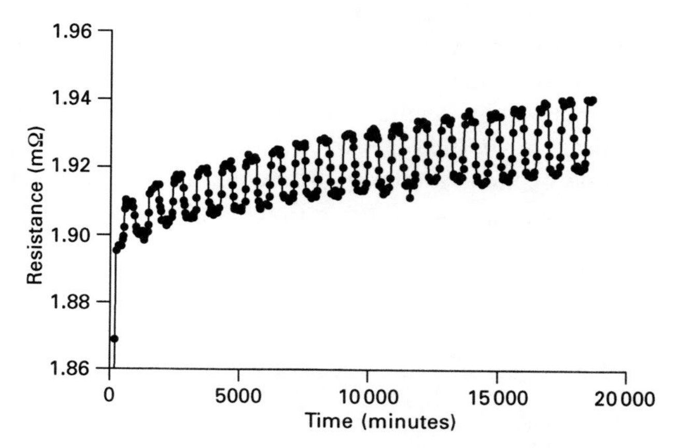

3.33 Resistance drift during cyclic humidity test (85 °C/30–85%RH).

experiment at 85 °C/85%RH and the average value when relative humidity is decreased to 30%RH during the first humidity cycle. Compared to the initial resistance criterion, the response criterion captures the effect of hygroscopic swelling, which is a reflection of moisture expansion coefficient and moisture saturation concentration of the adhesive, the effect of the bonding process as well as the coplanarity characteristics of the bump and pad. Together with the response, the hysteresis, being the difference in contact resistance level between the first and second cycle, provides a fingerprint of the adhesive interconnect. The potential of this method is illustrated in Table 3.6, showing hysteresis and response for FCOB interconnects to Au, Cu, and Al.

3.6 Case studies

3.6.1 Heat seal connector

A heat seal connector is an example of the use of an interconnection with a thermoplastic anisotropic conductive adhesive. The connector consists of a flex foil with a C-pattern screen printed on it. On top of this, the adhesive layer with conductive particles is applied. The interconnect is made by pressing the flex to the substrate using a hot bar (thermode). It is a low-cost option to connect, for example, an LCD with the printed circuit board containing the driver electronics. The aging kinetics of this type of adhesive interconnect were determined under static temperature loading. Heat seal connectors to a printed circuit board were aged at 60 °C, 70 °C, and 80 °C. To be able to monitor the contact resistance on-line, the triple track test structure as shown in Fig. 3.6 was used. Figure 3.34 shows the actual samples. The resistance of the inner row of contacts is monitored. The outer contacts are used to be able to electrically contact the flex during the accelerated test. The test structure allows one to monitor continuously the contact resistance drift of all the interconnects of the heat seal connector during the high temperature storage. Figure 3.35 shows the result for some interconnects at 80 °C. It shows that the resistance of the interconnect increases in the range of a few ohms to 50 ohms over 500 hrs testing at 80 °C. The interconnects with the highest drift in resistance are the outer ones (Track 1 and 2). The rounding at the thermode edge explains the large difference with the inner contacts in this example. Applying an arbitrary or functional failure criterion, e.g.

Table 3.6 Reponse and hysteresis of contact resistance to cyclic damp heat conditions

Metal system	Adhesive	Response	Hysteresis	Conditions
Au to Au	NCP	2–100 μΩ	< 3 μΩ	85 °C/30–85%RH
Au to Cu	ACP with Ni	≈ 17 mΩ	≈ 10 mΩ	60 °C/25–85%RH
Au to Al	ACP with Ni	≈ 1.2 Ω	≈ 2 Ω	60 °C/25–85%RH

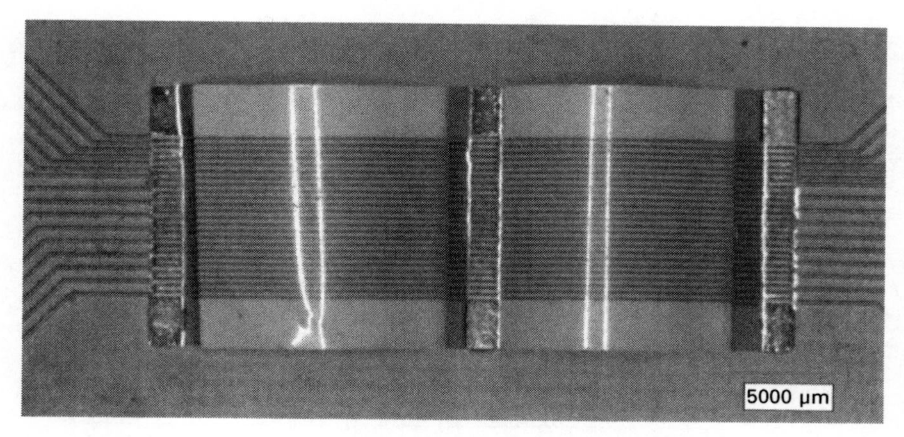

3.34 Top view of heat-seal connector in triple-bond test structure. See also Fig. 3.6.

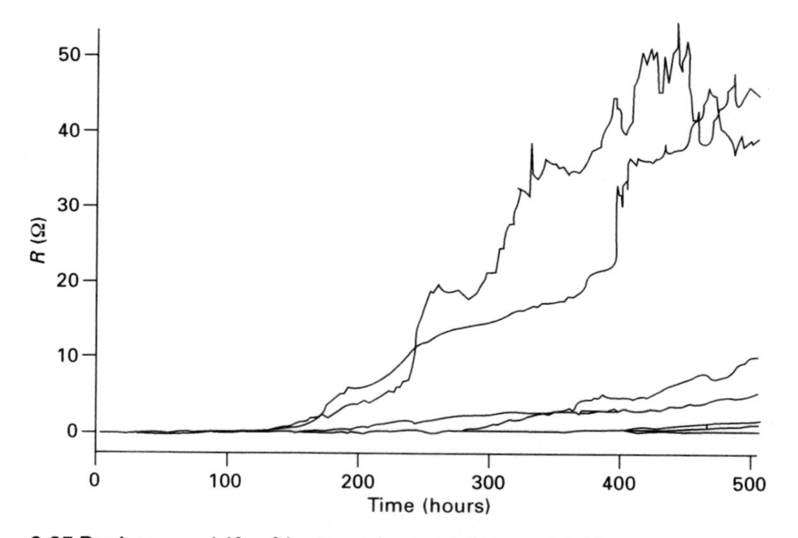

3.35 Resistance drift of heat-seal connectors at 80 °C.

of 10% resistance increase, allows one to determine times to failure for all the individual interconnections in the test and to make cumulative failure distributions for the three temperature levels. The cumulative failure distributions for the tests at 60 °C, 70 °C, and 80 °C, assuming a lognormal distribution, are shown in Fig. 3.36. If an Arrhenius T-dependence of the contact degradation is assumed, the activation energy for aging at static temperature loading can be estimated. For the type of HSC connectors used in this study, the estimated activation energy for dry aging, $Ea \approx 1$ eV. From the test results and from this activation energy, the acceleration transform and

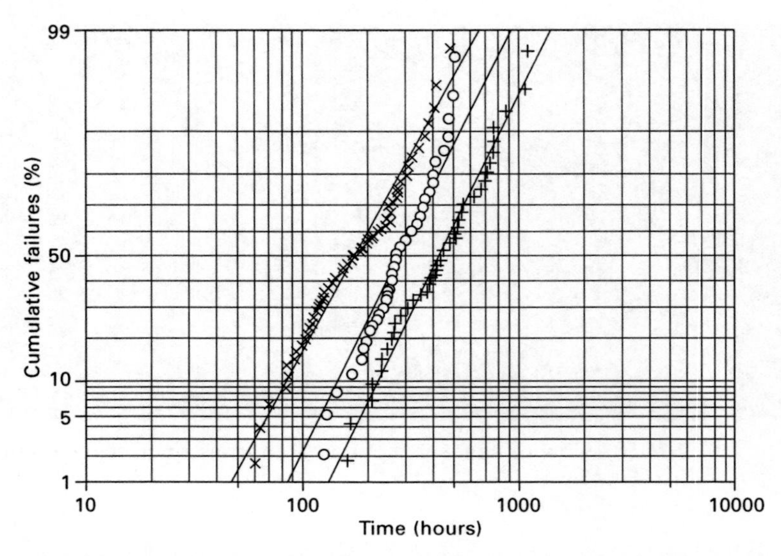

3.36 Log-normal failure distributions for heat-seal connectors in dry aging: 60 °C (+), 70 °C (○), 80 °C (×).

the expected life time (*TTF*) at any field temperature level can be calculated for this type of interconnect:

$$\frac{TTF_{T1}}{TTF_{T2}} = \exp\left[11628\left(\frac{1}{T_2} - \frac{1}{T_1}\right)\right] \qquad [3.4]$$

3.6.2 Ultra-thin fine pitch ball grid array package

In this section, the application of conductive adhesives as first level interconnects in actual flip-chip packages will be treated. This concerns an ultra-thin ball grid array package of about 300 μm thickness; Fig. 3.37 shows a schematic of such a device (in Fig. 3.5 the footprint is shown). Because of this so-called 'die down' layout, the thickness of the die must be in the range of 160 μm. This poses a limitation to the bonding pressure in order not to damage the thin die. A further requirement concerns the subsequent processing: the second level solder balls are made by stencil printing of solder paste followed by a reflow step. Finally, these packages are meant for soldering on a rigid or flexible substrate. Thus, the flip chip on flex construction has to withstand at least two reflow process steps.

For assessment of the moisture sensitivity level, the same procedure as described in Section 3.4.2 applies. This part of the investigation was done by mounting dies with a pitch of 100 μm to foil substrates. Two different anisotropic adhesives were used (S-1 and H-3, see Table 3.4). The die of

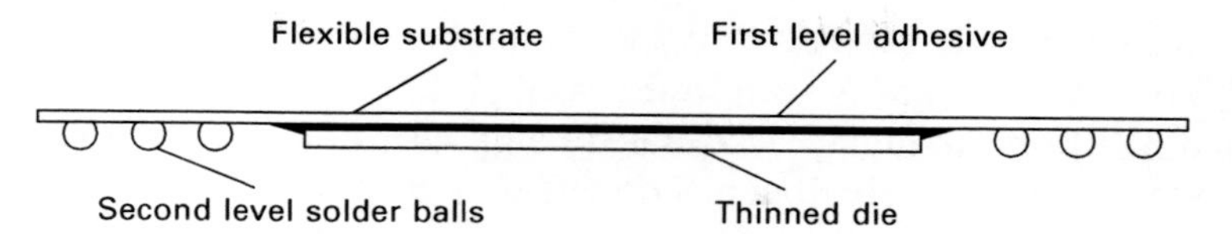

3.37 Ultra-thin ball grid array package with conductive adhesive as first level interconnect, and solder balls for board level interconnection.

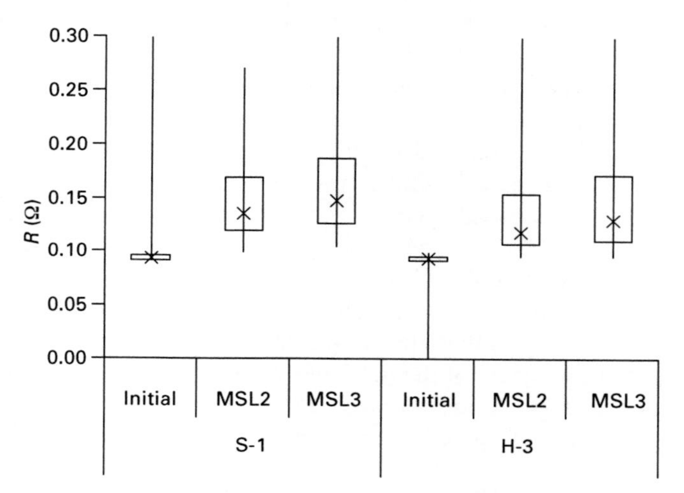

3.38 Single-bump resistance at indicated conditions (see text). Bonding pressure was 200 N/mm^2 bump area. S-1 and H-3 ACF on foil N-6 (see Table 3.2), 100 μm pitch, 0.16 mm die. Au-bumps of 15 μm height. Median (×), 2nd and 3rd quartiles (boxes), minimum and maximum values (lines).

160 μm thickness – as was mentioned in Section 3.3.2 (Fig. 3.12) – is more vulnerable than the standard type of 600 μm and therefore the bonding pressure was kept at a moderate level of 200 MPa. Further, preconditioning at one level lower (MSL3, see Table 3.5, accelerated equivalent condition) should be done as well. After the moisture–reflow test, the endurance performance was assessed by subjecting the assemblies to damp heat and thermal shock tests. The result of the moisture sensitivity experiment is presented in Fig. 3.38. Apart from a few outliers, between the two adhesives no significant differences occur. Of more importance is the fact that the two preconditioning levels do not lead to any significant difference in the distribution of the contact resistance. The yield of the contact, defined as the number of good contacts after the test relative to the initial number of contacts, was 100% at MSL2 and MSL3. Subjecting the same assemblies – after receiving the MSL3 reflow treatment – to stress tests, showed no difference between the two adhesives

in the damp heat test (85 °C/85%RH) and thermal shock test (–55 °C/+125 °C/20′–20′ cycle) as is seen in the Weibull failure distributions (Figs 3.39 and 3.40). In the highly accelerated damp heat test (110 °C/85%RH, Fig. 3.41), the Weibull distributions do differ in that the adhesive S-1 performs slightly better, but both distributions have become broader than in the other damp heat test. (Note that the data of Figs 3.39 and 3.41 are the same as those in Fig. 3.28.) This indicates, in principle, the possibility of another failure mechanism, but earlier experience as described in Section 3.4.3 has shown that the failure modes for the two types of humidity stress test are the same.

The next step brings us to assemblies with still smaller pitches of 60 μm and 40 μm (see de Vries *et al.*, 2005b). In this case also a non-conductive adhesive was applied. Cross-sections of typical examples of these assemblies are shown in Figs 3.42 and 3.43. One can see the very intimate contact between the bumps and the bond-pads. MSL3 was chosen for preconditioning. The result is given in Fig. 3.44. Of the 60 μm-pitch assemblies, 96 contacts were tested and 128 in the 40 μm-pitch version. A difference between anisotropic and non-conducting adhesives was observed: the samples made with the latter performed significantly better in the reflow test. Tentatively, this can be attributed to post curing of the adhesive, as has been reported elsewhere for assemblies on FR4 (Seppälä *et al.*, 2001) and on flexible foils (Yin *et al.*, 2003). The same difference between the two types of adhesive occured

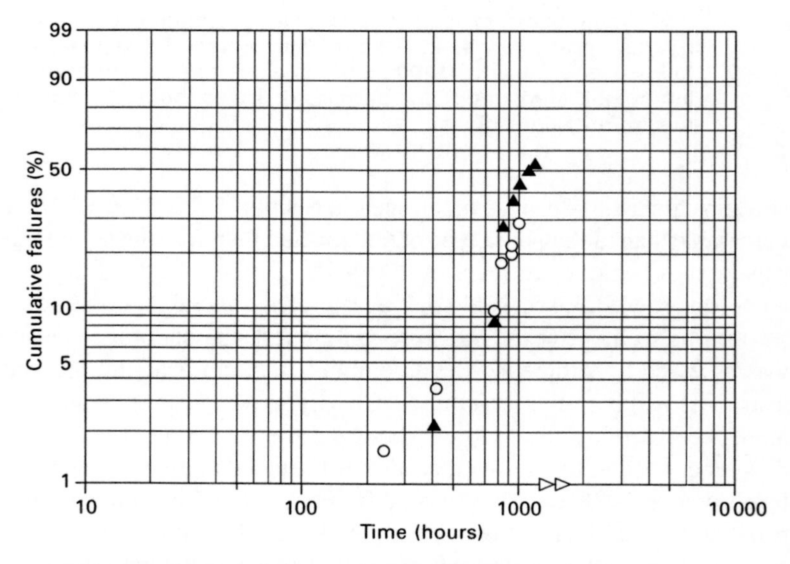

3.39 Weibull failure distributions of 100 μm pitch single-bump contacts in 85 °C/85%RH after MSL3, 0.16 mm die. Adhesive S-1 (○), H-3 (▲), suspended data (▷).

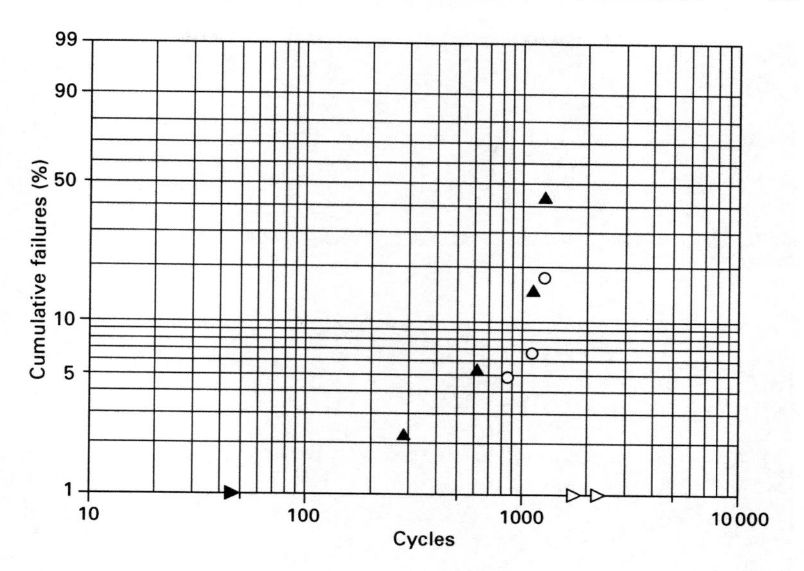

3.40 Weibull failure distributions of 100 μm pitch single-bump contacts in thermal shock test –55 °C/+125 °C/40′-cycle after MSL3, 0.16 mm die. Adhesive S-1 (○), H-3 (▲), suspended data (▷, ▶).

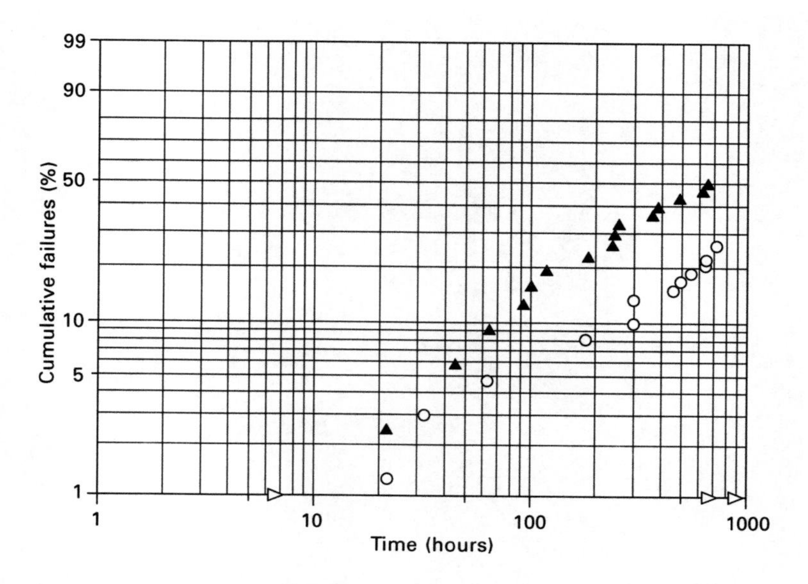

3.41 Weibull failure distributions of 100 μm pitch single-bump contacts in 110 °C/85%RH after MSL3, 0.16 mm die. Adhesive S-1 (○), H-3 (▲), suspended data (▷).

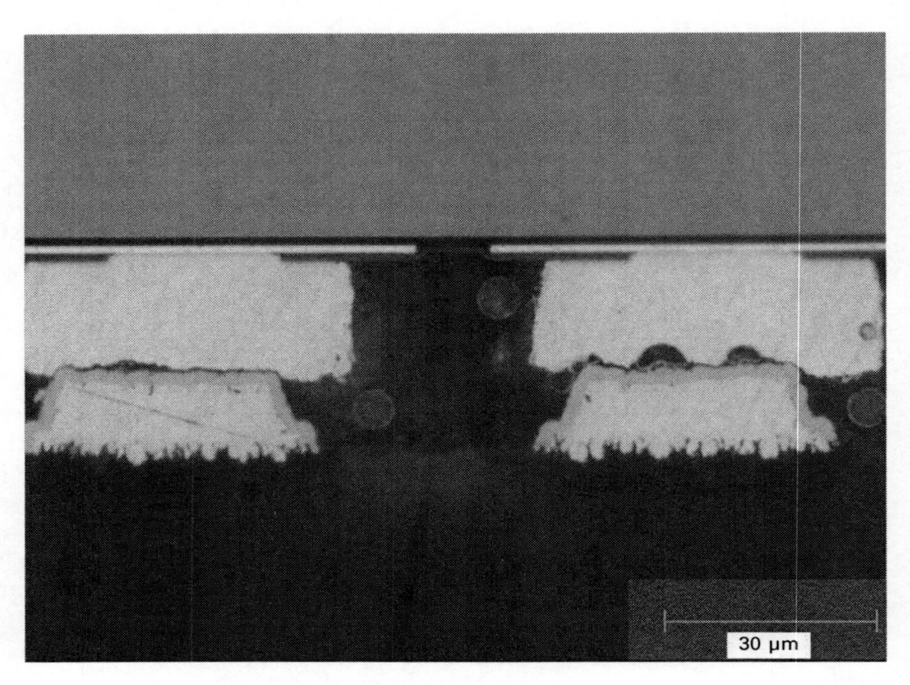

3.42 ACF-based assembly, 60 μm pitch.

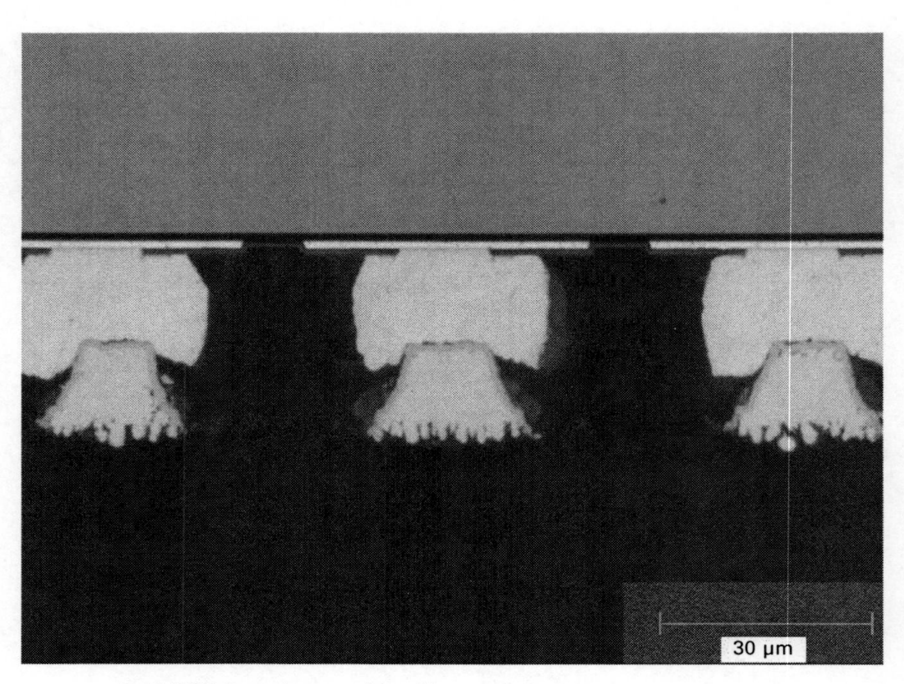

3.43 NCF-based assembly, 40 μm pitch.

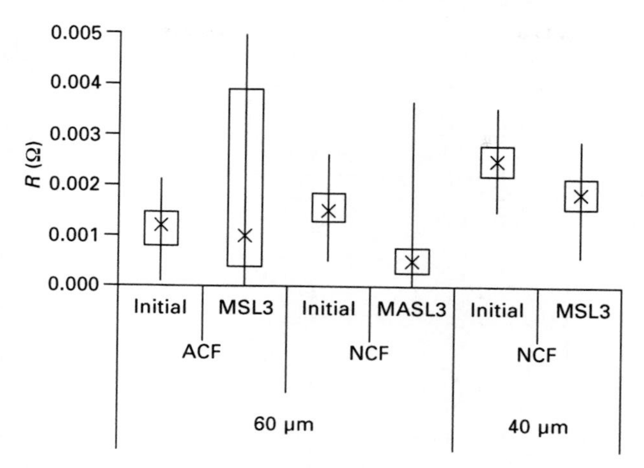

3.44 Single-bump resistance at indicated conditions (see text). Bonding pressure was 200 N/mm^2 bump area. Au-bumps of 15 μm height. ACF S-1 and NCF on foil N-6. Pitch 60 μm and 40 μm, 0.16 mm die. Median (×), 2nd and 3rd quartiles (boxes), minimum and maximum values (lines).

in the stress tests carried out after the reflow test. In the damp heat test at 110 °C/85%RH, the 40 μm-pitch assemblies made with non-conductive adhesive were even better than the 100 μm-pitch version based on conductive adhesive, as shown in Fig. 3.45. Similar observation can be made for the test at the milder condition of 85 °C/85%RH, where the results in Fig. 3.46 show that the ACF-samples with a pitch of 100 μm failed earlier than the 60 μm samples, but in the non-conductive adhesive assemblies of 60 μm-pitch only one failure was detected after 2400 hours in test. That very first result is represented in Fig. 3.46 by the probability line through this one data point assuming the same Weibull-slope ($\beta = 2.4$) as for the 100 μm version (dashed line in Fig. 3.46).

The longevity of fine-pitch anisotropic adhesive interconnections has thus been well established. In 80 μm pitch assemblies, Palm *et al.* (2001a) reported no failures for two types of adhesive after 2000 hours accelerated damp heat testing (85 °C/85%RH), while in one other type, three failures were detected. However, in the thermal shock test (–40 °C/+125 °C), more failures occurred in two of these adhesives while one passed the test that was terminated after 1880 cycles. Assemblies of similar pitch were made by Kim *et al.* (2004) and the resistance increased almost linearly with time in 1000 hours damp heat testing (80 °C/85%RH). In the accompanying thermal shock test (–15 °C/+100 °C), no increase of the resistance was found after 1000 cycles. The mechanical robustness of adhesively-bonded samples with a pitch of 80 μm was studied (Lu and Chen, 2008), together with the performance in the damp heat test (85 °C/85%RH). Under these circumstances the contact

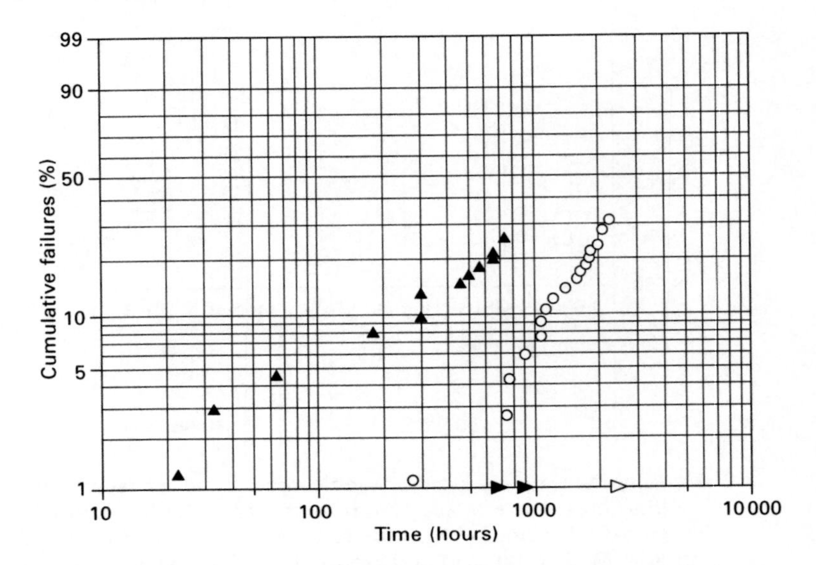

3.45 Weibull failure distributions of pitch single-bump contacts in 110 °C/85%RH after MSL3, 0.16 mm die. ACF S-1 100 μm pitch (▲), NCF 40 μm pitch (○), suspended data (▶, ▷).

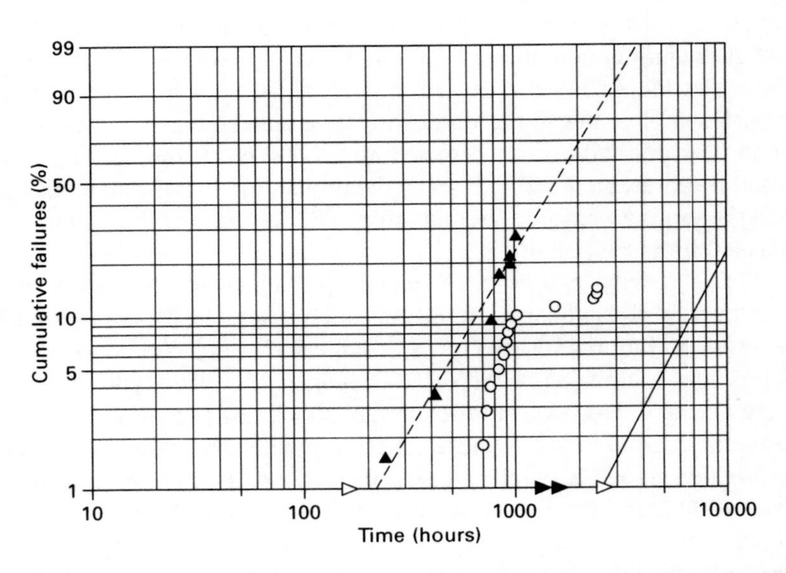

3.46 Weibull failure distributions of pitch single-bump contacts in 85 °C/85%RH after MSL3, 0.16 mm die. ACF S-1 100 μm pitch (▲, dashed line), same 60 μm pitch (○), NCF 40 μm pitch (solid line), suspended data (▶, ▷).

resistance increased sharply in the first 100 hours, and then remained constant depending on the bonding conditions. If the bonding process was done at low temperature (160 °C or lower) the resistance continuously increased. Bonding at temperatures up to 190 °C lead to resistance values that were constant until after 900 hours, when the resistance became slightly higher. All the samples reported in this paragraph were not subjected to a reflow process, as were those of the earlier mentioned study of Yin *et al.* (2003). In this latter case, the ~80 μm pitch assemblies survived at least 500 hours in the accelerated damp heat test (85 °C/85%RH) when the test was stopped. The resistance only moderately increased.

3.6.3 Anisotropic conducting film (ACF) versus non-conducting film (NCF)

In this final section of the case studies, the differences between these two types of anisotropic conductive adhesives is addressed. Remember that, in principle, NCA is also an anisotropic conductive adhesive.

In the first place, basic reasoning tells us that the conductivity is less for NCA than for ACA, but the electrical isolation is better. In particular, for very small pitches this is an important issue, because the gap between the bumps is perhaps just one half of the pitch. Typically, the diameter of the conductive filler particles is in the range of 5 μm, while the gap is 20 μm. Thus, a distinct risk for short circuiting exists, in particular if one realizes that the number of interconnections is extremely high. A brief statistical argumentation may be illustrative. Assume a two-dimensional situation. By means of the Poisson distribution, one can describe the probability that a bridge of conductive filler particles is formed between two adjacent bumps:

$$P(k) = e^{-\mu} \frac{\mu^k}{k!} \qquad [3.5]$$

The parameter k is the critical number of particles needed to form such bridge, and equals g/s, where g is the width of the gap between the two bumps, and s the diameter of the particles (see Fig. 3.47). Further, there are a number of possible bridges equal to L/s, where L is the length of the bump. The parameter μ is the average number of particles per surface area, which can be estimated as follows. Taking the particle density as 1.5×10^6 /mm^3 (De Vries *et al.*, 2005b), one calculates the number of particles in the space between the bumps (see Table 3.3). Here it is assumed that the particle size distribution is monodisperse. In the next step we state that the particles are evenly distributed over the height (h) of the bump, so that there are h/s layers with particles. This gives a value for the average μ. For various widths of the gap and diameters of the particles, the probability was calculated for a flip

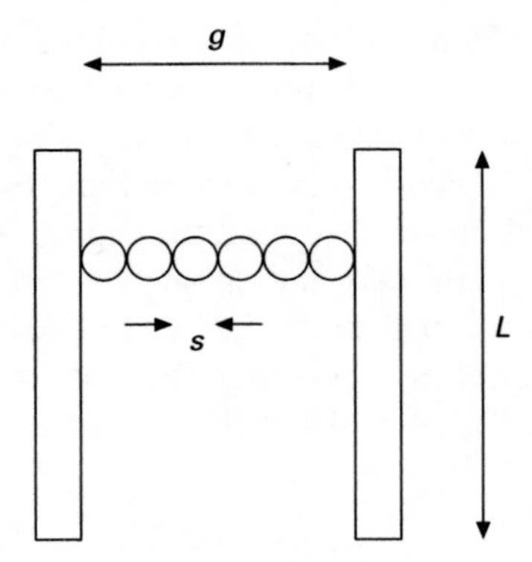

3.47 Model showing bridge of conducting particles (diameter *s*) between two adjacent bumps of length *L* at distance *g* from each other.

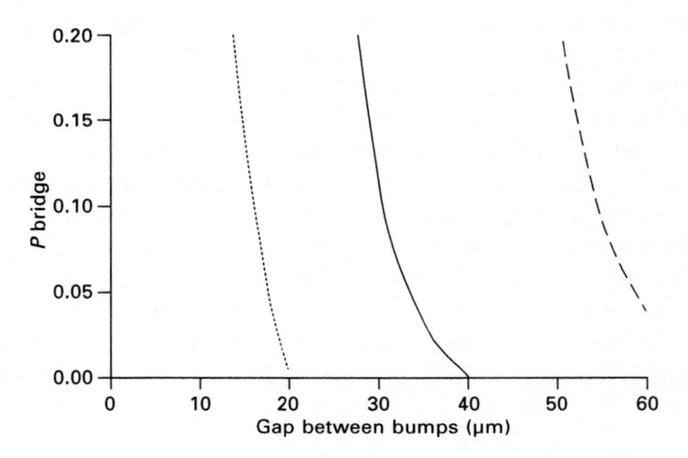

3.48 Probability of forming a bridge of conductive particles in flip chip with 400 bumps between two neighboring bumps with indicated gap. Particle diameter 3 μm (dotted line), 4 μm (solid line), and 5 μm (dashed line).

chip with 400 bumps that a conducting bridge occurs. In Fig. 3.48 the result makes clear that the combination of particle size and gap width is critical in the region below 40–50 μm. For the calculations the gap was taken as one half of the pitch. The authors emphasize that this statistical approach is meant only to illustrate the issue of conductive adhesives in a very small pitch, and that a handful of assumptions were made.

In the second place, we have presented various experimental results. The remarkable difference in performance between non-conductive adhesives and those with (Au-coated polymer) filler particles after the reflow test procedure needs further explanation. The filling factor is comparatively low, and both adhesives have the same base material. Still, a different failure mechanism can be conceived. Upon heating up, the adhesive passes through the glass transition point (which is in the range of 125–145 °C, see Table 3.2) and thus possesses a low elasticity and accordingly a higher coefficient of thermal expansion. In contrast, the metallized polymer particles with a much higher glass transition temperature (~300 °C) will retain their elasticity and relatively low thermal expansion coefficient. The connection between the bump and the track can thus become partly lost during heating up, and the stiffer filler particles, in the absence of an external pressure, will hamper restoration of the connection upon cooling down. The diagram of Fig. 3.2 gives circumstantial support for this theory. It shows the resistance of a chain of conducting adhesive interconnections after 2400 hours at 110 °C/85%RH, when the assembly was again put under gradually increasing pressure. Thus, part of the contact area between the bumps and the tracks might be permanently lost. In non-conductive adhesives, this mechanism is obviously not present.

Regarding the differences in the reliability performance between interconnections made from conductive and non-conductive adhesives, Liu and Lai (2002) were one of the first to report. They tested ACF- and NCF-based samples on FR4-substrate. The pitch was 100 μm. In general, the failure distributions of the anisotropic conductive adhesive joints were broader than those of the non-conductive adhesive. As was mentioned in Section 3.2, Kim *et al.* (2008a) investigated the resistance against reflow of ACF- and NCF-bonds in 200 μm-pitch samples. The ohmic behavior of the interconnections changed in the non-conductive adhesives after one or two reflow cycles, while in the anisotropic conductive version the resistance was ohmic up to two or three reflow cycles. On the other hand, the value of the resistance of the NCF-samples remained constant whereas in the ACF-samples it gradually increased. No endurance stress tests were done. Flip chip assemblies with a pitch of 60 μm were made (Chiang *et al.*, 2006). These were put to test in a pressure cooker (121 °C/100%RH). The resistance was measured at regular intervals and was found to increase by a factor of four after 192 hours. The same type of samples showed a factor of 2.5 higher resistance after 500 hours in high-temperature storage (150 °C).

Farley *et al.* (2009) observed that with non-conductive adhesive interconnects, the bonding conditions could result in a diffusion bond between the Au-bump and the Au-finish on the substrate. As a result, the bonding mechanism, and hence also the degradation mechanism, will become completely different. This observation could be an alternative explanation for the different behavior between ACAs and NCAs.

3.7 Conclusions

In this chapter, the use of conductive and non-conductive adhesives is described for heat seal connectors and fine-pitch flip chip on foil assemblies. Such adhesives have a different nature from that of soldered interconnections. Therefore, the degradation mechanisms also differ, and other stress factors apply. These latter are dominated by cyclic exposure to moisture and temperature. Still, given the right combination of materials with respect to the diffusivity of water and given the right processing, reliable interconnections can be produced. Such joints can withstand reflow treatments with peak temperatures up to 250 °C and subsequently endure several hundred thermal cycles (–55 °C/+125 °C) or several hundred hours accelerated damp heat testing (85 °C/85%RH). For very fine-pitch interconnections, down to 40 μm, non-conductive adhesives perform better: over 1000 hours accelerated damp heat test without failure. On-line measuring of the contact resistance is a powerful method for monitoring the degradation of adhesive interconnects. Effects such as post-curing during the accelerated tests are made visible. It can also be used to exploit the sensitivity of adhesive interconnects to a changing environment and use the response of the contact resistance to a varying humidity as a fingerprint of the adhesive and of the bonding process.

3.8 Acknowledgments

The authors thank their colleagues for co-operation during the investigations themselves and for support while preparing this chapter: Jan van Delft, Esther Janssen, Gerard Kums, Gerard Lijten, Cees Slob, Norbert Sweegers, Will van der Zanden, Xiujuan Zhao, Ee Hua Wong. Part of the work was carried out under the Brite-Euram projects ITERELCO, CUMULUS, and CIRRUS.

3.9 References

Caers J, Vries J de, Zhao X J, Wong E H (2003a), 'Some characteristics of anisotropic conductive and non-conductive adhesive flip chip on flex interconnection', *J. Semiconductor Techn. and Science*, 3, 122–131.

Caers J, Zhao X J, Lekens G, Dreesen R, Croes K, Wong E H (2003b), 'Moisture induced failures in flip chip on flex interconnections using anisotropic conductive adhesives', *Proc. Int. IEEE Conf. on the Business of Electronic Product Reliability and Liability*, 171–176.

Caers J, Zhao X J, Wong E H, Ong C K, Wu Z X, Ranjan R (2003c), 'Prediction of moisture induced failures in flip chip on flex interconnections with non-conductive adhesives', *53rd Electr. Comp. and Techn. Conf.*, 1176–1180.

Cao L, Li S, Lai Z, Liu J (2005), 'Formulation and characterization of anisotropic conductive adhesive paste for microelectronics packaging applications', *J. Electr. Mater.*, 34, 1420–1427.

Chan Y C, Tan S C, Lui N S M, Tan C W (2006), 'Electrical characterization of NCP-

and NCF-bonded fine-pitch flip-chip-on-flexible packages', *IEEE Trans. Adv. Pack.*, 29, 735–740.

Chang S M, Jou J H, Hsieh A, Chen T H, Chang C Y, Wang Y H, Huang C M (2001), 'Characteristic study of anisotropic-conductive film for chip-on-film packaging', *Microelectronics Reliability*, 41, 2001–2009.

Chiang W K, Chan Y C, Ralph B, Holland A (2006), 'Processability and reliability of non-conductive adhesives (NCAs) in fine-pitch chip-on-flex applications', *J. Electr. Mater.*, 35, 443–452.

Divigalpitiya R, Hogerton P (2003), 'Contact resistance of anisotropic conductive adhesives', *Proc. IMAPS*, 471–476.

Farley D, Kahnert T, Sinha K, Solares S, Dasgupta A, Caers J, Zhao X J (2009), 'Cold Welding: A New Factor Governing the Robustness of Adhesively Bonded Flip-Chip Interconnects', *59th Electr. Comp. and Techn. Conf.*, 67–73.

Frisk L, Cumini A (2009), 'Effect of substrate material and thickness on reliability of ACA bonded FC joints', *Soldering & Surface Mount Techn.*, 21(3), 16–23.

Frisk L, Ristolainen E (2005), 'Flip chip attachment on flexible LCP substrate using an ACF', *Microelectronics Relilability*, 45, 583–588.

Ikeda T, Kim W K, Miyazaki N (2006), 'Evaluation of the delamination in a flip chip using anisotropic conductive adhesive films under moisture/reflow sensitivity test', *IEEE Trans. Comp. Pack. Techn.*, 29, 551–559.

Jagt J (1999), 'Reliability of electrically conductive adhesive joints in surface mount applications', in Liu J (ed.) *Conductive Adhesives for Electronics Packaging*, Electrochemical Publications.

Jagt J, Buijsman A, Lijten G (2000), 'Flip chip technologies with adhesives – possibilities and challenges', *Adhesives in Electronics, 4th International Conference on Adhesive Joining & Coating Technology in Electronics Manufacturing*, 18–21 June 2000, Helsinki University of Technology, Espoo, Finland.

Janssen E, Kums G (2000), 'The development of an industrial technology for flip chip on flex circuitry', *Proc. Semicon.*, 179–184.

JEDEC (2008), 'Moisture/Reflow Sensitivity Classification for Nonhermetic Solid State Surface Mount Devices', *IPC/JEDEC J-STD-020D.1*.

Jokinen E, Ristolainen E (2001), 'Flip chip joining of thin chips on flexible PEN substrate', *Proc. 2001 Int. Symp. Microelectr.*, 600–604.

Jokinen E, Seppälä A, Pienimaa A, and Ristolainen E (2000), 'Reliability of ACF flip chip joint on flexible substrate', *37th IMAPS Nordic Conf.*, 192–196.

Kim J Y, Kwon S, Ihm D W (2004), 'Reliability and thermodynamic studies of an anisotropic conductive adhesive film (ACAF) prepared from epoxy/rubber resins', *J. Mater. Proc. Techn.*, 152, 357–362.

Kim J W, Jung S B (2007a), 'Behavior of anisotropic conductive film joints bonded with various forces under temperature fluctuation', *J. Electr. Mater.*, 36, 1199–1205.

Kim J W, Jung S B (2007b), 'Effect of bonding force on the reliability of the flip chip packages employing anisotropic conductive film with reflow process', *Mater. Sc. Eng.*, A452–453, 267–272.

Kim J W, Kim D G, Lee Y C, Jung S B (2008a), 'Analysis of failure mechanism in anisotropic conductive and non-conductive film interconnections', *IEEE Trans. Comp. Pack. Techn.*, 31, 65–73.

Kim J W, Koo J M, Lee C Y, Noh B I, Yoon J W, Kim D G, Park S K, Jung S B (2008b), 'Thermal degradation of anisotropic conductive film joints under temperature fluctuation', *Int. J. Adh. Adh.*, 28, 314–320.

Lai Z, Liu J (1996), 'Anisotropically conductive adhesive flip-chip bonding on rigid and flexible printed circuit systems', *IEEE Trans. Comp. Pack. Man. Techn.*, B19, 644–660.

Lam D, Yang F, Tong P (1999), 'Chemical kinetic model of interfacial degradation of adhesive joints', *IEEE Trans. Comp. Pack. Techn.*, 22, 215–220.

Lawrence Wu C M, Chau M L (2002), 'Degradation of flip chip on glass interconnection with ACF under high humidity and thermal aging', *Soldering & Surface Mount Techn.*, 14, 51–58.

Li L, Morris J, Liu J, Lai Z, Ljungkrona L, Lai C (1995), 'Reliability and failure mechanism of isotropically conductive adhesive joints', *Proc. 45th Electr. Comp. & Techn. Conf.*, 114–120.

Ling S, Binh L, Lew A, Nhan E (2001), 'A study on advanced flip-chip interconnect technologies for space application', *Proc. 2001 HD Int. Conf. High-dens. Interconnect and Syst. Pack.*, 131–136.

Liu J (1996), 'On the failure mechanism of anisotropically conductive adhesive joints on copper metallization', *Int. J. Adhesion and Adhesives,* 16, 285–287.

Liu J (1998), 'Recent advances in conductive adhesives for direct chip attach applications', *Microsystem Technologies,* 5, 72–80.

Liu J, Lai Z (1999), 'Reliability of anisotropically conductive adhesive joints on a flip-chip/FR4 substrate', *Advances in Electronic Packaging*, 26, 1691–1697.

Liu J, Lai Z (2002), 'Reliability of anisotropically conductive adhesive joints on a flip-chip/FR4 substrate', *J. Electr. Packaging*, 124, 240–245.

Lu S T, Chen W H (2008), 'Reliability of Ultra-thin Chip-on-Flex (UTCOF) with Anisotropic Conductive Adhesive (ACA) Joints', *58th Electr. Comp. and Techn. Conf*, 1287–1293.

Määttänen J, Palm P, De Maquillé Y, Bauduin N (2002), 'Development of fine pitch (54 μm) flip chip on flex interconnection process', *Proc. IEEE Polytronic*, 155–159.

Majeed B, Paul I, Razeeb K M, Barton J, Ó'Mathúna S C (2007), 'Effect of gold stud bump topology on reliability of flip chip on flex interconnections', *IEEE Trans. Adv. Pack.,* 30, 605–615.

Miessner R, Aschenbrenner R, Reichl H (1999), 'Reliability study of flip chip on FR4 interconnections with ACA', *45th Electr. Comp. and Techn. Conf.,* 595–601.

Nysaether J, Lai Z, Liu J (2000), 'Thermal cycling lifetime of flip chip on board circuits with solder bumps and isotropically conductive adhesive joints', *IEEE Trans. Adv. Pack.,* 23, 743–749.

Olliff D, Gaynes M, Kodnani R, Zubelewicz A (1997), 'Characterizing the failure envelope of a conductive adhesive', *Int. Symp. Adv. Pack. Mater.,* 124–126.

Palm P, Määttänen J, Tuominen A, and Ristolainen E (2001a), 'Reliability of 80 μm pitch flip chip attachment on flex', *Microelectronics Reliability*, 41, 633–638.

Palm P, Määttänen J, De Maquillé Y, Picault A, Vanfleteren J, Vandecasteele B (2001b), 'Reliability of different flex materials in high density flip chip on flex applications', *First Int. IEEE Conf. Polymers and Adhesives in Microelectronics and Photonics*, 224–229.

Palm P, Määttänen J, De Maquillé Y, Picault A, Vanfleteren J, Vandecasteele B (2003), 'Comparison of different flex materials in high density flip chip on flex applications', *Microelectronics Reliability,* 43, 445–451.

Rusanen O, Lenkkeri J (1999), 'Thermal stress induced failures in adhesive flip chip joints', *Int. J. Microcircuits and Electr. Pack.,* 22, 363–369.

Seppälä A, Pienimaa S, Ristolainen E (2001), 'Flip chip joining on FR4 substrate using ACFs', *Int. J. Microcircuits and Electr. Pack.*, 24, 148–159.

Vries J de (2004), 'Failure mechanism of anisotropic conductive adhesive interconnections in flip chip ICs on flexible substrates', *IEEE Trans. Comp. Pack. Techn.*, 27, 161–166.

Vries J de, Janssen E (2003), 'Humidity and reflow resistance of flip chip on foil assemblies with conductive adhesive joints', *IEEE Trans. Comp. Pack. Techn.*, 26(3), 563–568.

Vries J de, Delft J van, Slob C (2005a), '100 μm pitch flip chip on foil assemblies with adhesive interconnections', *Microelectronics Reliability*, 45, 527–534.

Vries J de, Delft J van, Slob C (2005b), 'SMT-compatibility of adhesive flip chip on foil interconnections with 40 μm pitch', *IEEE Trans. Comp. Pack. Techn.*, 28(3), 499–505.

Wu S X, Hu K X, Yeh C P (1999), 'Contact reliability modeling and material behavior of conductive adhesives under thermomechanical loads', in Liu J. (ed.) *Conductive Adhesives for Electronics Packaging*, Electrochemical Publications.

Wu Y P, Alam M O, Chan Y C, Wu B Y (2004), 'Dynamic strength of anisotropic conductive joints in flip chip on glass and flip chip on flex packages', *Microelectronics Reliability*, 44, 295–302.

Wuytswinkel G van, Dreezen G, Luijckx G (2002), 'The effects of temperature and humidity aging on the contact resistance of novel electrically conductive adhesives', *Proc. IEEE Polytronic*, 225–230.

Yim M J, Paik K W (2001), 'Effect of non-conducting filler additions on ACA properties and the reliability of ACA flip chip on organic substrates', *IEEE Trans. Comp. Pack. Techn.*, 24, 24–32.

Yin C Y, Alam M O, Chan Y C, Bailey C, Lu H (2003), 'The effect of reflow process on the contact resistance and reliability of anisotropic conductive film interconnection for flip chip on flex applications', *Microelectronics Reliability*, 43, 625–633.

3.10 Appendix: List of abbreviations

ACA anisotropic conducting adhesive
ACF anisotropic conducting film
ACP anisotropic conducting paste
BGA ball grid array (semiconductor package)
CTE coefficient of thermal expansion
DSC differential scanning calorimetry
ENIG electroless nickel immersion gold
FR flame retardant
HAST highly accelerated stress test
ICA isotropic conducting adhesive
ICP isotropic conducting paste
JEDEC Joint Electron Device Engineering Council(s)
MSLA moisture sensitivity level assessment
NCA non-conducting adhesive
NCF non-conducting film

NCP	non-conducting paste
PA	polyamide
PCB	printed circuit board
PI	polyimide
RH	relative humidity

4 Isotropic conductive adhesives in electronics

J. E. MORRIS, Portland State University, USA

Abstract: Isotropic conductive adhesives are metal/epoxy composites with high loading beyond the percolation threshold for electrical conduction. The primary reliability issues are impact resistance and galvanic corrosion of the contacts, among others. Nanoparticles and carbon nanotubes are beginning to be incorporated into the materials, for improved performance.

Key words: isotropic conductive adhesive (ICA), drop test, corrosion, immersion silver.

4.1 Introduction

There are two primary categories of electrically conductive adhesives (ECAs): isotropic conductive adhesive (ICA) and anisotropic conductive adhesive (ACA), with the latter dividing further into paste (ACP) and film (ACF) types. Tin–lead (Sn-Pb) solder toxicity and environmental issues triggered initial ECA interest, with the advantages of lower processing temperature, no-flux, no-clean, and simple processing (Zwolinski, 1996; Detert and Herzog, 1999), but it has been other advantages which continue to drive research for flip-chip (Rusanen *et al.*, 1997; Lohokare *et al.*, 2006), surface mount technology (SMT), optoelectronics, and MEMS packaging. ICAs are also used extensively in die-attach, for small passive chip attachment in automotive electronics, and in RFID tags for both antenna and chip connections. More recently, high-density SMT board via-fill has emerged as a major commercial application (Das *et al.*, 2006; Kisiel *et al.*, 2005a, b), and novel uses continue to be proposed, e.g. as conformal interconnect for a 3D IC stack (Robinson *et al.*, 2008). One disadvantage of ICAs is that the polymer cure time is inherently longer than solder reflow times, but it can be shortened by 'snap-cure' polymers (Moscicki *et al.*, 2005) which use fast catalysts, and fast variable frequency microwave curing (Wang *et al.*, 2001; Fu *et al.*, 2003).

Much of the research focus in the literature is on reliability failure modes, e.g. resistance stability and adhesion shear tests (Liu *et al.*, 1995), humidity effects (Rorgren and Liu, 1995) and other thermal testing (Gaynes *et al.*, 1995; Rusanen and Lenkkeri, 1995), usually with comparisons to solder's properties (Hvims, 1995).

The ICA consists of a two-phase mixture of metal, typically a bimodal distribution of silver (Ag) flakes and powder (Figs 4.1 and 4.2), in a polymer (epoxy) adhesive. ICA resistivity drops dramatically (Fig. 4.3) when the metallic content exceeds the 'percolation threshold'. Ag is popular because it is less expensive than gold (Au) with superior conductivity and chemical stability, and, unlike nickel (Ni), Ag oxides show high conductivity (Lu *et al.*, 1999c; Markley *et al.*, 1999; Shimada *et al.*, 2000; Kotthaus *et al.*, 1997; Lu and Wong, 2000f). Ag-coated copper (Cu) also shows promise (Nishikawa *et al.*, 2008), and low melting point alloys, Sn-coated Ag particles, etc. have been reported (Moon *et al.*, 2003; Lu and Wong, 2000c; Suzuki *et al.*, 2004; Yamashita and Suganuma, 2006b).

Regardless of the filler metal, the flakes require lubrication to resist the tendency to 'clump' together during processing, e.g. with stearic acid (soap). Lu *et al.*, reported on various lubricants, first on the chemistry of the lubricant layers and their interaction with the Ag flakes, and on their thermal behavior during heating (Lu and Wong, 2000b), and then on their thermal decomposition (Lu and Wong, 2000a). Wong has achieved reduction of overall resistance by replacing the traditional stearic acid with shorter chain alternatives (Wong and Li, 2004; Li *et al.*, 2004a,b, 2006a,b). Miragliotta *et al.* (2002) has shown that the lubricant breaks down during cure, and leaves a carbon residue on the flake surface, which is expected to contribute to inter-particle conductance.

Intuitively, the higher the metal content, the higher the conductivity, traded off against weaker adhesion. There have been several attempts to improve electrical conductivity at low metal concentrations, e.g. by magnetic alignment

4.1 ICA bi-modal filler distribution (Ag flakes and powder), with surface layering evident (Li, 1996).

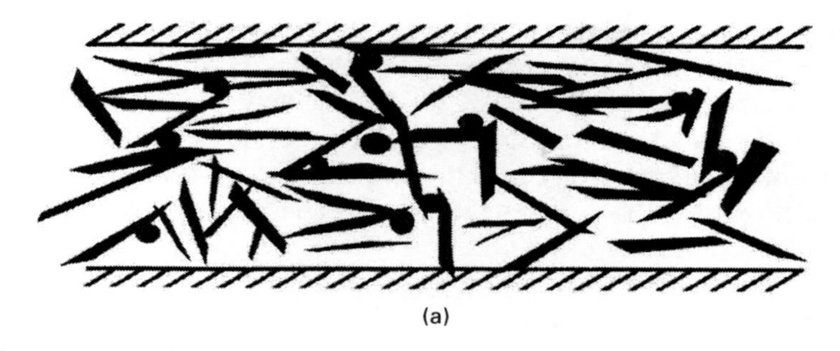

(a)

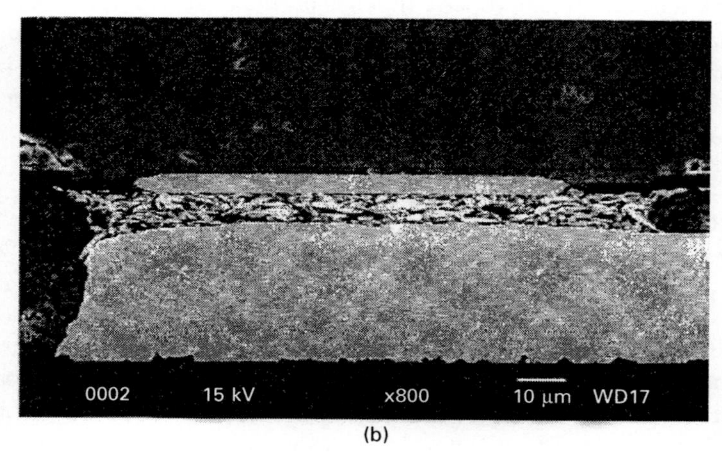

(b)

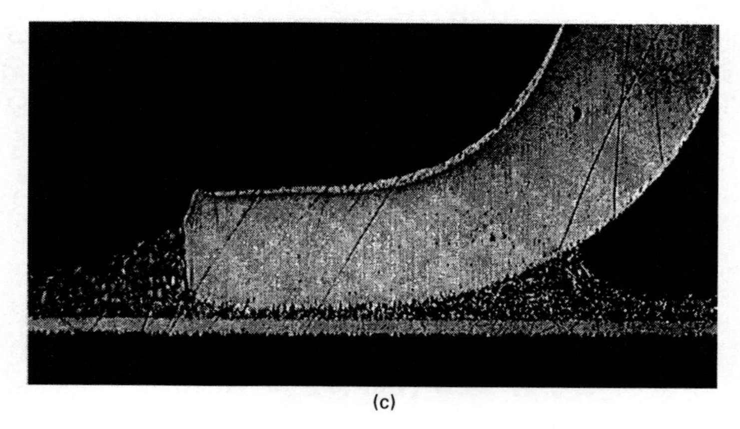

(c)

4.2 ICA contact joints: (a) schematic, (b) flip-chip on FR-4, (c) SMT on FR-4 (Kudtarkar and Morris, 2002; Morris, 2005).

of Ni filler rods (Sancaktar and Dilsiz, 1997) or particles (Ramakumar and Srihari, 2008), or by the use of polymer particles (Inada and Wong, 1998) or electric fields (Morris *et al.*, 1999) to force z-axis alignment of flakes. The coating of contacts by intrinsically conductive polymers has been investigated, but turns out to increase resistance (Lam *et al.*, 2006).

ICAs are typically dispensed by syringe, but may also be stencil- or screen-printed. Formulations are usually based on thermoset epoxy resins for the polymeric matrix, typically with thermoplastic added for rework capability. Although epoxies have excellent strength and adhesion capabilities, they tend to absorb moisture.

Self-alignment is a critical component to the success and reliability of solder attachment of area-array flip chips, and the absence of a similar surface-tension driven property in ICAs is widely seen as an impediment to their adoption for this role. Wu *et al.* (Wu *et al.*, 2001; Moon *et al.*, 2001) coaxed a minimal ICA self-alignment effect from LMPA content, but Kim *et al.* have shown that self-alignment is viable when driven by resin surface tension (Kim *et al.*, 2003).

4.2 General isotropic conductive adhesive (ICA) properties

4.2.1 Structure

The ICA resistivity only drops slightly as the metal content is increased until the 'percolation threshold' is reached (Fig. 4.3), when the first continuous metal path is established through the composite material. The primary concepts of percolation theory are well developed, especially for the elementary system of uniform conducting spheres (or cubes) in a perfectly insulating medium (Smilauer, 1991). Bi-modal particle distributions have been shown to reduce the percolation threshold (Kusy, 1977) with either flakes or powders used for the smaller particles. The electrical path includes the metallic resistance of the filler and the contact resistances between particles and at the contacts.

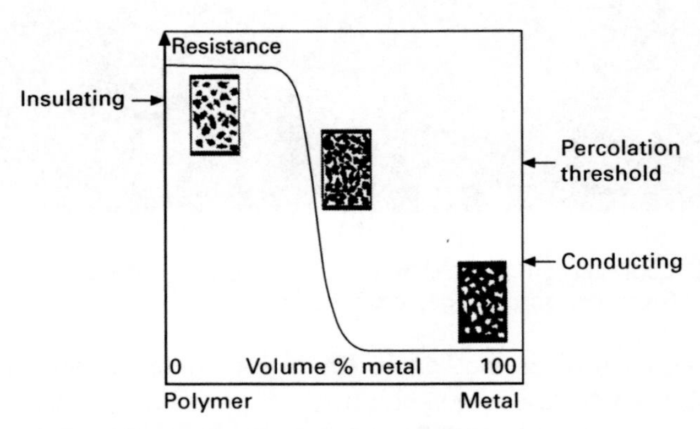

4.3 Percolation threshold (Morris, 1999).

Metallic resistance has been shown to dominate (Li *et al.*, 1993), in the absence of environmental effects.

At low thicknesses, the flakes are all layered, parallel to the substrate contact, but become more randomly distributed as thickness increases (Morris *et al.*, 2003), unless the application of pressure during or prior to cure forces all the flakes into alignment (Constable *et al.*, 1999). The effects of surface layering have been confirmed at a qualitative level (Li and Morris, 1997), and the consequential size effects demonstrated (Sancaktar and Dilsiz, 1998; Ruschau *et al.*, 1992), i.e. the increased/decreased conductance parallel/perpendicular to current flow. Morris *et al.* (2001) show the resistance of a z-axis contact as it is mechanically thinned, with a sharp drop in resistivity when the aligned surface layer is removed.

4.2.2 Electrical properties

It is necessary to separate contact resistance and the bulk composite resistivity in order to interpret physical effects on each, independently (see below), by the subtraction of the common four-terminal measurement, which removes test lead and contact effects, from a three-terminal measurement which includes one contact in the test current path (Morris *et al.*, 2001; Klosterman *et al.*, 1998).

Finite geometries can lead to errors if care is not taken. The ICA resistance under test is typically of the same order as PWB track resistances (Morris *et al.*, 2001; Kulkarni and Morris, 2003), and one should make some simple estimates of all resistances in the test set-up to determine contact thicknesses to avoid current crowding. Because ICA z-axis resistances are small, many experimental results in the literature have been obtained by x-axis measurements, i.e. by measuring ICA resistance along a long, thin printed track. Such x-axis measurements tend to give higher resistivities than the ideal random structures of the true bulk unless the thickness is much greater than the characteristic meandering percolation path length, and lower resistivities due to layering effects. The z-axis application of actual interest displays two opposite trends.

Li *et al.* (1993) showed that the ICA high-frequency behavior is fully attributable to skin effect in the metal filler, with no observation of the capacitive effects expected of tunneling contacts between filler particles. Their high-frequency ICA data have been extended by Wu *et al.* (1998), and into the GHz range by Dernevik *et al.* (1997), Sihlbom *et al.* (1997), and Otsuka and Akiyama (2007).

At frequencies where skin effect is dominant, the lower resistance advantage enjoyed by solder at DC disappears, as the effective cross-sectional area shrinks with the skin depth for both solder and ICA alike (Morris *et al.*, 1999), as supported by the observations of Hashimoto *et al.* (2008). For practical purposes, there is negligible high-frequency performance difference between ICAs and solder (Liong *et al.*, 2001).

4.2.3 Mechanical properties

ICA adhesive and shear strengths are similar to solder's, occasionally higher (Herzog *et al.*, 2004) although usually a little less (Suzuki *et al.*, 1998), but generally adequate (Luchs, 1996). Plasma cleaning of the adherent surfaces would seem to be a logical step to improve adhesion. Herzog *et al.* (2004) and Paproth *et al.* (2001) have shown that plasma treatments increase the polar component of surface energy, but preliminary data show no adhesion improvement with either argon or oxygen plasmas, despite removal of organic contaminants and oxides (Dernevik *et al.*, 1997; Morris *et al.*, 1999; Morris and Probsthain, 2000). Experimental studies consistently show that the mechanical component of adhesion dominates (Liong *et al.*, 2002; Chow *et al.*, 2002), with best results from surface roughening, (which may be accomplished by high-energy plasmas.) A conducting polymer interface layer can also promote ICA adhesion (Kuechenmeister and Meusel, 1997), and Keil *et al.* (2001) improved it by structuring the contact pad so that a proportion of the ICA contacts the FR-4 epoxy surface rather than the metal contact.

Wu *et al.* (1996) studied viscoelastic ICA properties, and concluded that stability requires a low-temperature cure, followed by a stabilization ramp to higher levels. Post-cure annealing decreases resistance further, even at less than the cure temperature (Inoue and Suganuma, 2005, 2006). At full cure conditions, however, the electrical resistance and the mechanical strength of conductive adhesives can be guaranteed (Liu *et al.*, 1997).

4.2.4 Thermal properties

The thermal performance of an adhesively assembled chip is of vital interest as power dissipation in the chip increases, and for die-attach or heat-sink bonding (Inoue and Suganuma, 2006). Sihlbom *et al.* (1998) have simulated power dissipations, and Kimura *et al.* (2003) and Inoue *et al.* (2006) have performed experimental studies, but this is an area that deserves much more attention.

4.2.5 Environmental properties

The environmental impact of ECAs has been studied by several research groups. Segerberg *et al.* (1997) compared the use of conductive adhesive joining with soldering for SMT applications and concluded that the relative environmental load of the conductive adhesives is dependent on the mining condition of Ag. Westphal (1998) concluded that conductive adhesives are generally better in terms of environmental loading compared to solder.

More work is needed to clarify environmental pros and cons, especially since environmental concerns have been an ICA technology driver, and Ag is not wholly environmentally benign. Curing agent toxicity, for example, is an overlooked issue (Yi *et al.*, 2006), and nano-Ag's anti-bacterial properties may become an environmental problem if concentrations rise (Luoma, 2008), threatening the bottom of the aquatic food chain.

4.3 Reliability

4.3.1 Impact resistance

Impact resistance is the primary impediment to more widespread ICA technology adoption. Early identification of drop-test failure as a significant ICA problem (Rusanen and Laitenen, 2004) led to the widespread adoption of the NCMS (National Center for Manufacturing Science) criterion (survival of six drops from 5 feet) as a *de facto* standard (Zwolinski, 1996). Survival rate correlates with the (imaginary) dissipation modulus instead of adhesive strength (Tong *et al.*, 1998a, b; Xu and Dillard, 2003), but the use of polymers with glass transition temperature T_g below room temperature (Luo and Wong, 2002; Yi *et al.*, 2008) would lead to other, greater problems. The addition of carbon fibers to the ICA seems to be helpful (Keil *et al.*, 2001), but the result is ambiguous without modulus measurements, because of a simultaneous change in the contact geometry to improve adhesion. Although the loss modulus is accepted to be the most significant parameter in impact resistance, the failure mode is adhesive rather than cohesive as expected. So shear adhesive strength should play a secondary role. Several sequences of drop test experiments, which have included low T_g and 'snap-cure' ICAs, are described by Morris and Lee (2008) with the conclusions outlined below.

It is well known that drop-test survival depends on the inertial mass of the component in question, as demonstrated by the variation in drop-test survival rate with component mass shown in Fig. 4.4 for dummy aluminum component blocks (Morris *et al.*, 2003). (The larger contact area available with the dummy blocks increases the number of drops to failure in comparison with 'real' components, and improves the statistical consistency of the results.) As a consequence, commercial products may use ICA attachment for small SMT passive devices, but solder or other attachment for larger processor chips, etc. In experiments with PWBs stocked with SMT devices of varied sizes and pin configurations, the larger components clearly fall off first (Kudtarkar and Morris, 2002).

A general trend was identified, that (within limits) under-cured samples survive more drops than those cured to specification, which in turn survive more than over-cured samples, in agreement with the loss modulus effect,

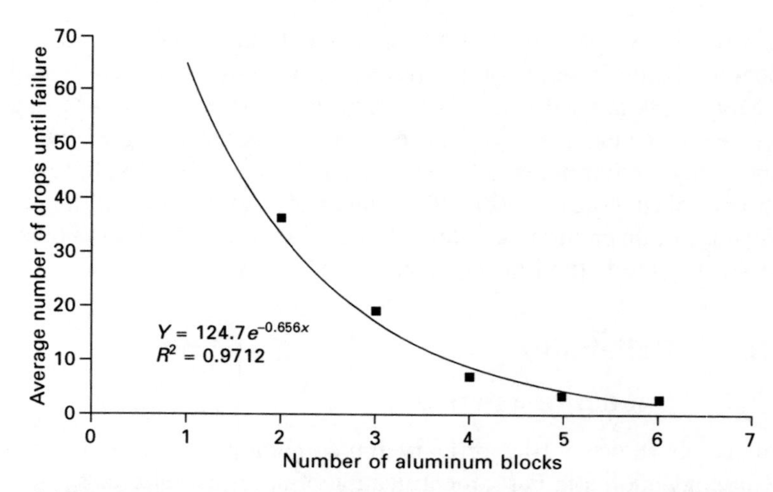

4.4 Drop-test survival using aluminum blocks as dummy chips (5-foot drops; 2.5cm × 2.5cm × 3mm aluminum blocks; 1cm² adhesive area; blocks stacked with cyanoacrylate 'super-glue'; single block survived 120 drops without failure).

T_g increasing with the degree of cure. An experimental ICA with low T_g (below room temperature) was also shown to be much more impact-resistant than others tested, and in this case the fully-cured low T_g adhesive performed better than the under-cured.

Dummy Al components also survive better than the PQFP components used, despite the higher mass, presumably due to greater adhesive strength to the rougher Al surfaces, a postulate supported by the fact that the point of failure for the Al samples was mainly at the substrate surface, whereas the PQFP failures usually occurred at the lead interfaces. It was also observed that PQFP leads bend under impact, and sometimes fail (fracture) before the ICA bond.

Figures 4.5 to 4.8 (Morris and Lee, 2008) summarize the key results for drop tests for a single ICA and two components of different sizes:

- Figs 4.5 and 4.6: small SO20GT devices
- Figs 4.7 and 4.8: larger 68-lead PLCCs

with two ideally equivalent ICA cure temperatures:

- Figs 4.5 and 4.7: 60 minutes at 150 °C
- Figs 4.6 and 4.8: 30 minutes at 175 °C

according to two profiles:

(i) simple ramp-up/hold/ramp-down specified by the manufacturer
(ii) with the addition of pre-heat/post-cool hold periods to the ramp up/down stages.

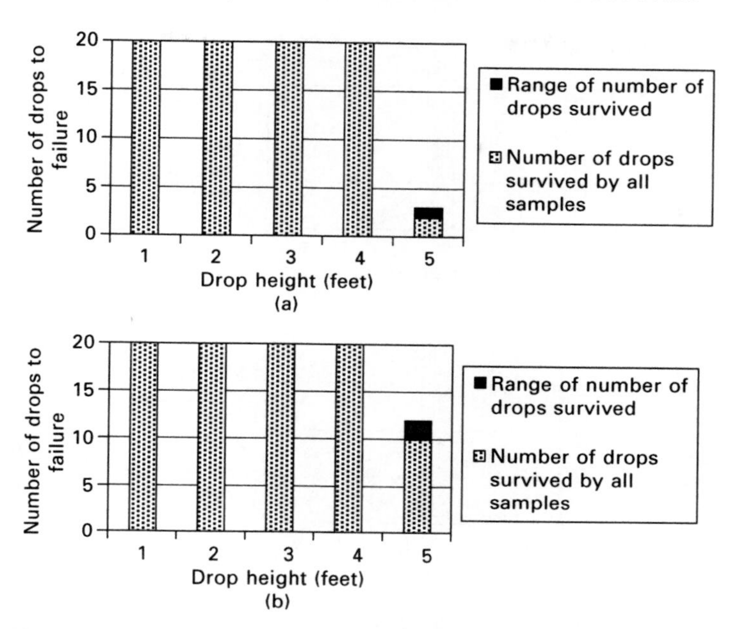

4.5 SO20GT cured at 150 °C: (a) ramp and hold, (b) pre-heat/post-cool.

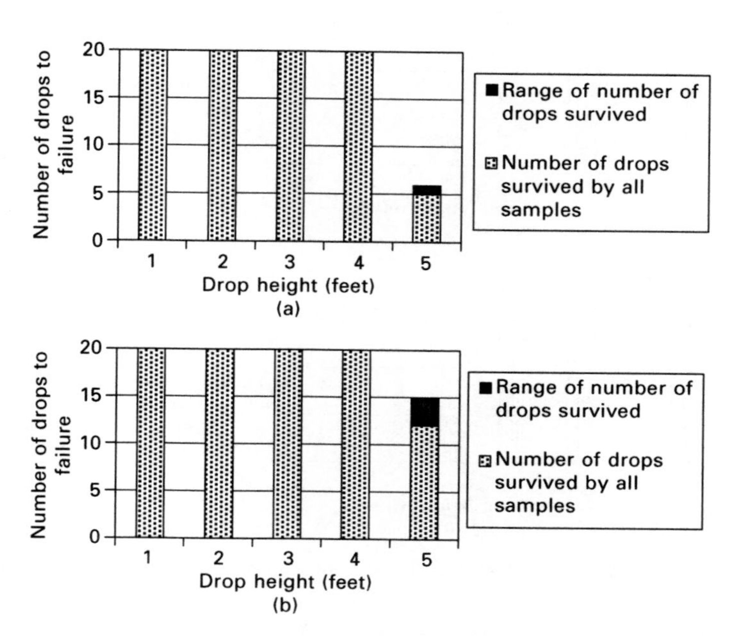

4.6 SO20GT cured at 175 °C: (a) ramp and hold, (b) pre-heat/post-cool.

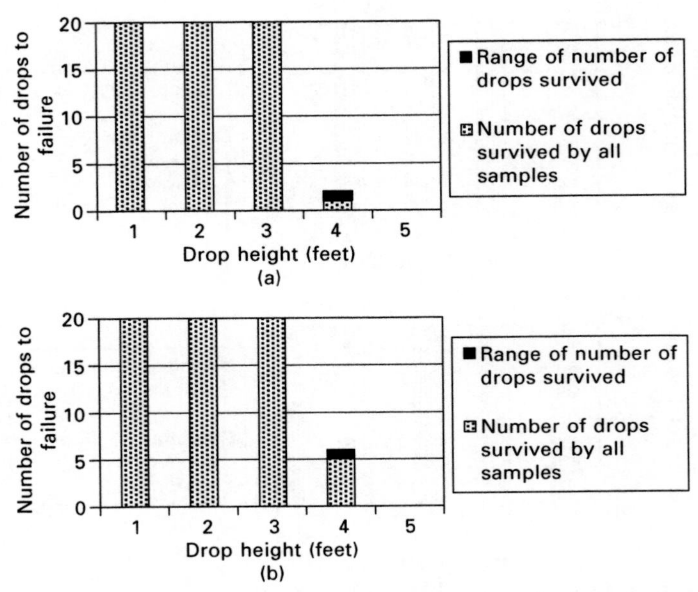

4.7 PLCC68 cured at 150 °C: (a) ramp and hold, (b) pre-heat/post-cool.

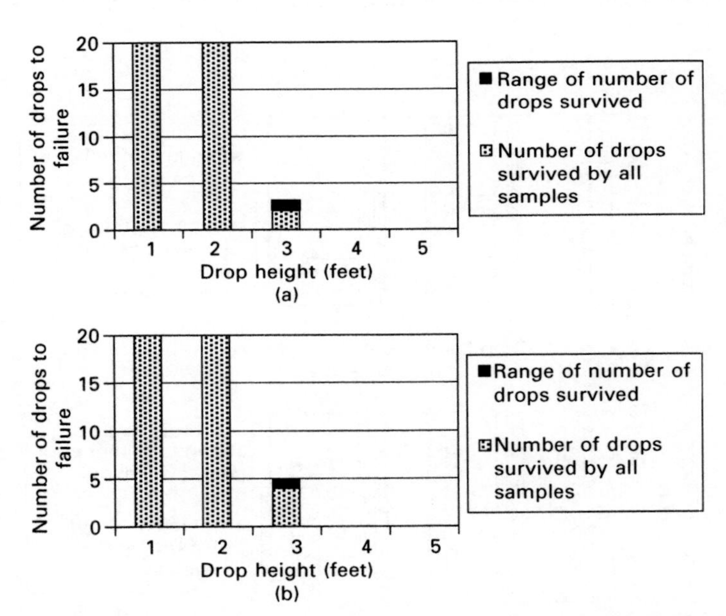

4.8 PLCC68 cured at 175 °C: (a) ramp and hold (b) pre-heat/post-cool.

Each sample, with one package per board, was dropped successively twenty times from each of 1, 2, 3, 4, and 5 feet heights, until at least one package lead detached from the board (which usually meant when all detached and the device fell off the board). Therefore, where the SO20GT samples fail after some variable number of drops from 5 feet in Figs 4.5 and 4.6, they have all previously survived twenty drops from each of 1, 2, 3, and 4 feet. The variation in the number of drops survived by multiple samples is indicated in each figure by the relatively small 'range.'

As expected, the larger PLCC68 components fall off sooner than the smaller SO20GTs. There is a small improvement in SO20GT impact resistance with the change in cure schedule from 60 minutes at 150 °C to 30 minutes at 175 °C, but a significant reduction in PLCC68 performance for both the simple ramp-and-hold and pre-heat/post-cool profiles. A possible explanation is that 175 °C results in an over-cure, to which the J-lead PLCC68 samples are more susceptible than the SO20GT gull wing configuration, due to the smaller contact area.

The primary result is that samples with pre-heating and post-cooling have better impact resistance than those without. From other experiments, it was previously determined that impact damage was cumulative, i.e. that samples which survived multiple small drops would then fail after fewer 5-foot drops than those that had not undergone the prior sequence of small drops.

When investigating plasma surface treatments, it was discovered that vacuum exposure of the ICA prior to cure, improved contact adhesion significantly, by elimination of bubbles from the material (Morris *et al.*, 2001). If cure proceeds too quickly, the organic solvents used to control viscosity for printing or dispensing cannot escape, and become entrapped in the epoxy as bubbles, and bubbles at the contact interface have also been correlated with weak ACF adhesion (Kim *et al.*, 2006). This could explain why impact failure is almost always interfacial rather than cohesive, despite the governing property being the bulk loss modulus, and is consistent with a cumulative damage model, with crack initiation at the bubbles. Bubble content can also be reduced or eliminated by a pre-cure heat soak at a temperature sufficiently high to drive out volatile components, but low enough for a negligibly small cure rate (Perichaud *et al.*, 1998).

4.3.2 Galvanic corrosion

Contact resistance reliability problems (Li *et al.*, 1995; Liu *et al.*, 1996; Botter *et al.*, 1998) are primarily due to galvanic corrosion between dissimilar metals at the interfaces (Lu *et al.*, 1999a,b; Lu and Wong, 1999b) in the presence of water, with the resulting oxide leading to micro-crack failure (Kim *et al.*, 2008). The resistance drift can be reduced or slowed by the addition of corrosion inhibitors, oxygen scavengers, and/or sacrificial anode material to the polymer matrix (Tong *et al.*, 1999; Lu and Wong, 1999a, 2000e; Takezawa *et al.*, 2002), and moisture can be minimized with anhydride-cured epoxies (Lu and Wong, 2000d). However, these techniques can only delay or reduce the effect, which requires the selection of compatible filler and contact materials.

Examples of resistance changes with time under 85/85 conditions (i.e. at 85 °C in 85% relative humidity) are shown in Fig. 4.9 (Li, 1996; Klosterman *et al.*, 1998), for Ag-filled ICA on Cu and Au contacts, where the contact and bulk resistances have been separated by combined three- and four-terminal measurements. The markedly different responses on Cu and Au correspond to their positions in the electrochemical series, relative to Ag.

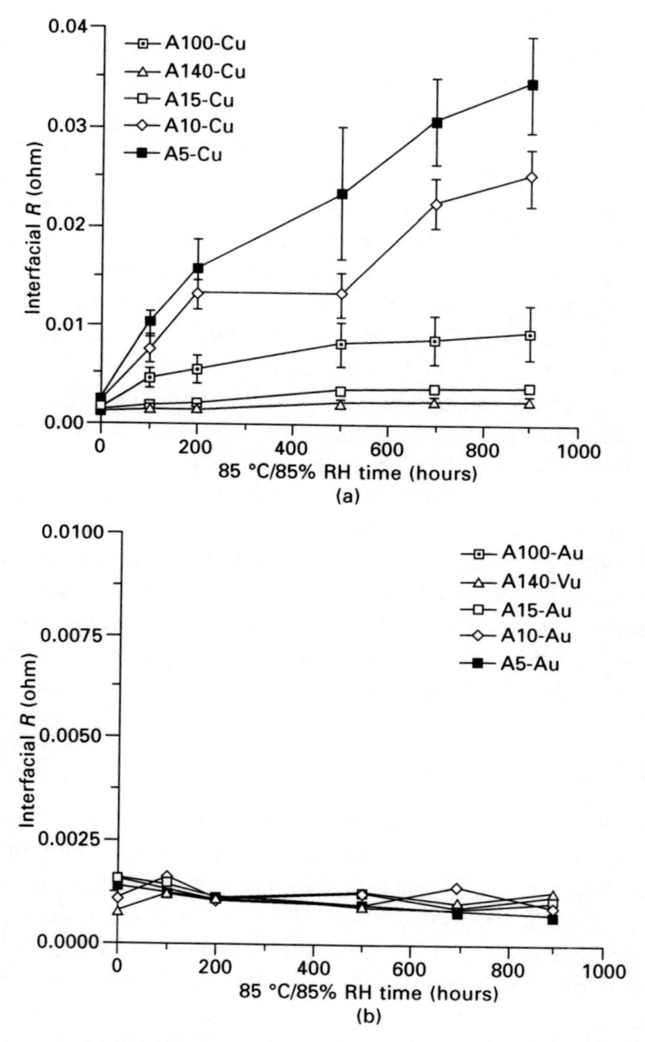

4.9 Ag ICA contact resistance changes at 85/85: (a) on Cu contacts, (b) on Au contacts, with (c) corresponding bulk resistivity variations.

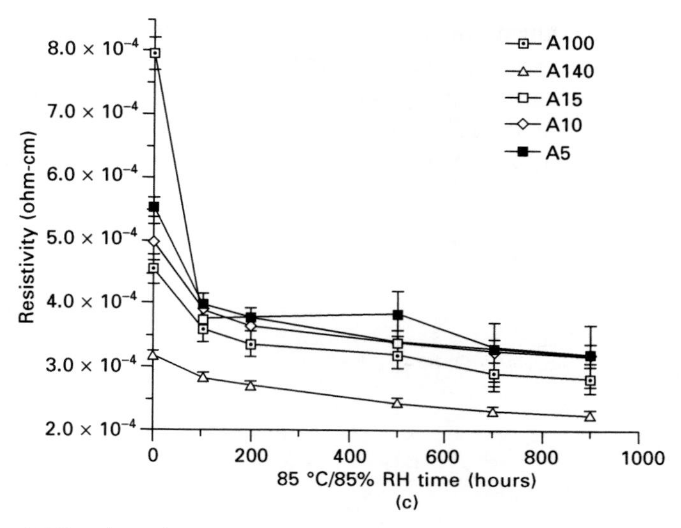

4.9 Continued.

The Cu/Ag contact resistance increases significantly (Fig. 4.9a), while there is essentially no change in the Au/Ag contact (Fig. 4.9b). The reduction in the bulk resistivity (Fig. 4.9c) is attributed to continued curing at 85 °C.

As part of the elimination of Pb from electronics, PWB metallic finishes must also change from hot-air-leveled SnPb solder to alternatives such as immersion-Ag, immersion-Sn, electroless-Ni/immersion-Au, or organic solderability preservative (OSP) (Pas, 2005). Immersion-Ag is especially becoming the Pb-free final finish choice for many OEMs in the telecommunications, computer, automotive, and consumer electronics industries. For Ag-filled ICAs, the use of an immersion-Ag PWB surface finish would completely eliminate the galvanic corrosion potential (Lee *et al.*, 2009) at the interface, leaving only the slower Ag oxidation and no insulating by-product.

The contact resistances of two different Ag-based ICAs, designated A and B in Fig. 4.10, were studied under 85/85 conditions as functions of time, with the data for contacts to Cu and immersion-Ag PWB tracks being shown as both absolute resistances and their ratios to initial values. The total measured resistance includes both the bulk and interfacial contact resistances, with the changes assumed to be dominated by the ICA/pad interface since the bulk resistivity has been shown to be relatively stable during aging, provided it is adequately cured. Two curing schedules were used: 27–30 minutes at 185 °C, and 22–24 minutes at 170 °C, with the initial degree of cure expected to be significantly greater for the former. The degree of cure does not seem to affect ICA-A resistance values, but ICA-B appears to be under-cured at 170 °C, with initial resistances about two orders higher than the 185 °C values.

The contact resistances are smaller for the immersion-Ag PWBs, which are also more stable, as expected. Nevertheless, there are small resistance increases observed for the Ag/Ag system, consistent with simple oxidation. Since the resistances of the 170 °C ICA-B samples also increase appropriately for the Cu or Ag contacts, the contact resistance is apparently increased by

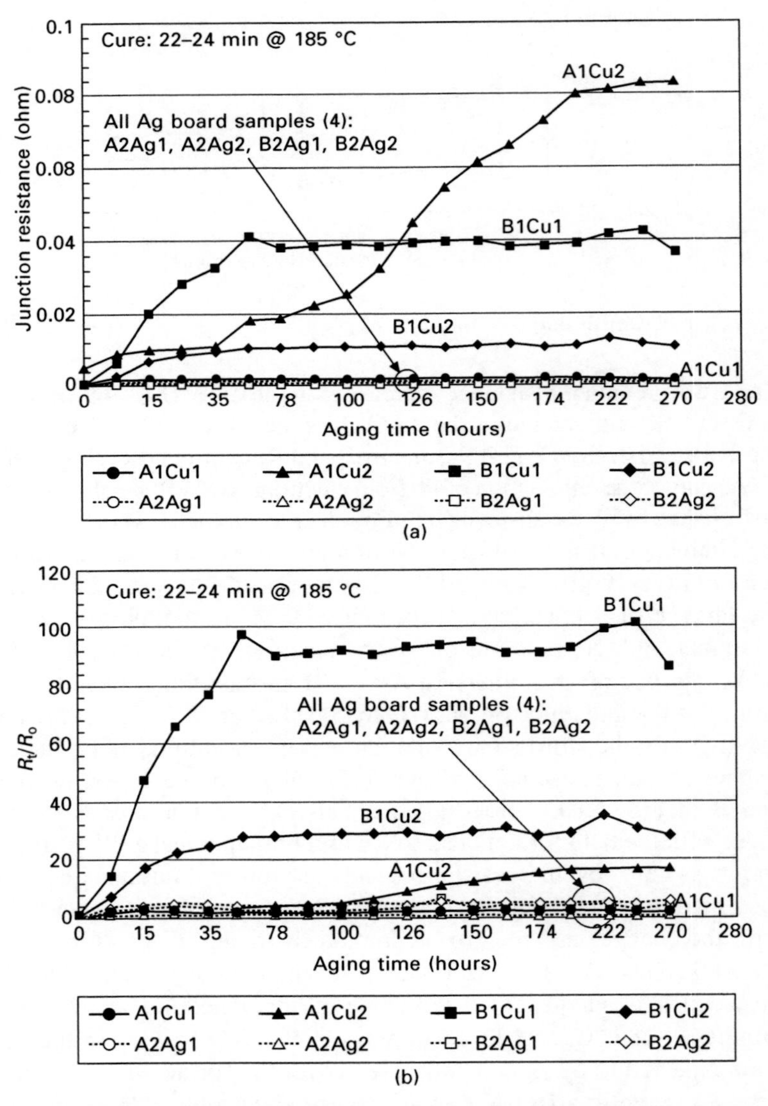

4.10 (a) and (c) 85/85 junction resistance shifts; (b) and (d) ratio of resistances to initial values (ICAs A and B, curing at 170 °C and 185 °C, Cu and immersion-Ag pad surfaces).

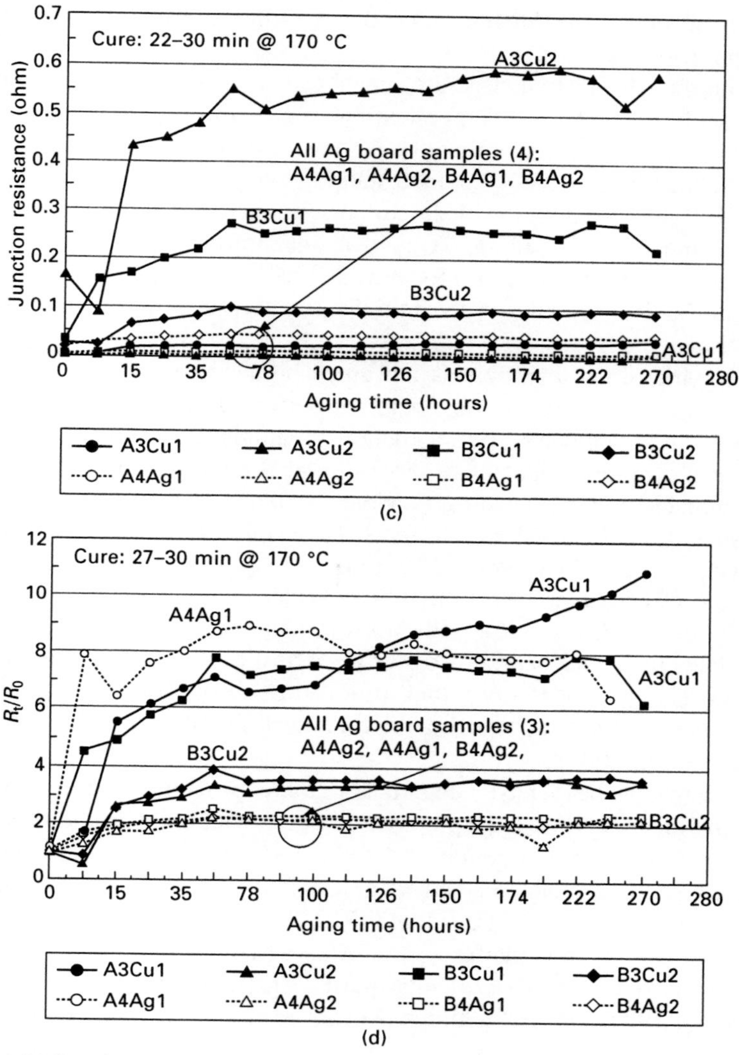

4.10 Continued.

under-curing, as well as the bulk resistivity. (Note that the resistance of sample A4Ag1 is anomalously low, which leads to the exaggerated plot for small changes in Fig. 4.10d. By contrast, the contact resistance of sample A3Cu2 is much larger than for any others, for unknown reasons.)

4.3.3 Other reliability problems

Apart from galvanic corrosion issues, the choice of surface contact metallization is also important for interdiffusion and intermetallic formation, e.g. Sn from Sn-Pb finishes diffuses into the Ag filler at 150° C (Yamashita and Suganuma, 2006a,c).

Interfacial and bulk fracture mechanisms have been studied by Gupta *et al.* (1999), but in SMT applications the thermoplastic properties of the polymer lead to the accumulation of plastic strain, which initiates cracking (Perichaud *et al.*, 1998). To understand the degradation mechanisms, Mo *et al.* (2002) focused on the electrical performance of a commercial ICA joint under mechanical loading. To gain insight into the electrical degradation mechanism, finite-element modeling (FEM) was executed, and the effects of mechanical loading on the initial intimate interaction among Ag fillers were analyzed.

Polymer creep coefficients are much higher than those of solders, and modeling predicts that ICAs would out-perform solder on room-temperature mechanical cycling tests by an order of magnitude (Rusanen, 2000). However, thermal cycling results do not show similar benefits (Nysaether *et al.*, 2000), presumably due to exceeding T_g. Note also that Ag/epoxy interfacial fracture is suggested by initial wear effects (Constable *et al.*, 1999), and has been directly observed (Chen *et al.*, 2006).

The effect of moisture on ICA polymer degradation has been studied by Khoo and Liu (1996), with moisture distribution within the joint being modeled by Dudek *et al.* (2005) for viscoelastic modeling of thermal cycling failure. Hygroscopic strain due to moisture absorption can be found from measured strain by subtracting the calculated thermomechanical component (Low *et al.*, 2007).

There has been past concern about the possible field-driven surface migration of Ag ions in the presence of water, which can lead to short circuits between adjacent ICA contacts (Suzuki *et al.*, 2004). Sancaktar *et al.* (2004) have recently correlated electromigration with Ag surface pitting, and there also appears to be a field threshold (Mo, 2005). Systematic study is required to establish the boundaries to the effect (Manepallis *et al.*, 1999), which is minimized by the use of high-purity Ag (with <10ppm Cu (Detert and Herzog, 1999), or Sn-Ag alloys (Suzuki *et al.*, 2004; Toida *et al.*, 2005) as Sn appears to inhibit migration. Ag migration is evident in un-cured material (Morris *et al.*, 1999), but it is not a problem in commercial products where it has been suggested that additives to the polymer seal the silver surface, defeating migration tendencies. Recent work (Yi and Wong, 2006a,b) has shown that a molecular self-assembled monolayer (SAM) of short chain di-carboxylic acid, for example, does just that. The addition of an appropriate SAM to particle surfaces also enables metal loading beyond the practical limit normally set by material viscosity (Jiang *et al.*, 2005).

At high current densities, electron momentum transfer drives mass electromigration within ultra-small flip-chip solder connections, and one might expect to see the same phenomenon in ICAs at the small contact points between Ag filler particles. Joule heating at these points of high resistance, however, causes localized polymer degradation, and this is the high current failure mode (Kotthaus *et al.*, 1998). The effect has been demonstrated for three commercial ICAs (Morris *et al.*, 1999), where surface temperature rises with the square of current. The sample temperature coefficient of resistance can be used to determine the internal temperature from the resistance increase with current, and failure is observed when the internal temperature reaches the polymer degradation temperature.

4.4 Modeling

4.4.1 Electrical modeling

While there have been some superficial efforts at structural modeling of ICAs, comparisons of electrical models with experiment are either strictly qualitative or fitted by parameter adjustment. For flakes of 10μm diameter by 1μm thick, and μm-sized smaller particles, the electron mean free path (mfp) is essentially bulk value, with no accounting required for size effects, other than for the constriction resistance at inter-particle contacts. (Note that the mfp is limited by particle dimensions in nano-particle ICAs (Kotthaus *et al.*, 1996).

Li and Morris (1997) developed electrical conduction models for Ag-filled ICAs, combining the microscopic resistance of the bulk Ag particles and the contact between Ag flakes with the macroscale resistor network calculation by percolation theory, confirming the effects of surface layering at a qualitative level, and bimodal size effects (Li and Morris, 1996; Sancaktar and Dilsiz, 1998). The model predicts the resistivity change as stress develops during the cure process, and variations with particle size distributions (Li and Morris, 1996). A dynamic model of compression effects demonstrates flake alignment quite dramatically (Mundlein and Nicolics, 2004), but as the structure fills up, the modeling process becomes more and more time-consuming. A potential energy technique has proved effective (McCluskey *et al.*, 1998) in reducing computation time, and compression algorithms can be applied to initially well-separated particles (Mustoe *et al.*, 1999; Mundlein *et al.*, 2002; Mundlein and Nicolics, 2004, 2005). Su and Qu (2004) extended this compression concept by modeling the curing process itself. The extension of structural modeling to electrical properties requires the assumption of intra-particle, inter-particle, and contact conduction processes, with the structural model itself providing the percolation component.

4.4.2 Cure modeling

Klosterman *et al.* (1998) focused on the influence of cure on resistivity, joint resistance and reliability, with novel analytical methods developed to define the cure conditions for optimum electrical properties and stability. The cure process seemed to have been modeled successfully by very simple mathematical expressions (Li and Morris, 1999), on the basis of a rapid resistance drop as the model 100%-cure point was approached. That success has been questioned, however, by the correlation of such resistance drops at ~20% degrees of cure (Inoue and Suganuma, 2006).

A review of polymer cure models used in microelectronics packaging applications reveals no clear consensus of the chemical rate constants for the cure reactions, or even of an effective model. The problem lies in the contrast between the actual cure process, which involves a sequence of distinct chemical reactions, and the models, which typically assume only one reaction (or two, with some restrictions on the independence of their characteristic constants). The standard techniques to determine the model parameters are based on differential scanning calorimetry (DSC), which cannot distinguish between the reactions (Morris *et al.*, 2009a), and hence yield results useful only under the same conditions, which completely misses the point of modeling. The obvious solution is for manufacturers to provide the modeling parameters, but failing that, an alternative experimental technique is required to determine individual reaction parameters, e.g. Fourier transform infra-red spectroscopy (FTIR).

Thermally cured epoxies and other polymers are extensively used in electronics packaging, as encapsulants, underfills, and adhesives, etc. The project which prompted this study was the microwave cure of a carbon-loaded epoxy encapsulant (Tilford *et al.*, 2007). The temperature rises more rapidly than in conventional isothermal or reflow oven curing systems, and the cure proceeds more uniformly within the material. Optimization of the microwave power level and application time cannot be readily accomplished experimentally, especially given the speed of the cure, so simulation is seen as the tool to sensible planning of the process development. A literature review was the obvious first step to establish the model, including the thermal dependences of the chemical reaction parameters. Prior experience suggests that the simplest first order model has proved to be effective in ICA applications, with the model parameters relatively easily obtained by DSC measurements (Klosterman *et al.*, 1998).

For α = degree of cure, the basic assumption of all models is that the reaction rate can be expressed as a function of reactant concentration, $f(\alpha)$, by

$$d\alpha/dt = K f(\alpha), \qquad [4.1]$$

with the temperature-dependent chemical rate constant $K = A \exp\text{-}(E/RT)$, for rate parameters A and E, and $R = 8.31$ J/K.mole. The models vary in the assumed form of $f(\alpha)$, as listed below.

(a) nth order model: $f(\alpha) = (1 - \alpha)^n$ [4.2]

- in this case, we can find α analytically (for constant T, i.e. isothermal cure) as:
  - 1st order: $d\alpha/dt = K(1 - \alpha)$, $\therefore \alpha = 1 - \exp(-Kt)$ [4.3]
  - 2nd order: $d\alpha/dt = K(1 - \alpha)^2$, $\therefore \alpha = 1 - [1 + Kt]^{-1}$ [4.4]
  - nth order: $d\alpha/dt = K(1 - \alpha)^n$, $\therefore \alpha = 1 - [1 + (n - 1)Kt]^{-1/(n-1)}$ [4.5]

(b) Auto-catalyzed model:

- single-step: $d\alpha/dt = K f(\alpha) = K \alpha^m (1 - \alpha)^n \ldots$

  but note $d\alpha/dt = 0$ for $\alpha = 0$ [4.6]
- double step (linear combination): $d\alpha/dt = (K_1 + K_2 \alpha^m)(1 - \alpha)^n$ [4.7]
- modified double step: $d\alpha/dt = K f(\alpha) = K(y_1 + y_2 \alpha^m)(1 - \alpha)^n$

  where $y_1 + y_2 = 1$ [4.8]

Note that only the double step auto-catalyzed model has more than a single chemical rate constant in the model, i.e. all others implicitly assume a single cure reaction, or at least a single rate-controlling reaction across the full temperature range of interest. In practice, for example, the cure of bisphenol-A diglycidyl ether (BADGE), a commonly employed epoxy in packaging applications and ICAs, requires two steps (with a third reaction occurring at high cure temperatures). So the double-step auto-catalyzed model is the *only* one with a realistic physical basis that can be expected to apply outside the measurement conditions used to determine A and E. This view is supported by published model data for a variety of polymers where the single-rate constant parameters, A and E, vary with temperature and/or degree of cure, making them applicable only within the range of measurement conditions, i.e. not useful as predictive tools. There are mathematical techniques available to extract the chemical rate parameters from isothermal and/or dynamic DSC data, but only the isothermal DSC scan can yield K_1, K_2, m, and n, (i) by plotting $d\alpha/dt$ versus α and varying the four parameters for best fit, or (ii) by plotting $d\alpha/dt$ versus $(1 - \alpha)^2$ with the assumption of $m = 1$, $n = 2$, (which seems to be supported by some data), or (iii) finding K_1 and K_2 independently from DSC peak values, but with m, n assumptions still required. Arhennius plots of K_1, K_2, then yield A_1, A_2, E_1, and E_2. Ko *et al.* (1994) have extracted consistent A_1 and E_1 values from DSC BADGE data, but m turns out to be temperature dependent, calling the A_2 and E_2 values derived from m into question.

The two BADGE reactions can be written as:

(i) primary amine + epoxide → secondary amine
(ii) secondary amine + epoxide → tertiary amine

neither of which totally dominates the rate control. The double step autocatalyzed model could conceivably be re-written as:

$$d\alpha/dt = [K_1(1 - \alpha_1)^{m1} + K_2(\alpha_1 - \alpha_2)^{m2}](1 - \alpha)^n \quad [4.9]$$

where α is the fraction of reacted epoxides, so $(1 - \alpha)$ is the fraction remaining, and the primary, tertiary, and secondary amine concentrations are $(1 - \alpha_1)$, α_2, and $(\alpha_1 - \alpha_2)$ respectively.

The obvious solution is for manufacturers to provide full and accurate modeling parameters in their material data sheets. However, FTIR studies of the cure process have been shown to be capable of distinguishing between the successive reactions. It is proposed that FTIR replace DSC in the determination of cure model parameters (Morris *et al.*, 2009a,b).

4.4.3 Flow modeling

There is a real need for a better understanding of the ICA dispensation process, whether by stencil or screen printing, or by syringe. Flow modeling of these processes would enable a scientific approach to the problem of flake layering, for example. As for all modeling, one must first have the real material parameters, and so the first step has now been taken with the work of Zhou and Sancaktar (2008a), who measured and modeled (fitted) the rheological properties of highly filled Ni/epoxy ICAs as functions of loading, shear rate, temperature, and time (degree of cure), providing input parameters for process flow modeling. In a separate study (Zhou and Sancaktar, 2008b), they identified stratification zones within a dispensation syringe, with Ni particles concentrating in a Ni-rich region at the tube circumference and the center/axis of the tube being epoxy-rich. As dispensation proceeds, pressure-induced 'filtration' occurs, with epoxy-rich material dispensed from the center, enriching the Ni content of the residue, which then requires increased syringe pressure to maintain the mass flow. Under certain circumstances, the pressure can oscillate as it increases under flow rate control. Process modeling can lead to greater understanding of these phenomena and their minimization.

4.5 Nanotechnologies in isotropic conductive adhesives

4.5.1 Nanoparticles

Kotthaus *et al.* (1997) reported on an ICA filled with porous aggregates of nano-size Ag particles, with the goal of decreasing the metal loading to improve adhesion. Total metal loading can be decreased with good electrical conductivity using a bimodal filler distribution (Fu *et al.*, 1999; Mach *et al.*, 2008), but the nanoparticles increase resistivity for given total filler content, due to mean free path limits and increased numbers of contacts (Wong *et al.*, 2005), although small decreases have been observed (Zhang *et al.*, 2008).

Moon *et al.* (2005) show the thermal behavior of silver nanoparticles with respect to the sintering reaction, which is critical to the effective use of nanoparticles to improve ICA performance (Bai *et al.*, 2005; Jiang *et al.*, 2005, 2006). Surface changes of the particles during sintering and crystal structure variation are also addressed in Moon *et al.* (2005). Nanoparticle shapes can be controlled during synthesis by inclusion of $AgNO_3$ into the epoxy resin (Pothukuchi *et al.*, 2004).

Ye *et al.* (1999) observed apparently sintered contacts of ~50nm diameter between Ag nanoparticles in an ICA, and similar contacts have been observed between micron-scaled ICA particles (Li and Morris, unpublished data, 1994). Sintering is clearly the key to the lower ICA resistivities appearing in the literature, and may explain the results of Yang *et al.* (2008) by a 'room temperature sintering' process (Wakuda *et al.*, 2008) of enhanced surface diffusion following removal of the protective surface layer.

Surface treatment of the flakes by a 0.2% 'reactive' solution in ethanol, following an ethanol rinse, dramatically lowers resistivity from $100\,\mu\Omega\cdot$cm to $6\,\mu\Omega\cdot$cm, presumably also by sintering, and doubles adhesion from 2.5MPa to 5MPa (Yang *et al.*, 2008).

4.5.2 Carbon nanotubes (CNTs)

The addition of carbon nanotubes (CNTs) to the Ag-flake/epoxy mix (Lin and Lin, 2004) lowers the percolation threshold, as expected, and may be more effective than Ag nano-particles. Ag nanoparticle/nanowire mixtures have also been demonstrated (Wu *et al.*, 2006a,b), as have CNT/epoxy composites (Li and Lumpp, 2006; Zhang *et al.*, 2008). However, the high resistance contacts between CNTs limit the conductivities that can be achieved with random-structure composites (Yan *et al.*, 2007). Nevertheless, CNTs will be used in the future as supplementary filler, between the Ag flakes, to both reduce resistivity and improve impact resistance (Heimann *et al.*, 2007, 2008; Wirts-Ruetters *et al.*, 2008).

4.6 Conclusions

Fundamental materials research has yielded greatly improved understanding of ICA reliability test results. However, the conventional wisdom remains that ICAs will continue as a niche technology where new no-Pb solders cannot be used. Impact strength remains the primary liability for more widespread adoption. Since the failure mode is usually adhesive rather than cohesive, even though failure correlates with the polymer loss modulus, greater attention to crack initiation at the contact interface should yield dividends. The other long-standing fundamental issue in the field is a detailed understanding of the inter-particulate interface, and its electrical properties (Morris, 2005, 2007).

Newcomers to the ICA field are encouraged to seek out other reviews. Li and Morris (1998) cover basic background principles, while Liu and Morris (1999) offer a more comprehensive technology review, updated in Morris and Liu (2006) and Morris *et al.* (2010). The most complete sources of information are dedicated books (Liu, 1999; Gomatan and Mitttal, 2008; Li *et al.*, 2010), with the last of these focused on nanotechnologies, and an on-line course (Morris and Liu, 2000 at www.cpmt.org). A new book (Morris, 2011) is in current development.

4.7 References

Bai, J. G., Zhang, Z. Z., Calata, J. N. and Lu, G.-Q., Characterization of low-temperature sintered nanoscale silver paste for attaching semiconductor devices, *Proc. 7th High Density Microsystem Design, Packaging and Component Failure Analysis (HDP'05)*, Shanghai, China (2005), pp. 272–276.

Botter, H., Van Der Plas, R. B. and Junai, A., Factors that influence the electrical contact resistance of isotropic conductive adhesive joints during climate chamber testing, *International Journal of Microelectronic Packaging*, **1**(3), (1998), pp. 177–186.

Chen, L. C., Lai, Z., Cheng, Z. and Liu, J., Reliability investigation for encapsulated isotropic conductive adhesives flip chip interconnection, *Journal of Electronic Packaging*, **128**, (2006), pp. 177–183.

Chow, L. L. W., Li, J. and Yuen, M. M. F., Development of low temperature processing thermoplastic intrinsically conductive polymer, *Proc. 8th International Symposium on Advanced Packaging Materials*, Braselton, GA (2002), pp. 127–131.

Constable, J., Kache, T., Teichmann, H., Mühle, S. and Gaynes, M., Continuous electrical resistance monitoring, pull strength, and fatigue life of isotropically conductive adhesive joints, IEEE *Transactions on Components and Packaging Technology*, **22**(2), (1999), pp. 191–199.

Das, R., Lauffer, J. and Egitto, F., Electrical conductivity and reliability of nano- and micro-filled conducting adhesives for z-axis interconnections, *Proc. 56th IEEE Electronic Component and Technology Conference*, San Diego CA (2006), pp. 112–118.

Dernevik, M., Sihlbom, R., Lai, Z., Starski, P. and Liu, J., High-frequency measurements and modelling of anisotropic, electrically conductive adhesive flip-chip joint, EEP-Vol. 19-1, *Advances in Electronics Packaging*, Vol. 1 ASME (1997), pp. 177–184.

Detert, M. and Herzog, T., ICAs for automotive applications in higher temperature ranges, *Proc 49th Electronic Components and Technology Conference*, San Diego, CA (1999) Paper s10p1.

Dudek, R., Berek, H., Fritsch, T. and Michel, B., Reliability investigations on conductive adhesive joints with emphasis on the mechanics of the conduction mechanism, *IEEE Transactions on Components and Packaging Technologies*, **23**(3), (2000), pp. 462–469.

Fu, Y., Liu, J. and Willander, M., Conduction modelling of a conductive adhesive with bimodal distribution of conducting element, *International Journal of Adhesion and Adhesives*, **19**(4), (1999), pp. 281–286.

Fu, Y., Wang, T. and Liu, J., Microwave-transmission, heat and temperature properties of electrically conductive adhesive, *IEEE Transactions on Components and Packaging Technology*, **26**(1), (2003), pp. 193–198.

Gaynes, M., Lewis, R., Saraf, R. and Roldan, J., Evaluation of contact resistance for isotropic electrically conductive adhesives, *IEEE Transactions on Components, Packaging and Manufacturing Technology Part B.*, **18**(2) (1995), pp. 299–304.

Gomatan, R. and Mittal, K. L., (editors), *Electrically Conductive Adhesives*, VSP, (2008) Leiden.

Gupta, S., Hydro, R. and Pearson, R., Fracture behavior of isotropically conductive adhesives, *IEEE Transactions on Components and Packaging Technology*, **22**(2), (1999), pp. 209–214.

Hashimoto, K., Akiyama, Y. and Otsuka, K., Transmission characteristics in GHz region at the conductive adhesive joints, *Proc. 58th Electronic Components and Technology Conference*, Orlando, FL (2008), pp. 2067–2072.

Heimann, M., Wirts-Ruetters, M., Boehme, B. and Wolter K.-J., Investigations of carbon nanotubes epoxy composites for electronics packaging, *Proc. 58th Electronic Components and Technology Conference*, Orlando, FL (2008), pp. 1731–1736.

Heimann, M., Lemm, J. and Wolter K.-J., Characterization of carbon nanotubes epoxy composites for electronics applications, *Proc. 30th International Spring Seminar on Electronics Technology*, Cluj-Napoca, Romania (2007), pp. 1–6.

Herzog, T., Koehler, M. and Wolter, K-J., Improvement of the adhesion of new memory packages by surface engineering, *Proc. 54th Electronic Components and Technology Conference*, Las Vegas (2004), pp. 1136–1141.

Hvims, H., Conductive adhesives for SMT and potential applications, *IEEE Transactions on Components, Packaging & Manufacturing Technology Part B.*, **18**(2), (1995), pp. 284–291.

Inada, T. and Wong, C. P., Fundamental study on adhesive strength of electrical conductive adhesives (ECAs), *Proc. 3rd International Conference on Adhesive Joining and Coating Technology in Electronics Manufacturing*, Binghamton, NY (1998), pp. 156–159.

Inoue, M. and Suganuma, K., Effect of curing conditions on the interconnect properties of isotropic conductive adhesives composed of an epoxy-based binder, *Proc. High Density Microsystem Design, Packaging and Component Failure Analysis Conference*, Shanghai (2005), pp. 128–133.

Inoue, M. and Suganuma, K., Effect of curing conditions on the electrical properties of isotropic conductive adhesives composed of an epoxy-based binder, *Soldering and Surface Mount Technology*, **18**, (2006), pp. 40–45.

Inoue, M., Muta, H., Maekawa, T., Yamanaka, S. and Suganuma, K., Thermal Conductivity of Isotropic Conductive Adhesives Composed of an Epoxy-based Binder, *Proceedings of High Density Microsystem Design, Packaging and Component Failure Analysis Conference*, Shanghai (2006), Paper A11.

Jiang, H., Moon, K. S., Zhu, L., Lu, J. and Wong, C. P., The role of Self-Assembled Monolayer (SAM) on Ag nanoparticles for conductive nanocomposite, *Proc. 10th International Symposium and Exhibition on Advanced Packaging Materials Processes*, Irvine, CA (2005).

Jiang, H., Moon, K. S., Yi, L. and Wong, C. P., Ultra high conductivity of isotropic conductive adhesives, *Proc. 56th Electronic Components and Technology Conference*, San Diego, CA (2006), pp. 485–490.

Keil, M., Bjarnason, B., Wickstrom, B. and Olsson, L., *Advanced Packaging*, September, 2001.

Khoo, C. G. L. and Liu, J., Moisture sorption in some popular conductive adhesives, *Circuit World*, **22**(4) (1996), pp. 9–15.

Kim, H. J., Chung, C. K., Yim, J., Hong, S. M., Jang, S. Y., Moon, Y. J., and Paik, K.-W., *Proc. 56th IEEE Electronic Components and Technology Conference*, San Diego, CA (2006), pp. 952–958.

Kim, J. M., Yasuda, K. and Fujimoto, K., Resin Self-alignment Processes for Assembly of Microelectronic and Optoelectronic Devices, *Proc. 3rd International IEEE Conference Polymers and Adhesives in Microelectronics and Photonics*, Montreux, Switzerland (2003), pp. 17–22.

Kim, S. S., Kim, K. S., Suganuma, K. and Tanaka, H., Degradation Mechanism of Ag–Epoxy Conductive Adhesive Joints by Heat and Humidity Exposure, *Proc. 2nd Electronics Systemintegration Technology Conference*, Greenwich, UK (2008), pp. 903–907.

Kimura, K. *et al.*, Thermal conductivity and RF signal transmission properties of Ag-filled epoxy resin, *Proc. 53rd IEEE Electronic Component and Technology Conference*, New Orleans, LA (2003), pp. 1383–1390.

Kisiel, R., Borecki, J., Felba, J. and Moscicki, A., Electrically conductive adhesives as vias fill in PCBs: The influence of fill shape and contact metallization on vias resistance stability, *Proc. 28th International Spring Seminar on Electronics Technology*, Wiener Neustadt, Austria (2005a) pp. 193–198.

Kisiel, R., Borecki, J., Koziol, G. and Felba, J., Conductive adhesives for through holes and blind vias metallization, *Microelectronics Reliability*, **45**, (2005b), pp. 1935–1940.

Klosterman, D., Li, L. and Morris, J. E., Materials characterization, conduction development, and curing effects on reliability of isotropically conductive adhesives, *IEEE Transactions on Components, Packaging and Manufacturing Technology, Part A*, **21**(1), (1998), pp. 23–31.

Ko, M. B., Kim, S. C. and Jo, W. H., Cure behavior of diglycidyl ether of bisphenol A/diaminodiphenylmethane in the presence of α, ω-methyl carboxylate–butadiene–acrylonitrile copolymer, *Korea Polymer Journal*, **2**(2), (1994), pp. 131–139.

Kotthaus, S., Haug, R., Schaefer, H., Guenther, B. and Schaefer, H., Investigation of isotropically conductive adhesives filled with aggregates of Nano-sized Ag-particles, *Proc. 2nd International Conference on Adhesive Joining and Coating Technology in Electronics Manufacturing*, Stockholm (1996), pp. 14–17.

Kotthaus, S., Guenther, B. H., Haug, R. and Schäfer, H., Study of isotropically conductive bondings filled with aggregates of nano-sited Ag-particles, *IEEE Transactions on Components, Packaging and Manufacturing Technology, Part A*, **20**(1), (1997), pp. 15–20.

Kotthaus, S., Haug, R., Schäfer, H. and Hennemann, O., Current-induced degradation of isotropically conductive adhesives, *IEEE Transactions on Components, Packaging, and Manufacturing Technology, Part A*, **21**(2), (1998), pp. 259–265.

Kudtarkar, S. A. and Morris, J. E., Reliability of electrically conductive adhesives, *Proc. 8th International Symposium on Advanced Packaging Materials*, Braselton, GA (2002), pp. 144–150.

Kuechenmeister, F. and Meusel, E., Polypyrrole as an interlayer for bonding conductive adhesives to activated aluminum bond pads, *IEEE Transactions on Components, Packaging and Manufacturing Technology Part A*, **20**(1), (1997), pp. 9–14.

Kulkarni, A. and Morris, J. E., Reliability lifetime studies on isotropic conductive adhesives,

Proc. 3rd International IEEE Conference Polymers and Adhesives in Microelectronics and Photonics, Montreux, Switzerland (2003), pp. 333–336.

Kusy, R. P., Influence of particle size ratio on the continuity of aggregates, *J. Appl. Phys.* **48**, (1977), pp. 5301–5305.

Lam, Y.-Z., Swingler, J. and McBride, J. W., The Contact Resistance Force Relationship of an Intrinsically Conducting Polymer Interface, *IEEE Transactions on Components and Packaging Technology*, **29**(2), (2006), pp. 294–302.

Lee, J., Cho, C. S. and Morris, J. E., Electrical and reliability properties of isotropic conductive adhesives on immersion silver printed circuit boards, *Microsystem Technologies*, **15**(1), (2009), pp. 145–149.

Li, J. and Lumpp, J.K., Electrical and mechanical characterization of carbon nanotube filled conductive adhesive, *Proc. IEEE Aerospace Conference*, BigSky, Montana, Paper #1519, (2006), pp. 1–6.

Li, L., *Basic and Applied Studies of Electrically Conductive Adhesives*, Ph.D. Thesis, State University of New York at Binghamton, 1996.

Li, L. and Morris, J. E., Electrical conduction models for isotropically conductive adhesives, *Journal of Electronics Manufacturing*, **5**(4), (1996), pp. 289–298.

Li, L. and Morris, J. E., Electrical conduction models for isotropically conductive adhesive joints, *IEEE Transactions on Components, Packaging and Manufacturing Technology, Part A* **20**(1) (1997), pp. 3–8.

Li, L. and Morris, J. E., An introduction to electrically conductive adhesives, *International Journal of Microelectronic Packaging*, **1**(3), (1998), 159–175.

Li, L. and Morris, J. E., 'Curing of isotropic electrically conductive adhesives' in *Conductive Adhesives for Electronics Packaging*, J. Liu, ed., Electrochemical Press, UK (1999), pp. 99–116.

Li, L., Kim, H., Lizzul, C., Sacolick, I. and Morris, J. E., Electrical, structural and processing properties of electrically conductive adhesives, *IEEE Transactions on Components, Packaging and Manufacturing Technology*, **16**, (1993), pp. 843–851.

Li, L., Morris, J. E., Liu, J., Lai, Z., Ljungkrona, L. and Li, C., Reliability and failure mechanism of isotropically conductive adhesives joints, *Proc. 45th Electronic Components and Technol. Conference*, Las Vegas, NV (1995), pp. 114–120.

Li, Y., Moon, K., Li, H. and Wong, C. P., Conductivity improvement of isotropic conductive adhesives with short-chain dicarboxylic acids, *Proc.55th Electronic Components and Technology Conference*., Las Vegas, NV (2004a), pp. 1959–1964.

Li, Y., Moon, K., Li, H. and Wong, C. P., Conductivity improvement of isotropically conductive adhesives, *Proc. 6th IEEE CPMT Conf. High Density Microsystem Design and Packaging and Component Failure Analysis (HDP'04)*, Shanghai (2004b), pp. 236–241.

Li, Y., Moon, K.-S. and Wong, C. P., Electrical property improvement of electrically conductive adhesives through *in situ* replacement by short-chain difunctional acids, *IEEE Transactions on Components and Packaging Technologies*, **29**(1), (2006a), pp. 173–178.

Li, Y., Moon, K.-S., Whitman, A. and Wong, C. P., Enhancement of electrical properties of electrically conductive adhesives (ECAs) by using novel aldehydes, *IEEE Transactions on Components and Packaging Technologies*, **29**(4), (2006b), pp. 758–763.

Li, Y., Lu, D. and Wong, C.-P., *Electrically Conductive Adhesives with Nanotechnologies*, Springer, 2010.

Lin, X. and Lin, F., The improvement of the properties of silver-containing conductive adhesives by the addition of carbon nanotube, *Proc. 6th IEEE CPMT Conference on*

High Density Microsystem Design and Packaging and Component Failure Analysis (HDP'04), Shanghai (2004), pp 382–384.

Liong, S., Zhang, Z. and Wong, C-P., High frequency measurement of isotropically conductive adhesives, *Proc. 51st Electronic Components and Technology Conference*, Orlando, FL (2001), s35p6.

Liong, S., Wong, C. P. and Burgoyne, W. Jr., Adhesion improvement of thermoplastic isotropically conductive adhesive, *Proc. 8th International Symposium on Advanced Packaging Materials*, Braselton, GA (2002), pp. 260–270.

Liu, J. (Ed.), *Conductive Adhesives for Electronics Packaging*, Electrochemical Publications Ltd (1999), UK.

Liu, J. and Morris, J. E., State of the art in electrically conductive adhesive joining, *Proc. Workshop Polymeric Materials for Microelectronics and Photonics Applications*, EEP-Vol. 27, ASME, (1999), pp. 259–281.

Liu, J., Ljungkrona, L. and Lai, Z., Development of conductive adhesive joining for surface-mounting electronics manufacturing, *IEEE Transactions on Components, Packaging and Manufacturing Technology Part B*, **18**(2), (1995), pp. 313–319.

Liu, J., Gustafsson, K., Lai, Z. and Li, C., Surface characteristics, reliability and failure mechanisms of tin, copper and gold metallisations, *Proc. 2nd International Conference on Adhesive Joining and Coating Technology in Electronics Manufacturing*, Stockholm (1996), pp. 141–153.

Liu, J., Lundström, P., Gustafsson, K. and Lai, Z., Conductive adhesive joint reliability under full-cure conditions, EEP-Vol. 19-1 *Advances in Electronics Packaging*, Vol. 1 ASME (1997), pp. 193–199.

Lohokare, S. K., Lu, Z., Scheutz, C. A. and Prather, D. W., Electrical characterization of flip-chip interconnects formed using a novel conductive-adhesive-based process, *IEEE Transactions on Advanced Packaging*, **29**(3), (2006), pp. 542–547.

Low, R. C., Miessner, R. and Wilde, J., Additional stresses of ECA joints due to moisture induced swelling, *Proc 16th European Microelectronics and Packaging Conference and Exhibition*, Oulu, Finland (2007) pp. 229–234.

Lu, D. and Wong C. P., High performance conductive adhesives, *IEEE Transactions on Electronics Packaging Manufacturing*, **22**(4), (1999a), pp. 324–330.

Lu, D. and Wong, C. P., Conductive adhesives with improved properties, *Proc. 2nd IEEE International Symposium on Polymeric Electronics Packaging*, Gothenburg, Sweden (1999b), pp. 1–8.

Lu, D. and Wong, C. P., Thermal decomposition of silver flake lubricants, *Journal of Thermal Analysis and Calorimetry*, **61**, (2000a), pp. 3–12.

Lu, D. and Wong, C. P., Characterization of Silver Flake Lubricants, *Journal of Thermal Analysis and Calorimetry*, **59**, (2000b), pp. 729–740.

Lu, D. and Wong C-P., Isotropic conductive adhesives filled with low-melting-point alloy fillers, *IEEE Transactions on Electronics Packaging Manufacturing*, **23**(3), (2000c), pp. 185–190.

Lu, D. and Wong, C-P., A study of contact resistance of conductive adhesives based on anhydride-cured epoxy systems, *IEEE Transactions on Components and Packaging Technologies*, **23**(3), (2000d), pp. 440–446.

Lu, D. and Wong, C. P., Development of conductive adhesives for solder replacement, *IEEE Transactions Components Packaging Technologies*, **23**(4), (2000e), pp. 620–626.

Lu, D. and Wong, C. P., Development of conductive adhesives filled with low-melting-point alloy fillers, *International Symposium on Advanced Packaging Materials*, Porterin, Isle of Man, British Isles (2000f), pp. 336–341.

Lu, D., Wong, C. P. and Tong, Q. K., Mechanisms underlying the unstable contact resistance of conductive adhesives, *Proc. 49th Electronic Components and Technology Conference*, San Diego (1999a) Paper s10p2.

Lu, D., Tong, Q. and Wong, C-P., Mechanisms underlying the unstable contact resistance of conductive adhesives, *IEEE Transactions on Electronics Packaging Manufacturing*, **22**(3), (1999b), pp. 228–232.

Lu, D., Tong, Q. K. and Wong C. P., A study of lubricants on silver flakes for microelectronics conductive adhesives, *IEEE Transactions on Components and Packaging Technologies*, **22**(3), (1999c), pp. 365–371.

Luchs, R., Application of electrically conductive adhesives in SMT, *Proceedings of the 2nd International Conference on Adhesive Joining and Coating Technology in Electronics Manufacturing*, Stockholm (1996), pp. 76–83.

Luo, S. and Wong, C. P., Thermo-mechanical properties of epoxy formulations with low glass transition temperatures, *Proc. 8th International Symposium on Advanced Packaging Materials*, Braselton, GA (2002), pp. 226–231.

Luoma, S. N., *Silver Nanotechnologies and the Environment*, Project on Emerging Nanotechnologies Report PEN-15, Woodrow Wilson International Center for Scholars and The Pew Charitable Trusts, September 2008.

Mach, P., Richter, L. and Pietrikova, A., Modification of electrically conductive adhesives for better mechanical and electrical properties, *Proc. 31st International Spring Seminar on Electronics Technology*, Budapest (2008) pp. 230–235.

Manepallis, R., Stepniak, F., Bidstrup-Allen, S. A. and Kohl, P., Silver metallization for advanced interconnects, *IEEE Transactions on Advanced Packaging*, **22**(1), (1999), pp. 4–8.

Markley, D. L., Tong, Q. K., Magliocca, D. J. and Hahn, T. D., Characterization of silver flakes utilized for isotropic conductive adhesives, *International Symposium on Advanced Packaging Materials, Processes, Properties and Interfaces*, Braselton, Georgia (1999), pp. 16–20.

McCluskey, P., Morris, J. E., Verneker, V., Kondracki, P. and Finello, D., Models of electrical conduction in nanoparticle filled polymers, *Proc. 3rd International Conference on Adhesive Joining and Coating Technology in Electronics Manufacturing*, Binghamton, NY (1998), pp. 84–89.

Miragliotta, J., Benson, R. C. and Phillips, T. E., Measurements of electrical resistivity and Raman scattering from conductive die attach adhesives, *Proc 8th International Symposium Advanced Packaging Materials*, Atlanta, GA (2002), pp. 132–138.

Mo, Z., Wang, X., Wang, T., Li, S., Lai, Z. and Liu, J., Electrical characterization of isotropic conductive adhesive under mechanical loading, *Journal of Electronic Materials*, **31**(9), (2002), pp. 916–920.

Mo, Z., Reliability and applications of adhesives for microsystem packaging, *Doktorsavhandlingar vid Chalmers Tekniska Hogskola*, (2005), p. 35.

Moon, K. Wu, J. and Wong, C.-P., Study on self-alignment capability of electrically conductive adhesives (ECAs) for flip-chip application, *Proc. International Symposium and Exhibition on Advanced Materials*, Braselton, GA (2001), 341–346.

Moon, K. S., Wu, J. and Wong, C. P., Improved stability of contact resistance of low melting point alloy incorporated isotropically conductive adhesives, *IEEE Transactions on Components and Packaging Technologies*, **26**(2), (2003), pp. 375–381.

Moon, K. S., Dong, H., Maric, R., Pothukuchi, S., Hunt, A., Li, Y. and Wong, C. P., Thermal behavior of silver nanoparticles for low-temperature interconnect applications, *Journal of Electronic Materials*, **34**(2), (2005), pp. 168–175.

Morris, J. E., Conduction mechanisms and microstructure development in isotropic, electrically conductive adhesives, in *Conductive Adhesives for Electronics Packaging*, J. Liu, ed., Electrochemical Press, UK (1999) pp. 37–77.

Morris, J. E., Isotropic conductive adhesives: Future trends, possibilities, and risks, *Proc. 5th International IEEE Conference on Polymers and Adhesives in Microelectronics and Photonics (Polytronic)*, Wroclaw (2005), pp. 233–234.

Morris, J. E., Isotropic conductive adhesives: Future trends, possibilities, and risks, *Microelectronics Reliability*, **47**(2/3), (2007) pp. 328–330.

Morris, J. E., *Electrically Conductive Adhesives*, Springer, 2011 (in development).

Morris, J. E. and Lee, J., Drop test performance of isotropic electrically conductive adhesives, *Journal of Adhesion Science and Technology*, **22**, (2008), 1699–1716.

Morris, J. E. and Liu, J., An Internet course on conductive adhesives for electronics packaging, *Proc. 50th Electronic Components and Technology Conf.*, Las Vegas (2000) pp. 1016–1020.

Morris, J. E. and Liu, J., Electrically Conductive Adhesives (ECAs) in *Micro- and Opto-Electronic Materials and Structures: Physics, Mechanics, Design, Reliability, and Packaging*, E. Suhir, Y. C. Li, and C-P. Wong (editors), Springer, US, 2006.

Morris, J. E. and Probsthain, S., Investigations of plasma cleaning on the reliability of electrically conductive adhesives, *Proc. 4th International Conference on Adhesive Joining and Coating Technology in Electronics Manufacturing*, Espoo, Finland (2000), pp. 41–45.

Morris, J. E., Cook, C., Armann, M., Kleye, A. and Fruehauf, P., Recent results of ICA testing, *Proc. 2nd International Symposium Polymeric Electronics Packaging (PEP'99)*, Gothenburg, Sweden (1999), pp. 15–25.

Morris, J. E., Anderssohn, F., Kudtarkar, S. and Loos, E., Reliability studies of an isotropic electrically conductive adhesive, *Proc. 1st International Conference Polymers and Adhesives in Microelectronics and Photonics*, Potsdam, Germany (2001), pp. 61–69.

Morris, J. E., Anderssohn, F., Loos, E. and Liu, J., Low-tech studies of isotropic electrically conductive adhesive, *Proc. 26th International Spring Seminar on Electronics Technology*, Stara, Lesna, Slovak Republic (2003), pp. 90–94.

Morris, J. E., Tilford, T., Bailey, C., Sinclair, K. I. and Desmulliez, M. P. Y., Polymer cure modeling for microelectronics packaging, *Proc. 32nd International Spring Seminar on Electronics Technology*, Brno (2009a).

Morris, J. E., Tilford, T., Bailey, C., Ferents, M., Sinclair, K. I. and Desmulliez, M. P. Y., Critical Analysis of Polymer Cure Modeling for Microelectronics Applications, *Proc. IMAPS-Nordic Conference*, Tonsberg, Norway (2009b).

Morris, J. E., Lee, J. and Liu, J., Isotropic conductive adhesive interconnect technology in electronics packaging applications, in *Surfaces and Interfaces in Nanomaterials for Interconnects*, C. Liu (editor), Springer (2010).

Moscicki, A., Felba, J., Sobierajski, T. and Kudzia, J., Snap-curing electrically conductive formulation for solder replacement applications, *Journal of Electronic Packaging, Transactions of the ASME*, **127**, (2005), pp. 91–95.

Mundlein, M. and Nicolics, J., Modeling of particle arrangement in an isotropically conductive adhesive joint, *Proc. 4th International Conference. Polymers and Adhesives in Microelectronics and Photonics*, Portland, Oregon (2004), pp-3.

Mundlein, M. and Nicolics, J., Electrical resistance modeling of isotropically conductive adhesive joints, *Proc. 28th International Spring Seminar on Electronics Technology*, Wiener Neustadt, Austria (2005), pp. 128–133.

Mundlein, M., Hanreich, G. and Nicolics, J., Simulation of the aging behavior of isotropic conductive adhesives, *Proc. 2nd International IEEE Conference Polymers and Adhesives in Microelectronics and Photonics*, Zalaegerszeg, Hungary (2002) pp. 68–72.

Mustoe, G. G. W., Nakagawa, M., Lin, X. and Iwamoto, N., Simulation of particle compaction for conductive adhesives using discrete element modeling, *Proc. 49th IEEE Electronic Components and Technol. Conference,* San Diego (1999), Paper s10p4.

Nishikawa, H., Mikami, S., Terada, N., Miyake, K., Aoki, A. and Takemoto, T., Electrical property of conductive adhesives using silver-coated copper filler, *Proc. 2nd Electronics Systemintegration Technology Conference*, Greenwich, UK (2008), pp. 825–828.

Nysaether, J., Lai, Z. and Liu, J., Thermal cycling lifetime of flip chip on board circuits with solder bumps and isotropically conductive adhesive joints, *IEEE Transactions on Advanced Packaging*, **23**(4), (2000), pp. 743–749.

Otsuka, K. and Akiyama, Y., The GHz region characteristic of the BGA-printed wired board interconnection by conductive adhesives, *Proc. 6th International Conference on Polymers and Adhesives in Microelectronics and Photonics*, Japan (2007), pp. 277–279.

Paproth, A., Wolter, K-J., Herzog, T. and Zerna, T., Influence of plasma treatment on the improvement of surface energy, *Proc. 24th International Spring Seminar Electronics Technology*, Romania (2001), pp. 37–41.

Pas, F.V.D., New, highest reliable generation of PWB surface finishes for lead-free soldering and future applications, *European Institute of Printed Circuits Summer Conference*, presentation, 2005.

Perichaud, M. G., Deletage, J. Y., Carboni, D., Fremont, H., Danto, Y. and Faure, C., Thermomechanical behaviour of adhesive jointed SMT components, *Proc. 3rd International Conference on Adhesive Joining and Coating Technology in Electronics Manufacturing*, Binghamton, NY (1998), pp. 55–61.

Pothukuchi, S., Yi, L. and Wong, C. P., Shape controlled synthesis of nanoparticles and their incorporation into polymers, *Proc. 54th Electronic Components and Technology Conference*, Atlanta, GA (2004), pp. 1965–1967.

Ramakumar, S. M. and Srihari, K., Influence of process parameters on component assembly and drop test performance using a novel anisotropic conductive adhesive for lead-free surface mount assembly, *Proc. 58th Electronic Components and Technology Conference*, Orlando, FL (2008), pp. 225–233.

Robinson, M., Leal, J. and Andrews, L., Conformal polymer interconnect for high density chip scale packages, *Advancing Microelectronics*, **35**(6), (2008), pp. 20–23.

Rörgren, R. and Liu, J., Reliability assessment of isotropically conductive adhesive joints in surface mount applications, *IEEE Transactions on Components, Packaging and Manufacturing Technology Part B.*, **18**(2), (1995), pp. 305–312.

Rusanen, O., Modelling of ICA creep properties, *Proc. 4th International Conference on Adhesive Joining and Coating Technology in Electronics Manufacturing*, Espoo, Finland, (2000), pp. 194–198.

Rusanen, O. and Laitinen, J. R., Reasons for using lead-free solders rather than isotropically conductive adhesives in mobile phone manufacturingt, *Proc. 4th International Conference. Polymers and Adhesives in Microelectronics and Photonics*, Portland, Oregon (2004), pp. 89–91.

Rusanen, O. and Lenkkeri, J., Reliability issues of replacing solder with conductive adhesives in power modules, *IEEE Transactions on Components, Packaging and Manufacturing Technology Part B.*, **18**(2), (1995), pp. 320–325.

Rusanen, O., Keraenen, Blomberg, M. and Lehto, A., Adhesive fli-chip bonding in a miniaturised spectrometer, *Proc. 1st IEEE Sympos. Polymeric Electronics Packaging Conference (PEP'97)*, (1997), pp. 95–100.

Ruschau, G. R., Yoshikawa, S. and Newnham, R. E., Percolation constraints in the use of conductor-filled polymers for interconnects, *Proc. 42nd Electronic Components and Technology Conference*, (1992), Atlanta, GA, pp. 481.

Sancaktar, E. and Dilsiz, N., Anisotropic alignment of nickel particles in a magnetic field for electronically conductive adhesives applications, *Journal of Adhesion Science and Technology*, **11**(2), (1997), pp. 155–166.

Sancaktar, E. and Dilsiz, N., Thickness dependent conduction behavior of various particles for conductive adhesive applications, *Proc. 3rd International Conference on Adhesive Joining and Coating Technology in Electronics Manufacturing*, Binghamton, NY (1998), pp. 90–95.

Sancaktar, E., Rajput, P. and Khanolkar, A., Correlation of silver migration to the pull out strength of silver wire embedded in an adhesive matrix, *Proc. 4th International Confer. Polymers and Adhesives in Microelectronics and Photonics*, Portland, OR (2004), RT2-1.

Segerberg, T., *Life Cycle Analysis – A Comparison Between Conductive Adhesives and Lead Containing Solder for Surface Mount Application*, IVF report, (1997).

Shimada, Y., Lu, D. and Wong, C. P., Electrical characterizations and considerations of electrically conductive adhesives (ECAs), *6th international Symposium on Advanced Packaging Materials*, Braselton, GA (2000), pp. 336–342.

Sihlbom, R., Dernevik, M., Lai, Z., Starski, P. and Liu, J., Conductive Adhesives for High-frequency Application, *Proc. Polymeric Electronics Packaging Conference*, Norrköping, Sweden (1997), pp. 123–130.

Sihlbom, A. and Liu, J., Thermal characterisation of electrically conductive adhesive flip-chip joints, *Proc. Electronic Packaging Technology Conference*, December 8–10, (1998), Singapore, pp. 251–257.

Smilauer, P., Thin metal films and percolation theory, *Contemporary Physics*, Vol. 32, (1991), pp. 89–102.

Su, B. and Qu, J., A micromechanics model for electrical conduction in isotropically conductive adhesives during curing, *Proc. 54th IEEE Electronic Component and Technology Conference*, Las Vegas, NV (2004), pp. 1766–1771.

Suzuki, K., Suzuki, O. and Komagata, M., Conductive adhesive materials for lead solder replacement, *IEEE Transactions on Components, Packaging, and Manufacturing Technology Part A*, **21**(2), (1998), pp. 252–258.

Suzuki, K., Shirai, Y., Mizumura, N. and Konagata, M., Conductive adhesives containing Ag-Sn alloys as conductive filler, *Proc. 4th International Confer. Polymers and Adhesives in Microelectronics and Photonics*, Portland, OR (2004), MP3-1.

Takezawa, H., Mitani, T., Kitae, T., Sogo, H., Kobayashi, S. and Bessho, Y., Effects of zinc on the reliability of conductive adhesives, *Proc. 8th International Symposium on Advanced Packaging Materials*, Braselton, GA (2002), pp. 139–143.

Tilford, T. *et al.*, Multiphysics simulation of microwave curing in micro-electronics packaging applications, *Journal of Soldering and Surface Mount Technology*, **19**, (2007), pp. 26–33.

Toida, G., Shirai, Y., Mizumura, N., Komagata, M. and Suzuki, K., Conductive adhesives containing Ag-Sn alloys as conductive filler, *Proc. 5th International Conference on Polymers and Adhesives in Microelectronics and Photonics*, Wroclaw, Poland (2005), pp. 7–12.

Tong, Q. Liu, D. and Wong, C. P., A Fundamental study on silver flakes for conductive adhesives, *Proc. 4th International Symposium and Exhibition on Advanced Packaging Materials, Processes, Properties and Interfaces*, Braselton, Georgia (1998a), pp. 256–260.

Tong, Q., Vona, S., Kuder, R. and Shenfield, D., The recent advances in surface mount conductive adhesives, *Proc. 3rd International Conference on Adhesive Joining and Coating Technology in Electronics Manufacturing*, Binghamton, NY (1998b), pp. 272–277.

Tong, Q. K., Markley, D., Fredrickson, G. and Kuder, R., Conductive adhesives with stable contact resistance and superior impact performance, *Proc. 49th Electronic Components and Technology Conference*, San Diego (1999), Paper s10p3.

Wakuda, D., Kim, C.-J., Kim, K.-S. and Suganuma, K., Room temperature sintering mechanism of Ag nanoparticle paste, *Proc. 2nd Electronics Systemintegration Technology Conference*, Greenwich, UK (2008), pp. 909–913.

Wang, T., Fu, Y., Becker, M. and Liu, J., Microwave cure of metal-filled electrically conductive adhesive, *Proc. 51st Electronic Components and Technology Conference*, Orlando, FL (2001). pp. 593–597.

Westphal, H., Health and environmental aspects of conductive adhesives – the use of lead based alloys compared with adhesives, in *Conductive Adhesives for Electronics Packaging*, edited by Liu, J., Electrochemical Publications Ltd, UK, (1998), pp. 415–424.

Wirts-Ruetters, M., Heimann, and Wolter K.-J., Carbon nanotube (CNT) filled adhesives for microelectronic packaging, *Proc. 2nd Electronics Systemintegration Technology Conference*, Greenwich, UK (2008), pp. 1057–1062.

Wong, C. P. and Li, Y., Recent advances on electrical conductive adhesives (ECAs), *Proc. 4th International Conf. Polymers and Adhesives in Microelectronics and Photonics*, Portland, OR (2004), PL-1.

Wong, C. P., Xu, J., Zhu, L., Li, Y., Jiang, H., Sun, Y., Lu, J. and Dong, H., Recent advances on polymers and polymer nanocomposites for advanced electronic packaging applications, *Proc. High Density Microsystem Design and Packaging, Component Failure Analysis*, Shanghai, China (2005), pp. 1–16.

Wu, J., Moon, K. and Wong, C-P., Self-alignment feasibility study and contact resistance improvement of electrically conductive adhesives (ECAs), *Proc. 51st Electronic Components and Technology Conference*, Orlando, FL (2001), Paper c17p2.

Wu, H., Wu, X., Liu, J., Zhang, G., Wang, Y., Zeng, Y. and Jing, J., Development of a novel isotropic conductive adhesive filled with silver nanowires, *Journal of Composite Materials*, **40**, (2006a), pp. 1961–1969.

Wu, H. P., Liu, J. F., Wu, X. J., Ge, M. Y., Wang, Y. W., Zhang, G. Q. and Jiang, J. Z., High conductivity of isotropic conductive adhesives filled with silver nanowires, *International Journal of Adhesion and Adhesives*, **26**, (2006b), pp. 617–621.

Wu, S., Mei, Y., Yeh, C. and Wyatt, K., Process induced residual stresses in isotropically conductive adhesive joints, *IEEE Transactions on Components, Packaging and Manufacturing Technology Part C.*, **19**(4), (1996), pp. 251–256.

Wu, S., Hu, K. and Yeh, C.-P., Contact reliability modelling and material behaviour of conductive adhesives under thermomechanical loads, in *Conductive Adhesives for Electronics Packaging* (edited by J. Liu), Electrochemical Publications Ltd, UK, (1998), pp. 117–150.

Xu, S. and Dillard, D., Determining the impact resistance of electrically conductive adhesives using a falling wedge test, *IEEE Transactions on Components and Packaging Technologies*, **26**(3), (2003), pp. 554–562.

Yamashita, M. and Suganuma, K., Differences in heat exposure degradation of Sn alloy platings joined with Ag–epoxy conductive adhesive, *Journal of Materials Science*, **41**, (2006a), pp. 583–585.

Yamashita, M. and Suganuma, K., Improvement in high-temperature degradation by isotropic conductive adhesives including Ag-Sn alloy fillers, *Microelectronics Reliability*, **46**, (2006b), pp. 850–858.

Yamashita, M. and Suganuma, K., Degradation by Sn diffusion applied to surface mounting with Ag–epoxy conductive adhesive with joining pressure, *Microelectronics Reliability*, **46**, (2006c), pp. 1113–1118.

Yan, Y. K., Xue, Q. Z., Zheng, Q. B. and Hao, L. Z., The interface effect of the effective electrical conductivity of carbon nanotube composites, *Nanotechnology*, **18**(25), (2007), 255705.

Yang, C., Yuen, M. M. F. and Bing Xu, Using novel materials to enhance the efficiency of conductive polymer; *Proc. 58th Electronic Components and Technology Conference*, Orlando FL (2008), pp. 213–218.

Ye, L., Lai, Z., Liu, J. and Thölén, A., Effect of Ag particle size on electrical conductivity of isotropically conductive adhesives, *IEEE Transactions on Electronics Packaging Manufactureing*, **22**(4), (1999), pp. 299–302.

Yi, L. and Wong, C. P., A novel non-migration nano-Ag conductive adhesive with enhanced electrical and thermal properties via self-assembled monolayers modification, *Proc. 56th Electronic Components and Technology Conference*, San Diego, CA (2006a), pp. 924–931.

Yi, L. and Wong, C. P., Silver migration control in electrically conductive adhesives, *Proc. High Density Microsystem Design, Packaging and Component Failure Analysis*, Shanghai (2006b), Paper B06.

Yi, L., Xiao, F., Moon, K.-S. and Wong, C. P., A novel environmentally friendly and biocompatible curing agent for lead free electronics, *Proc. 56th Electronic Components and Technology Conference*, San Diego, CA (2006), pp. 1639–1644.

Yi, L., Yim, M. J., Moon, K., Zhang, R. W. and Wong, C. P., Development of novel, flexible, electrically conductive adhesives for next-generation microelectronics interconnect applications, *Proc. 58th Electronic Components and Technology Conference*, Orlando, FL (2008), pp. 1272–1276.

Zhang, Z., Jiang, S., Liu, J. and Inoue, M., Development of high temperature stable isotropic conductive adhesives, *Proc. International Conference on Electronics Packaging Technology/High Density Packaging*, Shanghai (2008), E03-04, pp. 1–5.

Zhou J. and Sancaktar, E., Chemorheology of epoxy/nickel conductive adhesives during processing and cure, *Journal of Adhesion Science Technology*, **22**, (2008a), pp. 957–981.

Zhou, J. and Sancaktar, E., Stable and unstable capillary flows of highly-filled epoxy/nickel suspensions, *Journal of Adhesion Science Technology*, **22**, (2008b), pp. 983–1002.

Zwolinski, M., Electrically conductive adhesives for surface mount solder replacement, *IEEE Transactions on Components, Packaging and Manufacturing Technology, Part C*, **19**(4), (1996), pp. 241–250.

5
Underfill adhesive materials for flip chip applications

Q. K. TONG, Henkel Corporation, USA

Abstract: The advance of the semiconductor industry requires smaller and lighter micro-electronic packages with faster and more reliable electronic performance. Accordingly, flip chip technology, a direct chip attachment (DCA) technology that directly connects the circuitries between the silicon chip and the substrate, has been the fastest growing packaging in the semiconductor industry. Over the last decade it has also developed into many technology platforms, such as chip scale packaging (CSP) and multi-chip packaging. The reliability challenge of this new packaging technology was realized almost immediately and organic reinforcing materials were proposed, to be applied between the chip and the substrate ('underfill'). To improve the productivity and reduce the cost of this new packaging technology, many forms of processes and their corresponding underfill materials, including the original capillary underfill, no-flow underfill, and wafer-level underfill, are currently actively being explored.

Key words: flip chip, direct chip attachment, underfill, advanced packaging, reliability, electronic materials.

5.1 Introduction: flip chip and direct chip attachment technology

As the semiconductor industry advances, smaller and lighter micro-electronic packages with faster and more reliable electronic performance have become the general trend for the micro-electronic packaging industry. Direct chip attachment (DCA) technologies, which connect the electronic circuitry on the silicon chip directly to the circuitry on the substrate, have gained wide acceptance in the semiconductor industry. In contrast to the conventional wire-bonding packages, direct chip attachment connects the chip 'face down' directly to the substrate. Therefore, it gained the popular name 'flip chip' technology early in its development stage. Over the years, direct chip attachment technology developed into many technology platforms, including flip chip, chip scale packaging (CSP), and multi-chip packaging or 3D packaging, etc. Correspondingly, so-called 'underfill' packaging technology, which fills the re-enforcement materials into the gap between the chip and the substrate, has also been developed to ensure the reliability of these new types of package.

IBM first developed the C4 technology (Controlled Collapse Chip Connection) in the early 1960s for its mainframe computers.[1] This became the first generation of flip chip packages. At its early stage, the chip was connected to ceramic substrates. In the late 1960s, Delco Electronics was the first company to use flip chip technology for automotive applications The technology did not gain broad industrial acceptance, however, until mid 1990s. Since then, flip chip on organic substrates has gained popularity in the semiconductor industry. The flip chip technology also enabled many forms of DCA technologies, such as CSP, which usually refers to package sizes of less than 120% of the chip size, and multi-chip packaging (3D packaging). Currently, DCA is the fastest growing sector in the semiconductor packaging industry with a projected > 50% annual growth rate. Most mobile semiconductor devices such as cellular phones, pagers, laptops, as well as high-speed microprocessors are currently assembled with direct chip attachment, mostly flip chip technology.

5.2 Advantages of direct chip attachment technology

DCA technology, possesses many advantages in size, performance, flexibility, reliability, and cost over conventional wire-bonding packaging technology.

5.2.1 Smaller, thinner, and lighter packages

DCA technology directly connects the circuitry on the chip and the circuitry on the substrate by metal bumps in between, making the height (or thickness) of the package much less when compared with the use of conventional wire-bonding technology. In addition, conventional wire-bonding technology allows the wire connections only around the perimeter of the chip. Flip chip technology, however, uses the whole area of the chip and hence reduces the required size of the package by as much as 95%. Finally, the smaller and thinner flip chip package also significantly reduces the weight of the package. In some cases, the weight can be as little as 5% of the conventional wire-bonding package.

5.2.2 High performance

Since the electrical connections are made by directly connecting the chip and the substrate, flip chip technology offers the highest electrical speed performance among the available assembly technologies. Research has indicated that eliminating bond wires reduces the delaying inductance and capacitance of the connection by a factor of 10, and shortens the path by a factor of 25 to 100.[2] As discussed earlier, flip chip technology can make electrical connection using the whole chip area, and thus achieves the highest

I/O density of all assembly technologies. In addition, it also offers flexibility of I/O connectivity and allows more advanced assembly technology, such as 3D packaging.

5.3 Reliability challenge of flip chip technology

Challenges in the reliability of this new technology were recognized almost immediately when the flip chip technology was introduced to broad applications in the semiconductor industry. In the mid 1990s, organic substrates were broadly used in the semiconductor industry. Unlike the ceramic substrates that IBM used in its early C4 applications, organic substrates have a much higher coefficient of thermal expansion (CTE) than the silicon chip, as well as the solder bump connections. Therefore, the primary reliability challenge of this new type of microelectronic package was its reliability under thermal stress. During the thermal cycling reliability test (experiencing from – 40 °C to 150 °C repeatedly for as much as a few thousand cycles), the metal bumps (usually Sn/Pb eutectic alloy) connecting the circuitries of the chip and the substrate may crack and result in failure of the assembly. The root cause of this reliability failure is apparently the thermal mismatch of the low CTE of the silicon chip (3 ppm) and the high CTE of the organic substrate (usually 50–80 ppm). Encapsulant materials were soon introduced to reinforce the solder bumps and to enhance the solder interconnect reliability.[3] Since the flip chip and the substrate are connected by the solder bumps, the encapsulant material needs to be flowed into the gaps between the chip and the substrate. Therefore, this encapsulation process earned the name 'underfill process' and the corresponding encapsulant material was referred to as the 'flip chip underfill' material.

5.4 Advances in the flip chip underfill process and encapsulant materials

Since the flip chip technology gained broad acceptance in the semiconductor industry, scientists and engineers have been working collaboratively to further advance this technology and develop corresponding encapsulant materials. In order to improve the productivity and reduce the overall cost of this technology, many processes to underfill the chip assembly have been explored, which include the original capillary underfill process, so-called 'no-flow' or 'pre-applied' underfill processes, and the 'wafer level' pre-applied underfill process. Each of these processes requires different material properties of the encapsulant; in many cases, joint process development and material development programs have been carried out. The joint programs have involved universities and affiliated research centers, major semiconductor manufacturing companies, and major chemical companies that supply the underfill encapsulant materials.

5.4.1 Capillary underfill process

The first attempt to apply organic underfill encapsulant materials to reinforce the solder bump interconnects consisted of allowing the encapsulating materials to flow into the gap between the chip and the substrate by capillary force. Figure 5.1 illustrates the current 'capillary underfill' process in flip chip and CSP technologies. First, a chip is attached to a substrate during the 'reflow' process. After the solder interconnects are established, a low viscosity encapsulant material is applied to fill the gap between the chip and the substrate. Since the encapsulant material is flowed into the gap by capillary force, this is called the 'capillary underfill' process. In this process, each package is underfilled individually, which limits productivity. Although this process is currently widely employed in the industry and demonstrates enhanced reliability of the packages, it is slow since the encapsulant material has to flow gradually to fill the gap and each package has to be underfilled individually. Requirements on material flow properties are also stringent in order to shorten the flow time. The viscosity of the encapsulant must be low, and sometimes heat has to be applied to ensure that the encapsulant flows properly. Therefore, the slow underfilling and resulting relatively high cost become the bottlenecks for this capillary underfill process.

The underfill encapsulant material usually contains up to 50% by weight silica fillers to reduce its CTE from 50–90 ppm to 20–30 ppm and match the solder bump's CTE for maximum reinforcement. The primary material design challenge for the capillary underfill process is the requirement for low viscosity of the underfill material with a high percentage of silica fillers. If the flow property of underfill material is not designed properly, the flow may not be complete, or it may leave many voids in the package, which

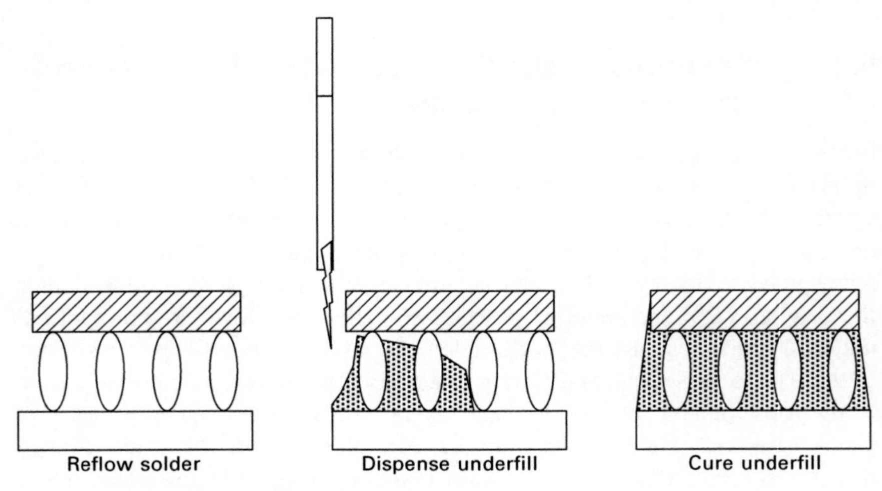

5.1 Capillary underfill process.

causes serious reliability problems. Another issue is filler settling – silica fillers gradually settle during the flow process resulting in non-homogeneity of the underfill encapsulant, which also affects reliability. Finally, it is required that the underfill encapsulant form a good fillet around the periphery of the die, which is essential for high package reliability. As the chip size increases, these challenges become more and more significant. In addition, low viscosity organics have to be cured into solid form and without generating volatiles. It has been reported that any voids generated under the chip may cause reliability failure of the flip chip assembly. Finally, the cure process of the low viscosity liquid underfill encapsulant has to be fast to ensure high productivity and reduce cost. In the meantime, in order to fulfill the maximum package reinforcement, the cured underfill encapsulant should be a high modulus solid, with low moisture absorption and good adhesion to the chip, solder bumps and the organic substrate.

Despite the above challenges, over the last 15 years, many material manufacturers have successfully developed a number of commercial underfill encapsulants. Many reports demonstrate that the underfill encapsulant has improved the reliability of flip chip packages by 10 to 100 times. The capillary underfill process is still by for the most dominant process for direct chip attachment.

5.4.2 'Known Good Die' issue and reworkability

During the development of DCA technology, the issue of reworkability of the assembly has been raised. With conventional wire bonding packaging technology, a failed package can be easily removed by melting the solder connection and replacing it by re-soldering a good package. After the underfill process with the solid encapsulant materials surrounding the solder bumps, it is no longer straightforward to replace the damaged chip. Over the years, both the flip chip rework process and reworkable encapsulant materials have been developed by several researchers. The proposed flip chip rework process is illustrated in Fig. 5.2 and includes:

- localized chip removal by heating
- underfill removal and site preparation
- new chip placement and solder reflow at the rework site
- application of underfill and subsequent curing

Solid thermoplastics are formed by polymerizing liquid small molecules into linear thermoplastic polymer. Heat can make the thermoplastics flowable for their removal, or, alternatively they can be removed by using a suitable solvent. Thermosets are solidified by 'curing' liquid small molecules and form a cross-linked network structure. Heat cannot make the thermosets flowable and solvents cannot dissolve the cross-linked network structure,

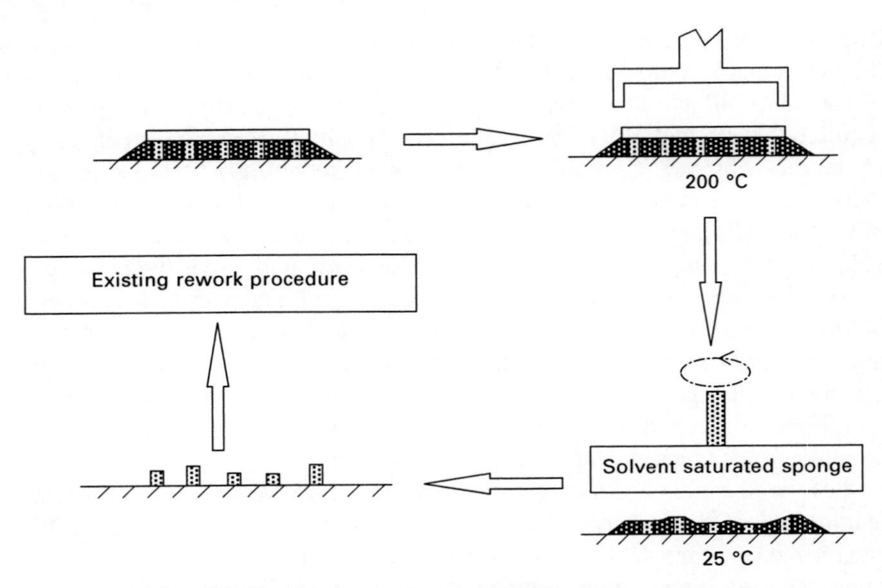

5.2 Localized chip removal and underfill clean-up procedure.

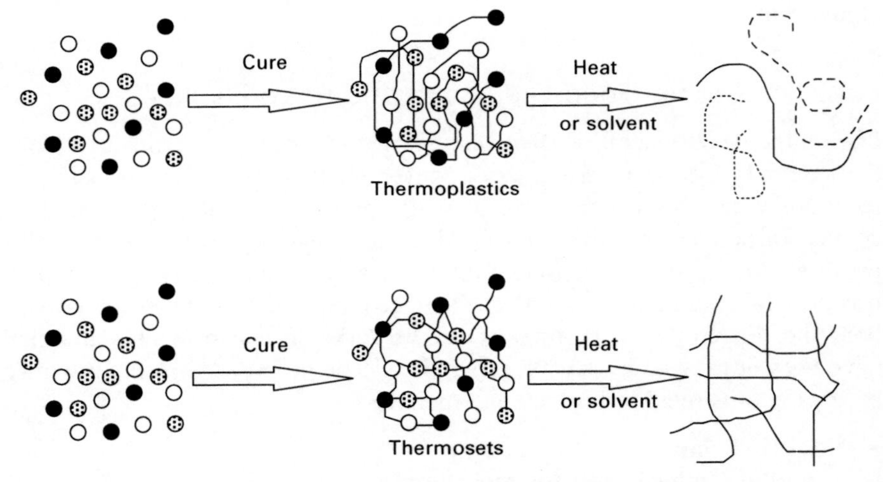

5.3 Comparison between the reactive thermoplastic and thermosetting chemistry.

either. Thermoset materials have to go through a degradation process to break down the network structure for their removal.

Reworkable underfill encapsulant materials have been developed in parallel. One of the approaches was 'reactive thermoplastic' and this is summarized in Fig. 5.3.[4]

Another approach has been to use thermally-degradable thermosetting materials. Upon heating, the thermosetting materials decompose and form low molecular weight monomers and oligomers. This allows the underfill encapsulant material to be easily removed during the rework process. The representative chemistry is the cleavable ester epoxy material developed by Professor Ober at Cornell University.[5] This approach uses a-terp-epoxide as the basic material for the underfill encapsulant. After curing, the material forms a thermoset solid. Upon heating during the rework process, its tertiary ester sites break down and the encapsulant becomes easily removable with common organic solvents.

In recent years, more reworkable encapsulant chemistries have been developed and explored by other universities and chemical companies. However, the actual application of the reworkable underfill technology has been limited, primarily due to cost considerations.

5.5 New material challenges to lead-free solder

Tin–lead (Sn-Pb) eutectic solder has been used for the semiconductor industry as the primary connecting materials for many decades. In recent years, the Pb toxicity issue has attracted public attention and, correspondingly, many environmental legislations have been imposed to limit Pb consumption in semiconductor industry. Alternative solder materials have been developed in recent years and most of them are metal alloys containing Sn, Cu, Ag, Zn, and Bi. So far, most of these newly developed Pb-free solder alloys have higher melting points than the conventional eutectic Sn-Pb solder. For example, the most commonly used Sn-Ag-Cu triple alloy has a melting point of 217 °C. The high melting points of these Pb-free alloys also imply high reflow temperatures during the flip chip assembly process, which in turn generates higher thermal stress on the package. In addition, the underfill encapsulant developed for eutectic solder applications faces a compatibility challenge with the Pb-free solder. The wetting behavior of the underfill on the new alloy solder bumps is usually not as good as on the eutectic solder bumps. The wetting behavior of an organic material is commonly believed to be essential to its adhesive performance, and good adhesion to the solder bumps is required for good reliability; this has become the new challenge for underfill encapsulant materials.

5.6 The 'no-flow' pre-applied underfill process

To accelerate the slow process of the capillary underfill, a so-called 'no-flow' process has been proposed and is under active development by many institutions and companies. Figure 5.4 depicts such a process. The main steps involve dispensing underfill on a substrate, aligning and placing the flip chip on the substrate, and then reflowing the assembly through a typical reflow

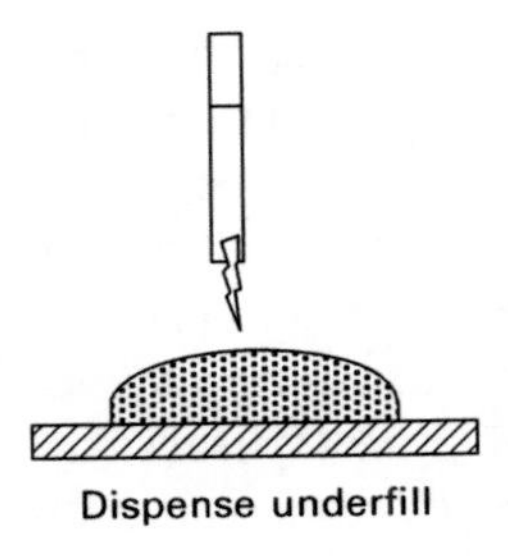

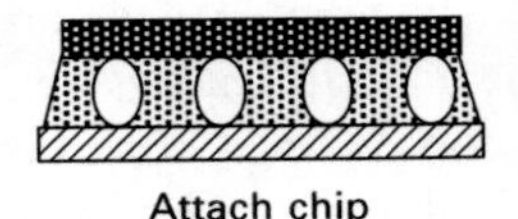

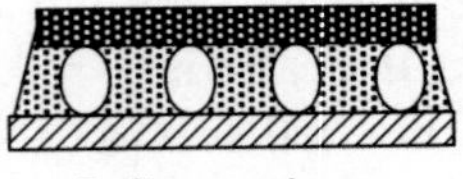

5.4 No-flow packaging process.

process. Different from the capillary underfill process, the no-flow underfill process does not require that the encapsulant materials flow slowly under the chip. The encapsulant materials can be quickly dispensed onto the substrates. After aligning and placing the chip on the substrate, the solder interconnect formation and the encapsulant curing occur simultaneously during the reflow process. Therefore, the no-flow process has the potential to accelerate the process speed and therefore reduce the overall process cost.

During the assembly process, however, there is a concern that air bubbles could be trapped during the fast chip attachment step, particularly for high I/O area array chips. In addition, filler particles in the encapsulant could be trapped between the solder balls and the substrate circuitry, which may affect the interconnect conductivity. Therefore, in most cases, inorganic fillers are not incorporated in current no-flow underfill encapsulants, which may limit the effectiveness of the underfill encapsulant on reliability enhancement due to the high CTE of the underfill encapsulant.

Active research and development effort has been launched in many institutions and companies. Equipment manufacturers, such as Asymtech, have even developed prototype no-flow underfill instrumentation. Many material suppliers have also developed corresponding no-flow underfill encapsulant materials. Unlike the capillary flow encapsulant materials, the viscosity of no-flow underfill materials needs to be relatively high, so that they do not flow freely to contaminate the surrounding area. During the solder bump reflow process after the chip is in place, the underfill encapsulant material should be solidified only after the solder bumps are reflowed. Ideally, the underfill encapsulant materials for this no-flow process should possesses cure 'latency', i.e. the underfill should not solidify immediately during the solder reflow. As soon as the solder reflows, however, the underfill encapsulant should be cured and solidify relatively fast to achieve its final properties. This cure latency requirement apparently raises a material design challenge. In the conventional solder reflow process, a flux is applied and solder reflow occurs in a separate process step. However, no-flow underfill encapsulant materials should also have a fluxing function to ensure reliable solder metal connections, since the solder reflow and underfill cure are achieved in one step. This is also a new challenge for the material developers. During the development of the no-flow underfill encapsulant materials, void formation became a major concern of this technology. It has been demonstrated in many research reports that voids in the package are a major source of reliability failure. Besides the typical origin of the voids formation – volatiles generated from the underfill encapsulant during the curing process – the no-flow process itself generates another potential voids formation mechanism. During the process, when the chip is placed on top of the substrate with the underfill on it, trapped air bubbles become another source for the undesirable void formation.

Filler entrapment is another hurdle for this technology. Without incorporation of silica filler, the material cannot achieve the desired low CTE. The high CTE of the encapsulant material may limit the underfill's reinforcement efficiency. However, if silica filler is used, the possibility of trapping filler particles between the solder bumps and the substrate circuitry endangers the interconnect reliability. Understanding these technical challenges, intensive development work has been carried out in many institutions and material supply companies.[6,7] Most of them use epoxy–anhydride based materials, but other chemistries have also been explored.[8]

5.7 The wafer level pre-applied underfill process

To further accelerate production speed and exploit process cost savings, the next logical step is to bring the underfill encapsulation process to 'wafer level'. The wafer level process is the most advanced and challenging flip chip and CSP option, and potentially the most efficient and cost-effective one, too.[9,10] This approach breaks the traditional 'front-end' and 'back-end' fields in the semiconductor industry and is the first attempt to merge the two separate sectors. As expected, however, this approach also encounters the most technical challenges.

As illustrated in Fig. 5.5, the process starts with the deposition of underfill encapsulant on a bumped wafer. This is followed by solidification of the underfill (or B-stage), and dicing the processed wafer into individual chips. During the assembly process, it is proposed that the individual chip is aligned and placed on the substrate, followed by an in-line reflow process. Ideally, the solder reflow and the curing of encapsulant are completed in one reflow step. By dispensing the underfill encapsulant onto a wafer instead of onto each individual chip, it is potentially a more efficient packaging process.

The wafer level process eliminates several steps in current flip chip/CSP packaging processes. However, new requirements for encapsulant materials have to be met to ensure the process is a feasible one. A non-tacky solid encapsulant at room temperature on the wafer is essential to meet both dicing and storage requirements. However, this non-tacky solid encapsulant

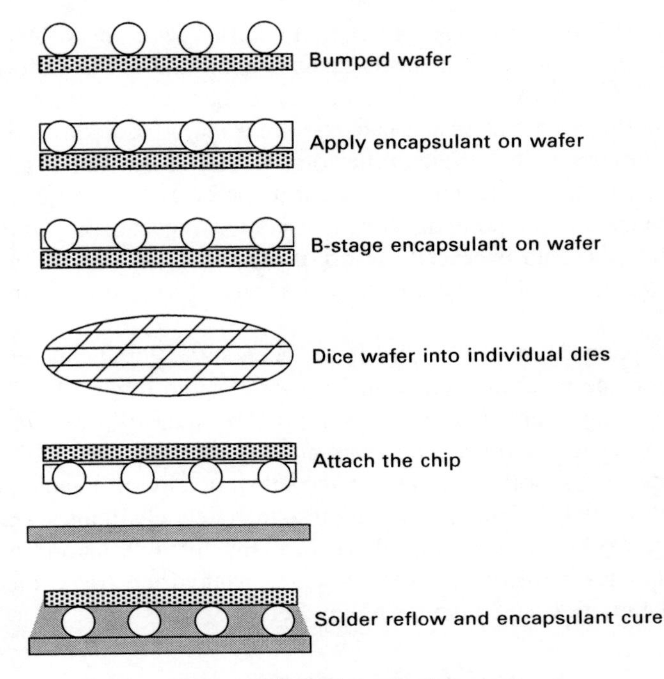

5.5. Wafer level flip chip/CSP packaging process.

also brings a challenge to attach the chip on the substrate during the solder reflow step. Poor wetting to the substrate and lack of fillet formation may result in poor end-use reliability of the package. Furthermore, full coverage of the solder bumps with filled encapsulant materials causes filler trapping concerns and the unevenness of the encapsulant surface may trap air voids during the attachment step. Besides the above concerns, there is another issue that needs to be considered in the final assembly process. This 'self-alignment' issue will be addressed in detail in Section 5.8. The wafer level dual encapsulation process.

5.7.1 Material requirements for the wafer level underfill

The performance requirements of wafer level underfill materials present material suppliers with many challenges they have never experienced before.[11] In the proposed wafer level flip chip processes, the underfill encapsulant material has to perform multiple functions at a series of processing steps, combining the roles of a number of materials that have been used independently up to now. The following is a step-by-step analysis of the material requirements and strategies in terms of underfill material development.

5.7.2 Depositing underfill onto the wafer

The method of depositing underfill on finished wafer will be the first step of this process, which used to be a 'front end' process. The challenge for the material supplier is to supply the material with the right rheological properties to facilitate deposition. Many processing methods have been explored, including stencil printing, screen printing, and spin coating. Depending on the application method, the underfill material rheology has to be adjusted accordingly. The residue of unreacted monomer or solvent should be avoided in any case, because any small amount of residual solvent or monomer may cause voids at the later high-temperature reflow stage and curing stage, affecting the reliability of the flip chip.

5.7.3 Solidifying the underfill

The wafer and package should be easy to handle under ambient conditions after the underfill is applied. Therefore, the underfill material should be in a solid, non-tacky form at room temperature for storage and handling. From the material development point of view, this means a solidification step, either by a chemical or a physical method, and this has to be carried out after the deposition step. This is referred by many people in this field as the 'B-stage' process in this process. The underfill encapsulant material on the wafer is in a solid, non-tacky form for storage and handling, as well as during the subsequent dicing step. In the meantime, the B-staged underfill encapsulant has to be able to flow again during the solder reflow process to allow the formation of the solder bump connections.

5.7.4 Dicing

During the dicing step, both the silicon wafer and the underfill will experience heat generated by the diamond saw. To protect the lifetime of the saw and the integrity of the underfill encapsulant, the B-stage solidified underfill material should have good mechanical properties under dicing conditions. Also, the material should be water resistant, so as not to absorb moisture from the cooling water during this process step. A high moisture content in the underfill material will be detrimental to the later processing stage as well as to the end-use reliability of the package. It has been reported that during dicing, the material may experience temperatures as high as 100 °C or more.[12]

5.7.5 Storage

After the dicing step, the diced wafer or the individual chips will be packed, shipped and stored for a long period of time. During this period, the properties of the underfill encapsulant should not change, so as not to affect the subsequent processing steps such as solder reflow and final cure. Shelf-life of the diced wafer or chips should meet industrial standards, typically from six months to a year at room temperature. If the underfill materials advances and further cures on the chip during storage, the resultant high viscosity may hinder the formation of the solder bumps during the next attachment and solder reflow steps.

5.7.6 Aligning and attaching

Pick and place equipment is currently available to place the flip chip with underfill on its substrate during the attachment process. 'Self-alignment' is an important phenomenon during the conventional reflow process, which enables the chip to align itself with maximum accuracy to its substrate circuitry. However, due to the existence of the underfill encapsulant on the chip, the self-alignment effect will be reduced or eliminated in the wafer level flip chip packaging process. Therefore, more accurate instrument alignment may be needed in this process.

5.7.7 Reflow of the solder and forming the interconnect

Compared with the conventional flip chip attaching process, this step represents the most drastic change. The metallurgical connection has to be made in the presence of the underfill medium without conventional fluxing agents, so there are multiple challenges to the underfill encapsulant material. First, the material should carry the fluxing capability, since the solder bumps are now being surrounded by the underfill materials. Second, the solder connection has to be made before the underfill undergoes significant cure because the high viscosity of a cured underfill will certainly hinder the flow of molten solder. That means that the B-staged solid underfill encapsulant has to be liquidized at the reflow temperature to allow the metallic interconnect formation. Finally, after the metal solder bumps reflow and form the interconnects, the underfill encapsulant should fully cure relatively fast to ensure high efficiency of the process. Therefore, a carefully designed material with desired cure latency is the primary challenge to the material developers.

5.7.8 Curing of the underfill encapsulant

As stated earlier, the goal of the wafer level flip chip process is to make the solder reflow and underfill cure as a single step. Solder reflow also dictates that the underfill cure should happen after the metallurgical interconnection is made. A simple analysis of the common reflow profile, such as the one illustrated in Fig. 5.6, tells us that the assembly stays above 150 °C for two

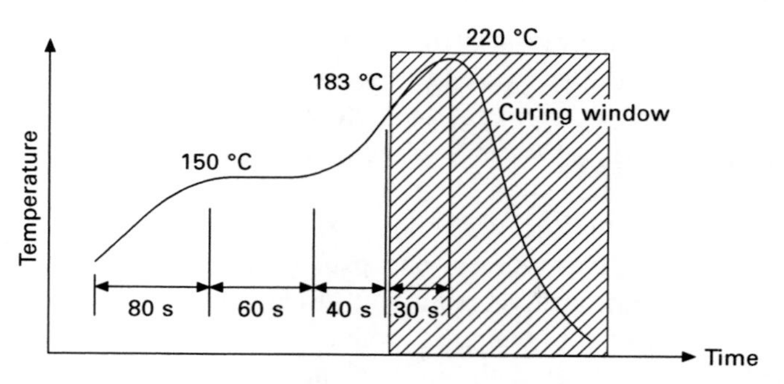

5.6 Eutectic solder temperature–time reflow profile.

minutes before it reaches the maximum reflow temperature. During this period, the underfill material should remain as a low viscosity liquid to allow the formation of the metal interconnections. After the peak solder melting temperature at 220 °C, the device stays above 150 °C for less than 2 minutes. It is in this temperature window that the cure of the underfill encapsulant should be completed. Thus, the development of a novel curing mechanism is essential for the success of the wafer level underfill encapsulant material. Currently, the state of art in this area is achieved in no-flow underfill applications, in which a post-cure is required after the regular reflow temperature profile is completed. However, the post-cure process definitely slows the overall process and increases the cost.

5.7.9 Reliability of the wafer level underfill

The reliability requirements for wafer level underfill are identical to those for conventional underfill. The after-cure properties of the material should include high glass transition temperature, low CTE, high modulus, and good adhesion to all surfaces.

5.8 The wafer level dual encapsulation process

In current surface mount processes, liquid flux is applied onto the solder bumps to ensure the proper solder interconnect formation. In addition, the liquid flux also holds the chip in place during the in-line reflow process, which is important to ensure the alignment of the package. In the wafer level packaging process, the solder bumps are surrounded by wafer-applied encapsulant and self-alignment is no longer effective. In addition, applying conventional liquid flux is not an option for the wafer level process. Since the metal interconnect formation and the reinforcement encapsulant filling are achieved in the same reflow step, the flux application and flux residue cleaning are practically impossible. Currently widely used 'no-clean' fluxes, which rely on the evaporation of their organic ingredients, may generate voids during the reflow process. In order to solve this difficulty and also make the wafer level packaging a completely compatible process to current industry infrastructure, a 'wafer level dual encapsulation' approach is here proposed and explored.

Figure 5.7 illustrates the wafer level dual encapsulation process. As implied by the name, two types of encapsulant materials are necessary for this process. Type-I encapsulant is a highly-filled, high modulus, and high Tg material. The encapsulant is applied on the wafer before the wafer is diced into individual chips. It goes through all the wafer level processing steps such as B-staging and dicing. Type-II encapsulant is a low-viscosity liquid material without inorganic filler. It is capable of fluxing the solder bumps during the solder reflow process. During the assembly process, either the chip is dipped into the liquid Type-II material, or a droplet of the material is dispensed onto the substrate before the chip is placed on the substrate. The low-viscosity liquid Type-II encapsulant holds the chip in place during the in-line reflow process and therefore the chip can be properly aligned.

The wafer level dual encapsulation process takes advantage of both the 'no-flow' and the 'wafer-level' packaging approaches. It still shares all the advantages of the two processes, such as high production output and low

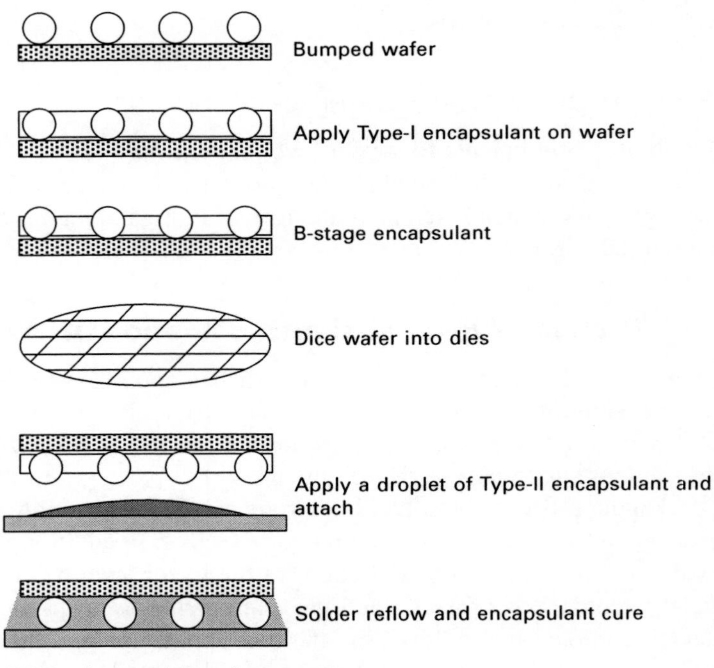

5.7 Dual-material wafer process.

processing cost. However, highly-filled Type-I underfill encapsulant with low CTE can be used in this process and this translates into better end-use reliability. It has been shown in the published literature that the most vulnerable point of the package, is at the die-bump interface due to the thermal stress generated from the CTE mis-match. Therefore, it is critical that the chip side is reinforced by a highly filled, high Tg, and low CTE Type-I encapsulant to enhance the reliability of flip chip/CSP packages. During the attachment step, a low-viscosity liquid Type-II encapsulant is applied to hold the chip in place. In addition, it provides better wetting and adhesion to the substrate. The low viscosity Type-II liquid encapsulant also flows easily during the attachment process and forms a nice 'fan-out' fillet around the chip after the reflow process. Both the better adhesion and the better fillet formation will contribute positively to the end-use reliability of the packages. Finally, by replacing solder flux with a liquid Type-II encapsulant, the proposed wafer level dual encapsulation process is compatible with the industry infrastructure of current surface mount technology (SMT). Therefore, this process and the corresponding encapsulant materials will be a feasible wafer packaging option for flip chip packaging and CSP.

In further detail, Type-I materials are highly-filled with silica filler, up to 60% by weight. Both solvent-containing formulations and solvent-free formulations are developed for this encapsulant. Depending on the application method, the viscosity of the formulation can be adjusted properly for stenciling or spinning coating processes. Fluxing agents are optional in the Type-I materials, although most of the formulations contain fluxing agents.

After the paste Type-I material is applied on the wafer, it is B-staged to form a smooth, void-free, and non-tacky solid coating on the wafer. The B-stage conditions need to be optimized to ensure the complete removal of the solvents and volatile ingredients in the formulation. Residual solvent and volatile ingredients have been shown to generate voids during the assembly process. On the other hand, it is desirable that the B-staged encapsulant softens during the solder reflow process. Although this wafer level dual encapsulation process obviates the stringent flow requirement for the Type-I material during the assembly step, experimental results have suggested that premature cure of the encapsulant should still be prevented.

The Type-II material is designed to be applied between the chip (with Type-I encapsulant on it) and the substrate during the assembly process. It is a low-viscosity liquid to allow easy dispensing and good spreading during the process, and to avoid the filler trapping problem, no filler is added. This material contains sufficient fluxing agent to flux the solder balls and ensure adequate interconnect formation. Premature cure of the encapsulant before the solder interconnect formation should be prevented – delayed cure is required for this material. To control the curing process, a proprietary curing agent has been developed. Finally, since void generation is a big concern during the assembly step at the high reflow temperature (up to 260 °C), the formulation should be free of solvents and volatile ingredients.

An experiment simulating the assembly process has been performed to demonstrate the feasibility of this wafer dual encapsulation process.[13] A droplet of Type-II material was placed on a piece of Cu finished FR4 board. The glass slide specimen coated with B-staged Type-I encapsulant was placed face down on the droplet to form the assembly. The assembly was then heated on a hot plate which was preheated at 240 °C. Through the glass slide, the fluxing of the solder balls, the interconnect formation, and the flow and curing behavior of the encapsulant materials were observed. Within 30 seconds, it was observed that the size of the solder balls was enlarged and the glass slide collapsed on the substrate. This phenomenon demonstrated the fluxing of the solder balls and the formation of the metal interconnects. It also confirmed that the encapsulants were not prematurely cured and did not prevent the wetting and spreading of the solder balls. In addition, it was also observed that the liquid Type-II material at the bottom flowed and formed good fillets around the glass cover slide. Finally, after heating on the hot plate for two minutes, the encapsulant materials were fully cured and the glass slide was solidly bonded onto the FR4 substrate. Good fillet formation was also reported. The packages processed with both the Type-I and Type-II encapsulant materials formed fillets around the chip boundary. As can be seen in Fig. 5.8, the low-viscosity liquid Type-II material flowed well, and formed a complete fan out fillet. Closer examination of the fillet from the side view shows that the Type-II encapsulant climbed up along the chip edges. It is common knowledge in the industry that fillet formation is necessary to achieve the required reliability of the flip chip and CSP packages.

5.9 Conclusions

Flip chip technology, or in a more generic term, direct chip attachment (DCA) technology, is still in its infant stage. Due to its advantages, such as small footprint, thin package, light weight, high I/O density, and fast electronic performance, it is gaining more and more acceptance within the semiconductor industry. To enhance the reliability of these assemblies, underfill encapsulant materials are usually used. It has been demonstrated that, with the underfill encapsulant reinforcement, thermal reliability of the packages usually increases by 3 to 10 times. To improve production efficiency and reduce costs, many underfill processes have been developed and corresponding underfill encapsulant materials have been explored. Although currently the original capillary underfill process still dominates, these other approaches are actively being explored and developed.

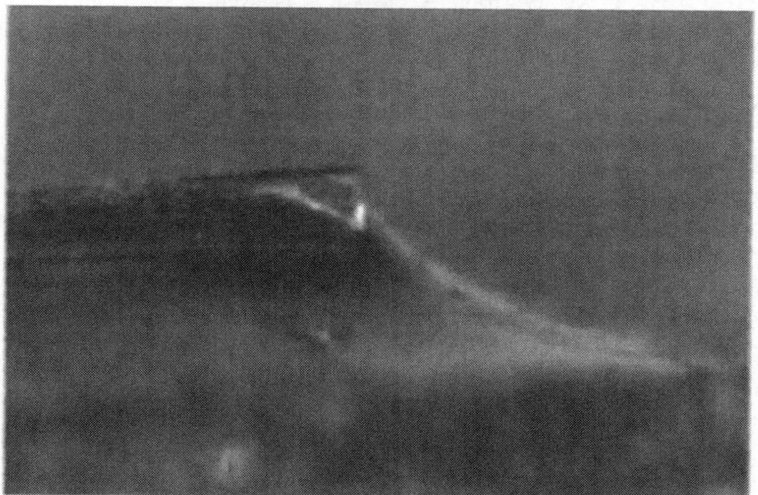

5.8 The liquid Type-II encapsulant flowed and climbed up on the edges of the chip and formed a good 'fan out' fillet.

5.10 References

1 E. Davis, W. Harding, R. Schwartz, and J. Coring, Solid Logic Technology: Versatile High Performance Microelectronics. *IBM Journal of Research & Development*, **8**, 102, 1964.

2 G. A. Riley, Introduction to Flip Chip: What, Why, How. FlipChips.com, October 2000.

3 F. Nakano, T. Soga, and S. Amagi, Resin-insertion Effect on Thermal Cycle Resistivity of Flip Chip Mounted LSI Devices. *Proceedings of the International Society of Hybrid Microelectronics Conference*, p. 536, 1987.

4 B. Ma, Q. Tong, A. Savoca, and T. DeBarrosl, Novel Fast Cure and Reworkable

Underfill Materials. *Proceedings of 3rd International Conference on the Adhesive Joining and Coating Technology in Electronics Manufacturing*, pp. 252–255, 1998.

5 J. Chen, C. Ober, and M. Poliks, Reworkable Thermosets: The Decomposition Mechanism and Development Network Breakdown of Epoxies with Tertiary Ester Links. *Polymer Materials: Science and Engineering*, **82**, 357–358, 2000.

6 C. P. Wong, S. H. Shi, G. Jefferson, High Performance No-Flow Underfills for Low-Cost Flip-Chip Applications, *Proceedings of 47th Electronic Component and Technology Conference*, pp. 850–858, 1997.

7 C. P. Wong, D. Baldwin, M. B. Vincent, B. Fennell, L. J. Wang, S. H. Shi, Characterization of a No-flow Underfill Encapsulant During Solder Reflow Process, *Proceedings of 48th Electronic Component and Technology Conference*, pp. 1253–1259, 1998.

8 A. Xiao, Q. Tong, S. Jayesh, and P. Morganelli, A Novel No-flow Flux Underfill Material for Advanced Flip Chip Packaging, *Proceedings of 52nd Electronic Component and Technology Conference*, pp. 1396–1401, 2002.

9 Q. Tong, B. Ma, E. Zhang, A. Savoca, L. Nguyen, C. Quentin, S. Lou, H. Li, L. Fan, and C. Wong, Recent Advances on a Wafer-level Flip Chip Packaging Process, *Proceedings of 50th Electronic Component and Technology Conference*, pp. 101–106, 2000.

10 C. Feger *et al.*, Wafer-level Underfill Materials and Processes for Lead-free Flip Chip Applications, *35th International Symposium on Microelectronics, International Microelectronics and Packaging Society (IMAPS)*, 2002.

11 B. Ma, E. Zhang, S. Hong, Q. Tong, and A. Savoca, Material Challenges for Wafer Level Packaging, *International Symposium on Advanced Packaging Materials: Processes, Properties, and Interfaces*, pp. 68–73, 2000.

12 L. Nguyen, H. Nguyen, V. Patwardhan, N. Kelkar, and S. Mostafazadeh, Method and Apparatus for Forming an Underfill Adhesive Layer, *US Patent 7253078*.

13 Q. Tong, A. Xiao, G. Dutt and S. Hong, Novel Materials and Processes for Wafer Pre-apply Flip Chip and Chip Scale Packaging, *Proceedings of 52nd Electronic Component and Technology Conference*, pp. 411–416, 2002.

Part II

Processing and properties

6

Structural integrity of metal–polymer adhesive interfaces in microelectronics

M. INOUE, Osaka University, Japan

Abstract: Realization of reliable bonding is the immutable requirement for adhesives. The interfacial adhesive strength is basically determined by the intermolecular interactions at interfaces, such as van der Waals and hydrogen bonds. When coupling agents are utilized for improving interfacial adhesion, it is essential to consider the chemistry of the adhesive molecules and the functional groups of the coupling agents. In addition to the intermolecular interactions, mechanical factors in meso-scale of adhesive layers can also contribute to the bonding strength of adhesive joints. In this chapter, the theoretical background for structural integrity of adhesive joints in microelectronics is discussed.

Key words: adhesion, bonding strength, intermolecular interactions, surface free energy, internal stress, viscoelastic behavior of adhesives, anchoring effect, environmental factors, conductive adhesives.

6.1 Introduction

Adhesion techniques are some of the most important elements in electronics packaging technologies. Advanced adhesives for electronics packaging have recently become diversified, in conjunction with the remarkable progress that is being made in microsystems. The adhesives are required to exhibit functions such as electrical and thermal conductivities in some cases, but realization of reliable bonding is the immutable requirement. In this chapter, the principal factors that determine the bonding strength of adhesive joints are identified and explained in detail, with the aim of establishing appropriate guidelines for developing advanced adhesives from the viewpoint of reliable bonding. Although the chemical composition of adhesive compounds seems complex, their design stems from the basic theory of adhesion and bonding, which is briefly introduced in this chapter.

6.2 Theoretical considerations of work of fracture and bonding strength of adhesive joints

6.2.1 Thermodynamic work of adhesion

Figure 6.1 shows schematically the concept of the thermodynamic work of adhesion;[1,2] the surfaces A and B are assumed to be perfectly flat. The interfacial free energy of A–B and the surface free energies of A and B are

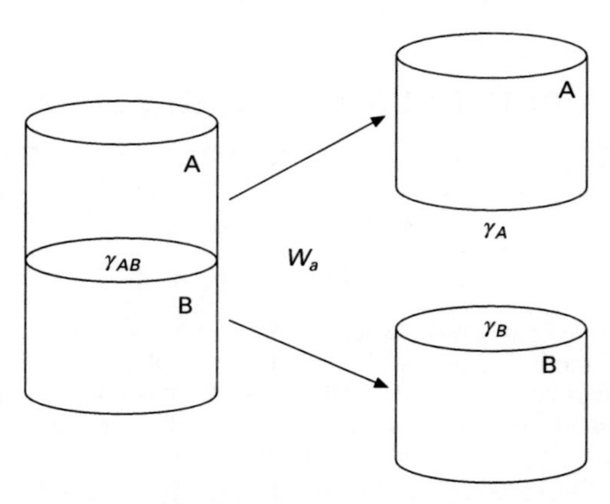

6.1 Concept of the thermodynamic work of adhesion.

represented by γ_{AB}, γ_A and γ_B, respectively. γ_{AB} is the interfacial free energy when the atoms on the surfaces of A and B are located at their equilibrium inter-atomic distances. The work of adhesion (W_a) can be expressed by Equation 6.1 (Dupré's equation):

$$W_a = \gamma_A + \gamma_B - \gamma_{AB} \qquad [6.1]$$

W_a can be estimated if the interfacial and surface free energies (γ_{AB}, γ_A, γ_B) are known. However, the surface energy of solids is difficult to measure and is consequently often estimated indirectly using the wettability of the surfaces of solids by liquids. The estimation method for the surface free energy of solids is explained in detail in Section 6.3.3.

When a liquid drop is placed on a solid surface, the equilibrium shape of the drop is determined by the balance between the solid/liquid interfacial tension (interfacial free energy; γ_{SL}), and the surface tensions (surface free energies) of the solid and liquid (γ_S and γ_L), as shown in Fig. 6.2. Thus, the surface free energy of the solid is represented by Equation 6.2 (Young's equation):

$$\gamma_S = \gamma_{SL} + \gamma_L \cos\theta \qquad [6.2]$$

where θ is the contact angle of the liquid on the solid surface. In this situation, we can express W_a between the solid and liquid using Equations 6.1 and 6.2 (Young–Dupré equation):

$$W_a = \gamma_S + \gamma_L - \gamma_{SL} = \gamma_L (1 + \cos\theta) \qquad [6.3]$$

In addition, it is assumed that γ_{SL} can be represented by Fowkes's equation (Equation 6.4)[3,4] and extended Fowkes's equations (Equations 6.5 and 6.6),[5–8] depending on the definition of chemical interaction at the interface:

$$\gamma_{SL} = \gamma_S + \gamma_L - 2(\gamma_L^d \gamma_S^d)^{1/2} \qquad [6.4]$$

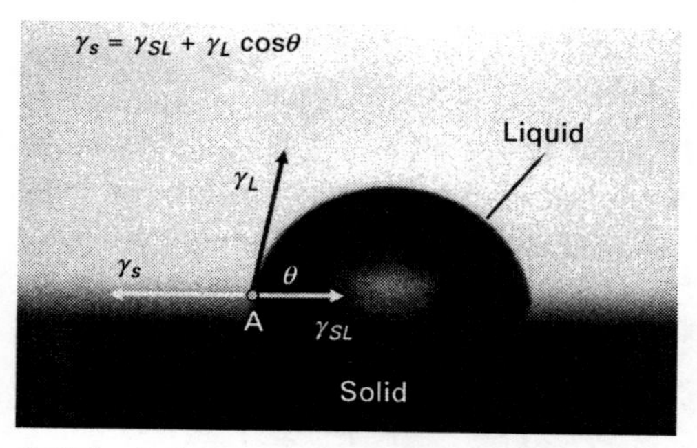

6.2 The balance between interfacial tension (γ_{SL}) and surface tensions (γ_S and γ_L) for a liquid drop on a solid surface.

$$\gamma_{SL} = \gamma_S + \gamma_L - 2\{(\gamma_L^d \gamma_S^d)^{1/2} + (\gamma_L^p \gamma_S^p)^{1/2}\} \qquad [6.5]$$

$$\gamma_{SL} = \gamma_S + \gamma_L - 2\{(\gamma_L^d \gamma_S^d)^{1/2} + (\gamma_L^p \gamma_S^p)^{1/2} + (\gamma_L^h \gamma_S^h)^{1/2}\} \qquad [6.6]$$

where γ^d, γ^p and γ^h are the dispersive, polar and hydrogen bonding components, respectively, of the surface free energy. By contrast, γ_{SL} is formulated as Equation 6.7 in the acid–base theory:[9]

$$\gamma_{SL} = \gamma_S + \gamma_L - 2\{(\gamma_L^{LW} \gamma_S^{LW})^{1/2} + (\gamma_L^- \gamma_S^+)^{1/2} + (\gamma_L^+ \gamma_S^-)^{1/2}\} \qquad [6.7]$$

where γ^{LW}, γ^+ and γ^- are the Lifshitz–van der Waals, Lewis acid (electron acceptor) and Lewis base (electron donor) components, respectively. The details (premises and limits of application) of these equations are explained in Section 6.3.

We adopt Equation 6.5 to continue discussion of W_a. By using Equations 6.2, 6.3 and 6.5, W_a can be expressed by

$$W_a = 2\{(\gamma_L^d \gamma_S^d)^{1/2} + (\gamma_L^p \gamma_S^p)^{1/2}\} \qquad [6.8]$$

In Section 6.2.2, we will estimate ideal adhesive strength between adhesive and substrate using Equation 6.8.

6.2.2 Interfacial interaction energy estimated from *ab initio* simulations

The interfacial interaction energy caused by the intermolecular interactions between adhesive molecules and substrate surfaces has been recently studied using *ab initio* simulation techniques. In these simulations, the molecular structure and intermolecular distance between models of adhesive molecules and substrate surfaces are optimized to estimate the total energy of the optimum structure (E_{AS}). The interaction energy (ΔE) is described by

$$\Delta E = E_{AS} - E_A - E_S \quad [6.9]$$

where E_A and E_S are the total energy of the models of adhesive molecules and surface, respectively. The concept of ΔE is similar to that of the thermodynamic work of adhesion represented by Equation 6.1. Furthermore, ΔE can be decomposed into several interaction components using energy decomposition analyses. For example, Kitaura and Morokuma[10] have proposed a methodology of energy decomposition analysis represented by:

$$\Delta E = E_{ES} + E_{PL} + E_{CT} + E_{EX} \quad [6.10]$$

where E_{ES}, E_{PL}, E_{CT} and E_{EX} are electrostatic, polarization, charge transfer and exchange energies, respectively. The energy decomposition analysis is conceptually similar to Equation 6.8. Minamizaki *et al.*[1,11] have analyzed the interaction energy between several adhesives and substrates using the energy decomposition method based on *ab initio* simulation.

6.2.3 Ideal adhesive strength and the bonding strength of adhesive joints

The interfacial potential ($V_{(r)}$) between A and B can be schematically drawn as shown in Fig. 6.3. The ideal adhesive strength is determined from the differential curve, $-dV/dr$, the maximum value of $-dV/dr$ being the ideal adhesive strength.[12] However, the interfacial potential is often difficult to obtain exactly for adhesive joints. Hence in the present study, the ideal adhesive strength is roughly estimated using W_a.[13]

The values of γ^d and γ^p for an epoxy-based adhesive and polyimide substrate are assumed as shown in Table 6.1. In this case, W_a between the epoxy and

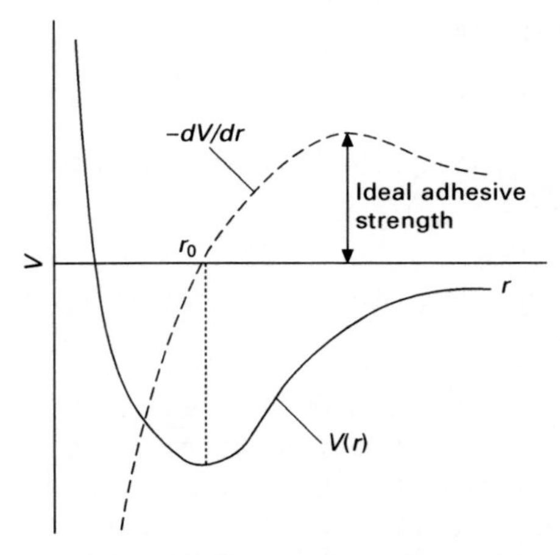

6.3 Schematic illustration of the interfacial potential and ideal adhesive strength between certain surfaces.

Table 6.1 **Magnitudes of the dispersive and polar components of the surface free energy of epoxy-based adhesive and polyimide that were assumed to roughly estimate the ideal adhesive strength using the thermodynamic work of adhesion**

	γ^d (mJ/m^2)	γ^p (mJ/m^2)	γ^{total} (mJ/m^2)
Epoxy adhesive	41.2	5.0	46.2
Polyimide	35.1	9.7	44.8

polyimide is estimated as 90 mJ m^{-2} using Equation 6.8. Furthermore, we assume that the bonds between epoxy adhesive and polyimide can be broken at a displacement of 0.5–1.0 nm from the equilibrium interatomic distance. The stress to break the chemical bonds is estimated as 90–180 MPa.

However, the ideal adhesive strength is always much larger than the bonding strength that can be experimentally measured using real adhesive joints. When we discuss the differences in the ideal adhesive strength and the bonding strength, the following two points should be taken into account. One is the imperfection of interfacial adhesion, including imperfect wetting of adhesives and the effect of internal stress generated by shrinkage of adhesives during the curing process. Imperfect wetting can provide a flaw that will be the origin of a fracture during bonding strength measurement. The internal stress generated in the adhesive layer significantly reduces adhesive strength (see Section 6.4.1). The other important point is the effects of fracture behavior during measurement of the bonding strength that depends on various factors such as geometry of joints, and loading conditions.[1] The viscoelastic deformation of adhesives can affect the bonding strength. Furthermore, the bonding strength often depends on the geometry of joints, even if the same adhesive and substrate materials are used for preparing the joints.

6.2.4 Relationship between work of adhesion and work of fracture of adhesive joints

The contribution of the thermodynamic work of adhesion to bonding strength measured by mechanical tests is usually obscured due to the effects of geometrical and loading factors. In the 1970s, Andrews and colleagues proposed a generalized fracture mechanics theory to analyze the fracture energy (Θ) of adhesive joints. According to Andrews and Kinloch,[14] Θ can be written:

$$\Theta = \Theta_0 \, \Phi \qquad [6.11]$$

where Φ is a loss function that depends on the debonding rate, temperature and strain level. Θ_0 is the true interfacial fracture energy.

Although Θ_0 is proportional to W_a, the experimental data often show that

Θ_0 is not equal to W_a. Thus, Kamyab and Andrews[15] proposed Equation 6.12 to analyze the relationship between Θ_0 and W_a:

$$\Theta_0 = W_a \Phi' \quad [6.12]$$

where Φ' is a loss function evaluated only over the process zone around the crack, and involves the appropriate parameters for the loss process concerned. Therefore, Θ can be rewritten as:

$$\Theta = W_a \Phi \, \Phi' = W_a \Phi_{total} \quad [6.13]$$

where Φ_{total} is the total loss function.

The experimental data showed that Φ_{total} is usually orders of magnitude greater than W_a. The stress relaxation due to viscoelastic deformation of adhesives during mechanical tests is one of the most effective factors in determining Φ_{total}. Thus the contribution of the viscoelastic behavior of adhesives is more dominant than the effects of interfacial adhesion during mechanical testing of adhesive joints. However, the effects of interfacial adhesion are still important, because it was established experimentally and theoretically that the total loss function (Φ_{total}) is proportional to W_a.[14] To realize significant stress relaxation due to viscoelastic deformation of adhesives during mechanical tests, sufficient interfacial adhesion between adhesives and substrates that is capable of withstanding stress is essential.

As mentioned above, the bonding strength of adhesive joints is not determined only by the interfacial adhesive strength, but also by the loss function. When we practically analyze the bonding strength, various factors such as defects at the interface, internal stress generated due to shrinkage of adhesives, geometrical and loading factors of mechanical tests, should be taken into account as well as viscoelastic behavior of adhesives. However, improvement of the interfacial adhesive strength is fundamentally important to increase the bonding strength of adhesive joints.

In the following sections, the influential factors determining the bond strength of adhesive joints are discussed in detail. The discussions will be useful for scientists and engineers who develop novel adhesives and packaging techniques, including nano-technology (nano-packaging).

6.3 Chemical and physical intermolecular interactions at interfaces

6.3.1 General view of the intermolecular interactions

Interfacial adhesive strength is basically determined by the intermolecular interactions at interfaces, although chemical reactions between the surface functional groups of substrates and the molecules of adhesives do not always occur (to form covalent bonds at the interface).

The intermolecular interactions are generally categorized into several types of bonds, such as van der Waals and hydrogen bonds. Although these categories of intermolecular interactions have been widely used, they are not always appropriate for interpreting adhesion phenomena in adhesive joints. The goal is to understand interfacial interactions through quantum chemical (*ab initio*) and molecular dynamics (MD) simulations. However, these simulation techniques are not yet generally applied for adhesive joints. The concepts of intermolecular interactions that are applied especially for adhesive joints will be introduced in the following sections.

6.3.2 Dispersive and polar components of intermolecular interactions

Van der Waals bonds

Van der Waals forces[16] originate from dipole–dipole, dipole–induced dipole, and induced dipole–induced dipole interactions. The intermolecular potential can be generally described as a function of intermolecular distance (r) using:

$$V(r) = -\frac{A}{r^6} + \frac{B}{r^n} \qquad [6.14]$$

The first and second terms on the right side of this equation represent the attractive and repulsive interactions. The exponent n in the repulsive term is a constant between 9 and 12. Equation 6.14 represents the Lennard–Jones potential when $n = 12$.

An attractive interaction is generated between two polar molecules (which have permanent dipoles) depending on their relative orientation. Although the interaction energy between two freely rotating polar molecules is zero, the molecules do not rotate completely freely. The average interaction energy (attractive interaction) of two rotating polar molecules is written as:

$$V(r) = -\frac{(2\mu_1^2\mu_2^2)/3(4\pi\varepsilon_0)^2\,kT}{r^6} \qquad [6.15]$$

where μ, ε_0, k, T are dipole moment, permittivity in vacuum, Boltzmann constant and temperature, respectively. Equation 6.15 describes the potential energy of dipole–dipole interaction (Keesom interaction).

The second contribution to van der Waals force is the dipole–induced dipole interaction. A polar molecule (with dipole moment μ_1) can induce a dipole in a neighboring polarizable molecule. These two molecules exhibit an attractive interaction between the permanent dipole of the first molecule and the induced dipole of the second molecule. In this case, the average interaction energy (attractive interaction) is written as a function of r:

$$V(r) = -\frac{(\alpha_2' \mu_1^2)/4\pi\varepsilon_0}{r^6} \quad [6.16]$$

where α_2' is the polarizability volume of molecule 2. The dipole–induced dipole interaction energy is independent of temperature.

An attractive interaction is also generated even between non-polar molecules. Transient dipoles are induced in non-polar molecules as well as polar molecules due to fluctuations in the instantaneous positions of electrons. The intermolecular interactions between the transient dipoles of non-polar molecules are called the dispersive (London) interactions. The dispersive interaction energy (attractive interaction) can be approximated by:

$$V(r) = -\frac{3/2\{\alpha_1'\alpha_2' I_1 I_2/(I_1 + I_2)\}}{r^6} \quad [6.17]$$

where I_1 and I_2 are the ionization energies of the two molecules.

Table 6.2 shows the coefficient of the attractive potential for dipole–dipole, dipole–induced dipole and dispersive interactions between several molecules.[1,12] Because the dispersive force can be generated in both non-polar and polar molecules, it is considered to be a universal intermolecular interaction. The dispersive interaction tends to increase with increasing molecular (atomic) size. Even in polar molecules, the dispersive interaction makes a significant contribution to the van der Waals force.

By contrast, permanent dipoles in molecules are necessary in order to generate dipole–dipole and dipole–induced dipole interactions. Consequently, these forces should be distinguished from the dispersive force when we analyze intermolecular interactions. Dipole–dipole and dipole–induced dipole interactions can be categorized as the polar component of intermolecular interactions.

Table 6.2 Coefficient of attractive potential for three types of van der Waals interactions between several molecules

	Dipole–dipole ($J\ m^6 \times 10^{-79}$)	Dipole–induced dipole ($J\ m^6 \times 10^{-79}$)	Dispersive ($J\ m^6 \times 10^{-79}$)
Ar			50
Kr			102
Xe			230
CH_4			97
CO_2			136
Na			5340
HI	6.2	4.0	191
HCl	18.6	5.4	106
CH_3Cl	109	23	284
H_2O	190	10	33
NH_3	77	9.7	61.5

Hydrogen bonds

The hydrogen bond is an attractive interaction of the form A–H···B, where A and B are highly electronegative elements such as N, O, F and anionic species (e.g. Cl^-).[16] In this case, a dominating attractive interaction is supposed to exist because the internuclear distance between formally non-bonded atoms becomes less than their van der Waals contact distance.

The result of an *ab initio* molecular orbital simulation (using a commercial code, Dmol[3]) for a primitive model composed two H_2O molecules (Fig. 6.4) is presented in order to understand the characteristics of hydrogen bonds. The *ab initio* simulation was carried out for intermolecular distances (interatomic distance between H(1) and O(2)) in the range of 0.1–0.7 nm. Figure 6.5 shows the interaction energy between the two H_2O molecules (at 0 K) obtained by the simulation. The interaction energy exhibits a minimum value at intermolecular distance 0.19 nm. Because the experimentally measured average length of hydrogen bonds between H_2O molecules is 0.177 nm, the simulation results are thought to be usable for interpreting the principle of the hydrogen bonding. Figure 6.6 shows the bond orders of H(1)–O(2) and O(1)–O(2) estimated by Mulliken's method. The population analysis indicates that a bonding orbital is formed between H(1) and O(2) when these atoms approach to within ~0.3 nm. Simultaneously, O(2) exhibits an anti-bonding interaction with O(1). The anti-bonding interaction between oxygen atoms increases drastically with decrease of the intermolecular distance below 0.2 nm. The interaction energy between H_2O molecules in the present models (shown in Fig. 6.4) is considered to be determined mainly by the balance of these bonding and anti-bonding interactions. Figure 6.7 shows the total electron density between the H_2O molecules in the model at an intermolecular

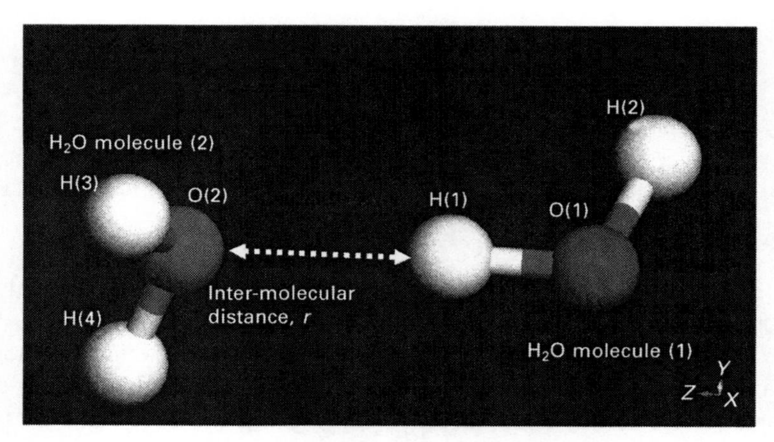

6.4 Primitive model for *ab initio* molecular orbital simulation for analysis of the hydrogen bond between H_2O molecules.

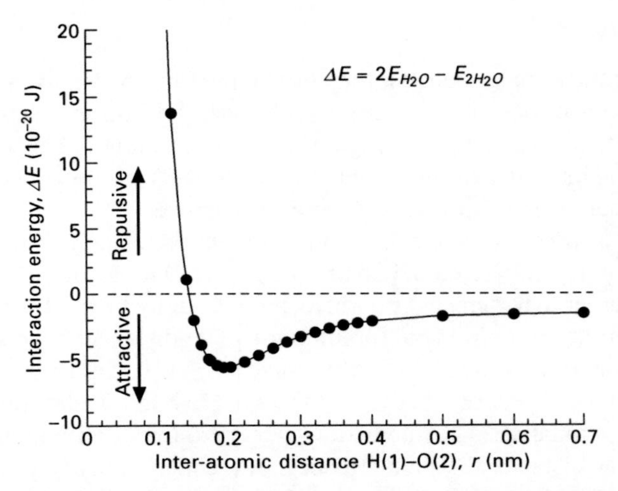

6.5 Interaction energy between H_2O molecules at 0 K that was simulated using the model shown in Fig. 6.4.

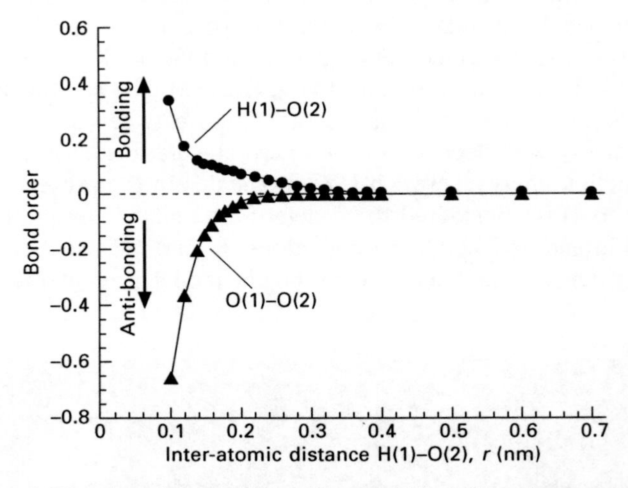

6.6 Bond orders of H(1)-O(2) and O(1)-O(2) estimated by Mulliken's method based on *ab initio* simulation results as a function of intermolecular distance.

distance of 0.19 nm. In this figure, the bond with a weak covalent nature formed between H(1) and O(2) corresponds to the hydrogen bond.

The mechanisms of the intermolecular interactions between H_2O molecules can be briefly summarized as follows. Because the van der Waals contact distance between H and O is 0.272 nm (the van der Waal radii of H and O atoms are 0.120 and 0.152 nm, respectively), H_2O molecules are considered to interact mainly by van der Waals forces at an intermolecular distance greater

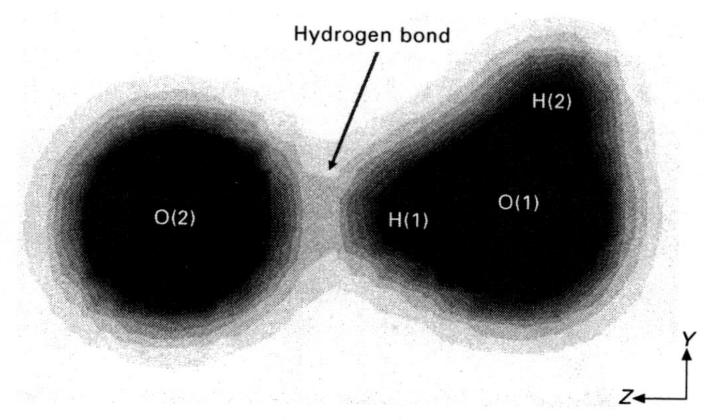

6.7 Contour diagram of total electron density between H_2O molecules at an intermolecular distance of 0.19 nm.

Table 6.3 Comparison of bonding energy for hydrogen bonds and other chemical and physical bonds

Bonding nature		Bonding energy (kJ/mol)
Chemical bonds	Ionic bonds	600–1500
	Covalent bonds	60–700
Hydrogen bonds	O–HO	25
	C–HO	8–13
	O–HN	17–30
	N–HO	8–13
	N–HF	21
	F–HF	30
van der Waals bonds	Dipole–dipole	4–20
	Dipole–induced dipole	< 2
	Dispersive	0.08–40

than ~0.3 nm. When the molecules approach more closely, hydrogen bonds with weak covalency are formed between neighboring H and O atoms. The bonding energy of hydrogen bonds is strongly limited by the anti-bonding interaction between O atoms of adjacent H_2O molecules. The weak covalency is the important characteristic of hydrogen bonds.

Table 6.3 shows typical values of hydrogen bonding energy compared with other chemical and physical bonding energies.[13] Hydrogen bonds generally exhibit a higher bonding energy than van der Waals bonds due to the covalency between H and electronegative species.

When hydrogen bonds are formed between the functional groups on a substrate surface and adhesive molecules, the contribution of the hydrogen bonds to adhesive energy is commonly included in the polar component.

By contrast, some researchers separate the effects of weak covalency in hydrogen bonds from the polar component. In this case, the intermolecular interactions are divided into three components including dispersive, polar and hydrogen bonding components.

6.3.3 Analysis of interfacial free energy using the intermolecular interaction concept

The discussion on the components of intermolecular interactions suggests that the interfacial free energy can be described as a function of these components.[17] To analyze the interfacial free energy, Fowkes[3,4] assumed that the surface free energy of substrate i is represented by Equation 6.18 using the dispersive component γ_i^d and the other components γ_i^x:

$$\gamma_i = \gamma_i^d + \gamma_i^x \qquad [6.18]$$

When material j that has only the dispersive component of surface free energy (γ_j^d) forms an interface with substrate i, the interaction energy between i and j is assumed to be described using the Berthelot relation based on the regular solution approximation. Under these conditions, Equation 6.19 can be used to describe the interfacial free energy between i and j (γ_{ij}):

$$\gamma_{ij} = \gamma_i + \gamma_j - 2(\gamma_i^d \gamma_j^d)^{1/2} \qquad [6.19]$$

Equations 6.4 and 6.19 are well known as the Fowkes's equation. However, the Fowkes's equation is applicable to interfacial interactions based solely on dispersive interaction.

Owen *et al.*[5,6] and Kaelble *et al.*[7] have taken the polar component into account as well as the dispersive component in their analyses of the interfacial free energy. According to their assumption, the interfacial free energy is written as Equation 6.5, using the additional term for the polar component. Furthermore, Kitazaki and Hata[8] separated the hydrogen bonding component from the polar component, and expressed the interfacial free energy by Equation 6.6, using a supplementary term for the effect of hydrogen bonds. The Berthelot geometric average may not strictly represent the hydrogen bonding component because of the covalency in hydrogen bonds. However, they introduced supplementary terms that deal with the Berthelot relation for the hydrogen bonding component, to establish Equation 6.6.[1] The components of the surface free energy for substrates and adhesives provide a useful guideline for the surface modification of substrates and the development of adhesives.

By contrast, intermolecular interactions at interfaces are expressed by another concept in the case of the acid–base theory,[9] as expressed by Equation 6.7. The Lifshitz–van der Waals component (γ^{LW}) includes the dispersive, dipole–dipole and dipole–induced dipole interactions. The Lewis acid (γ^+) and base (γ^-) components represent the interaction ability of surfaces as electron acceptor and electron donor, respectively. In the acid–base theory, the intermolecular interactions are classified from the viewpoint of charge transfer between the adhesive molecules and functional groups on substrate surfaces. In this classification, the hydrogen bonding component can be apparently distinguished from the polar component including dipole–dipole and dipole–induced dipole interactions. Fowkes[9] claimed that the acid–base interaction theory is adequate for describing the polarity or hydrophilicity of surfaces rather than the polar component introduced in the extended Fowkes's equations (6.5 and 6.6).

From the concept of interfacial interaction based on surface free energy, the maximum interfacial adhesive strength is considered to be ideally obtained when $\gamma_{SL} = 0$. In this ideal condition, the components of surface free energy are perfectly matched between adhesives and substrates (extended Fowkes's theories: $\gamma_S^d = \gamma_L^d$, $\gamma_S^p = \gamma_L^p$, $\gamma_S^h = \gamma_L^h$; acid–base theory: $\gamma_S^{LW} = \gamma_L^{LW}$, $\gamma_S^+ = \gamma_L^-$, $\gamma_S^- = \gamma_L^+$). Thus, mismatch in the intermolecular components of surface free energy between adhesives and substrates should be decreased in order to improve the interfacial adhesive strength.

The experimental technique for estimation of the surface free energy of solid surfaces is introduced next. When the surface free energy is assumed to be composed of dispersive and polar components, the surface free energy of solids can be estimated using Equations 6.2 and 6.5, which lead to:

$$\gamma_L(1 + \cos\theta) = 2\{(\gamma_L^d \gamma_S^d)^{1/2} + (\gamma_L^p \gamma_S^p)^{1/2}\} \qquad [6.20]$$

When we measure the contact angle for the solid surface using two different liquids that have known values for γ_L, γ_L^d and γ_L^p, the values for γ_S^d and γ_S^p can be estimated using Equation 6.20 (liquid drop method). The total value for the surface free energy (γ_S) is obtained by:

$$\gamma_S = \gamma_S^d + \gamma_S^p \qquad [6.21]$$

To estimate the surface free energy of solids composed of three components (dispersive, polar and hydrogen bonding components) using Equation 6.6, three different liquids are required for the contact angle measurement. The surface free energy can be similarly analyzed based on the acid–base theory (using Lifshitz–van der Waals, Lewis acid and Lewis base components) using Equation 6.7.

The results of surface free energy analysis by the liquid drop method for several polymer films are shown in Table 6.4. The surface free energy was separated into dispersive and polar components using H_2O and CH_2I_2 as the probe liquids, by means of Equations 6.20 and 6.21. PTFE film exhibited a small surface free energy due to extremely small dispersive and polar components. The surface free energy of polyimide film was increased

Table 6.4 Surface free energy of several polymer films estimated by the liquid drop method using H_2O and CH_2I_2

	γ^d (mJ/m^2)	γ^p (mJ/m^2)	γ^{total} (mJ/m^2)
Polyimide (as received)	35.1	9.7	44.8
Polyimide (after soaking in 5M-NaOH for 30 s)	28.0	42.8	70.8
PET	38.7	7.2	45.9
PTFE	15.4	2.7	18.1

Table 6.5 Variation of surface free energy of a Cu foil with surface finishing. The surface free energy was estimated by the liquid drop method using H_2O and CH_2I_2

	γ^d (mJ/m^2)	γ^p (mJ/m^2)	γ^{total} (mJ/m^2)
Cu foil on flex (as received)	29.4	0.4	29.8
After Ni plating (5 μm)	33.2	4.7	37.9
After Au flash plating (0.1 μm)	36.6	0.8	37.4

remarkably by soaking in 5 M NaOH aqueous solution for 30 s due to a significant increase in the polar component. Polar functional groups were effectively introduced on the surface of the polyimide film by the treatment with strongly alkaline solution (formation of carboxylic groups by breaking imide rings).[18] The analysis of surface free energy possibly provides useful information on the chemical properties of substrate surfaces.

The surface energies of metals and metalized surfaces can also be estimated by the liquid drop method. The experimentally estimated values of the surface energy of metals are strongly influenced by surface contamination, which can be categorized as inorganic or organic, depending on the nature of the adsorbate. O_2 and H_2O molecules are typical inorganic adsorbates. O_2 molecules adsorbed on metal surfaces (physisorption) are often decomposed to atomic oxygen (O) (chemisorption), and in the case of metals with high oxidation potential, an oxide layer is formed spontaneously. H_2O molecules can also chemisorb on metal surfaces to form hydroxyl groups on the surfaces. Physisorption of organic molecules on the surface of metals results in significant decrease of the polar component of the surface free energy.

Table 6.5 shows the variations in surface free energy of a Cu foil (used for conductive patterns on flexible substrates) depending on the surface finishing. The surface free energy was estimated by the liquid drop method using H_2O and CH_2I_2. The Cu foil surface was plated with Ni (5 μm thickness), then Au flash plated (0.1 μm thickness). Although the Cu surface exhibited a low value of the polar component, the magnitude of the polar component increased significantly due to the Ni plating. The Au flash plating increased the dispersive component and decreased the polar component. The variation in surface free energy is considered to be related to characteristics of the adsorbates on the surfaces. In general, the surface free energy of noble metals (such as Au, Ag, Cu, Pt and Pd) is easily affected by the adsorption of organic molecules, and the surface free energies of the original and the Ni/Au plated Cu foil are likely to have been influenced by surface contamination with organic molecules.[19, 20] By contrast, the effect of organic adsorbates on the surface free energy of transition metals such as Ni is relatively small.[19,20] Hence, the Ni-plated specimen exhibited a relatively higher value for the polar component.

Metals are considered to have intrinsically high surface free energies due to the large polar component. Table 6.6 shows the surface free energy of mechanically polished Ag, Cu and Ni sheets before and after exposure to air at ambient temperature for 48 h. The surface free energy of these sheets significantly decreased during exposure to air, due to decrease in the polar component. Hence, surface cleaning techniques using physical and chemical methods are effective in improving the adhesion strength of adhesives on metal surfaces. The ultimate bonding process utilizing surfaces cleaned in vacuum chambers is the surface activated bonding (SAB) process developed by Suga and colleagues.[21,22] This process enables direct metal/metal, metal/inorganic and metal/organic substances bonding without adhesives at low temperatures.

6.3.4 Solubility parameters for analyzing intermolecular interactions

When novel adhesives are developed, the solubility parameters[23] are often referred to in compounding the adhesives. Although solubility parameters do not directly relate to adhesive strength, they are useful for analyzing intermolecular interactions at interfaces as well as within adhesive compounds.

The cohesive energy density (c) of molecules (i) is defined by:

Table 6.6 Variation in surface free energy of mechanically polished Ag, Cu and Ni sheets due to exposure to air at ambient temperature for 48 h. The surface free energy was estimated by the liquid drop method using H_2O and CH_2I_2

	γ^d (mJ/m^2)	γ^p (mJ/m^2)	γ^{total} (mJ/m^2)
Ag (as polished)	35.1	24.5	59.6
Ag (after exposure to air)	37.8	2.2	40.0
Cu (as polished)	37.2	13.6	50.8
Cu (after exposure to air)	33.0	4.5	37.5
Ni (as polished)	31.8	24.1	55.9
Ni (after exposure to air)	32.5	7.7	40.2

$$c_{ii} = \frac{\Delta E_i^v}{V_i^o} \quad [6.22]$$

where ΔE_i^v and V_i^o are the energy of vaporization and molar volume of the molecule. Hildebrand proposed that the solubility parameter (δ) of a molecule can be written based on the regular solution approximation using the cohesive energy density, as:

$$\delta_i = (c_{ii})^{1/2} = \left(\frac{\Delta E_i^v}{V_i^o}\right)^{1/2} \quad [6.23]$$

The energy for mixing components 1 and 2 (ΔE_{mix}) is described by Eq. 6.24:

$$\Delta E_{mix} = \frac{n_1 V_1^o n_2 V_2^o}{n_1 V_1^o + n_2 V_2^o}(\delta_1 - \delta_2)^2 \quad [6.24]$$

The components exhibit good solubility when the magnitude of δ_1–δ_2 is small, because ΔE_{mix} consequently becomes small.

The solubility parameter (cohesive energy density) is related to intermolecular interactions in the same way as the surface free energy. Hansen[23] assumed that the solubility parameter is composed of dispersive (δ_d), polar (δ_p) and hydrogen bonding (δ_h) components, as represented by:

$$\delta = \delta_d^2 + \delta_p^2 + \delta_h^2 \quad [6.25]$$

In addition, some researchers have formalized the relationships between the surface free energy and the solubility parameter. Consequently, the solubility parameter is useful for designing adhesive compounds from the viewpoints of solubility of the components in adhesives, and the adhesive ability of the compounds on substrate surfaces.

The values of solubility parameter for many substances have been listed in the literatures.[23–25] If values of the solubility parameter for particular substances are not available, they can be estimated using the group contribution methods[23–25] that have been proposed by Hoy, van Kreveren and Hoftyzer, and others. The solubility parameters of the components of adhesives (main resin, diluents, curing agents and modifiers) should be determined using these methods before compounding the adhesives. The solubility parameter of a mixture is calculated by volume-wise summation of the solubility parameters of the individual components of the mixture.

6.3.5 Improvement of interfacial interactions using coupling agents

Adhesives usually adhere to substrate surfaces via intermolecular interactions such as dispersive, polar and hydrogen bonding interactions, as described previously. To improve the interfacial adhesion of adhesives to substrates and fillers, organometallic compounds containing Si, Ti, Al and Zr may be utilized as coupling agents. Typical molecular structures of coupling agents are illustrated schematically in Fig. 6.8. The molecules have functional groups both for hydrolytic reactions and for interactions with adhesive molecules. The groups for hydrolytic reactions transform to hydroxyl groups that can condense with hydroxyl groups on the surface of fillers and substrates. Consequently, the coupling agents are strongly fixed on the surfaces of fillers and substrates due to the condensation reactions.

There are two mechanisms for the interaction of adhesive molecules with substrate and filler surfaces modified by coupling agents. If the modified surface contains no functional groups that are reactive to the adhesive molecules, intermolecular interactions including dispersive, polar and hydrogen bonding interactions are induced between the molecules and modified surfaces. In this case, the interfacial interactions can be analyzed based on the concepts of the extended Fowkes's equations for interfacial free energy, and the Hansen's solubility parameters. Matching of the components between the adhesive molecules and the functional groups on the modified surfaces is the key to improving interfacial adhesion.

When the modified surfaces contain some functional groups that are reactive toward the adhesive molecules, chemical reactions can be promoted to form covalent bonds between the molecules and modified surfaces.[2] For the formation of covalent bonds, the extended Fowkes's equations are no longer valid. Here, the surface free energies of the modified surfaces and adhesive molecules are assumed to be given by Equation 6.21. The interfacial free energy with covalent bonding cannot be written as Equation 6.5.

The formation of covalent bonds completely deviates from the regular solution approximation that is the premise of the concept of solubility parameters, and consideration of interfacial interactions based on the solubility parameters is meaningless in this case. In fact, the bonding strength between adhesives and surfaces that are treated with silane coupling agents containing reactive functional groups is usually determined without regard to solubility parameters.

When coupling agents are utilized for improving interfacial adhesion, it is essential to consider the chemistry of the adhesive molecules and the functional

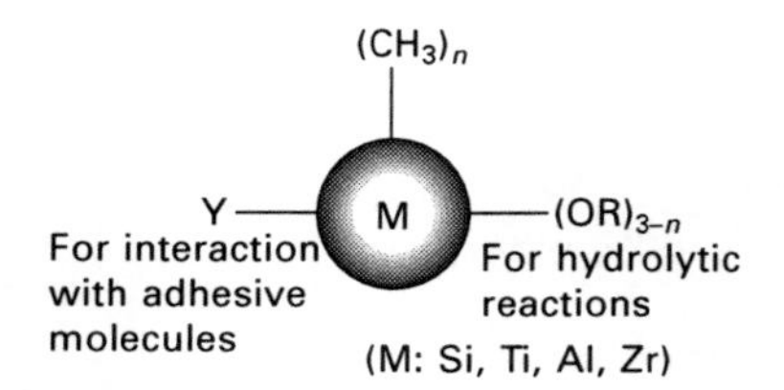

6.8 Schematic illustration for molecular structure of coupling agents.

groups of the coupling agents. If they do not exhibit a significant reactivity, the concepts of components (dispersive, polar and hydrogen bonding) of surface free energy and solubility parameter provide an effective guideline for improvement of interfacial adhesion. However, the reactivity between the adhesive molecules and the functional groups of coupling agents is the most important issue for selection of coupling agents when the interfacial adhesion is designed based on the formation of covalent bonds.

There are two methods of utilizing coupling agents. Fillers and substrates can be directly treated by dry and wet processes before mixing and bonding with adhesives, or the coupling agents can be pre-mixed into the adhesives (integral blend method). The integral-blended coupling agent molecules can diffuse in the adhesive paste to modify the surfaces of fillers and substrates. In this case, the coupling agents also contribute to modifying several properties of the adhesives. The integral blend method is often used for adhesives for electronics packaging.

6.3.6 Self-assembled monolayers (SAMs) for surface and interfacial modifications

In 1946, Zisman and colleagues[26] reported monolayer adsorption of a surfactant on a metal surface, although the possibility of self-assembly was not recognized. Since Nuzzo and Allara[27] demonstrated formation of SAMs of alkanethiolates on gold using dilute solutions (in 1983), various self-assembly systems have been investigated as shown in Table 6.7.

Figure 6.9 schematically illustrates the formation mechanism of SAMs on a substrate surface.[28,29] When surfactant molecules adsorb on the surface by chemical interactions (reactions) of surface-active functional groups, ordered molecular assemblies are realized if intermolecular interactions adequately occur between alkyl groups of the molecules. The function of a surface covered with SAMs is determined by the chemical properties of the terminal group of the surfactant molecules. To improve interfacial bonding, several

Table 6.7 Typical self-assembly systems

Surfactants	Substrates
R-SH, RS-SR′	Au, Ag, Cu, Pt, Pd
R-SCN	Fe, GaAs, InP
R-CN	Pt, Pd, Au, Ag
R-COOH	Al_2O_3, AgO, CuO
R-SiH_3	Au
R-$SiCl_3$	SiO_2, SnO_2, TiO_2
R-$SiOCH_3$, R-$SiOC_2H_5$	GeO_2, ZrO_2, Al_2O_3

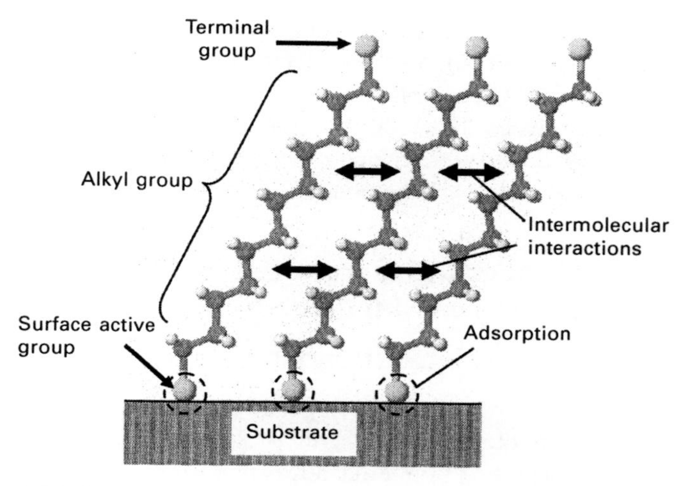

6.9 Schematic illustration of formation mechanism of self-assembled monolayers (SAMs) on a solid surface.

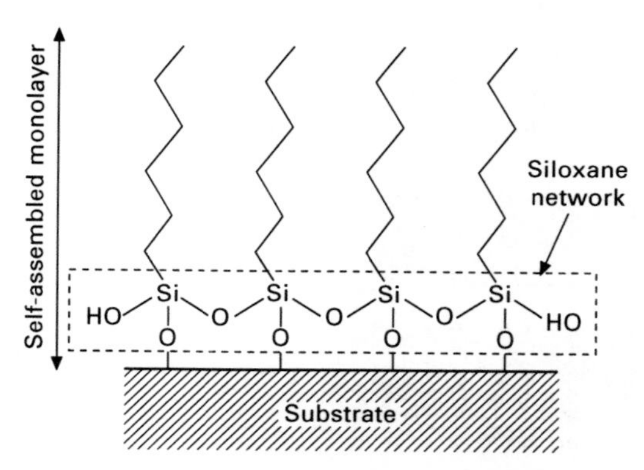

6.10 Schematic illustration of molecular structure of SAMs formed using organosilanes ($R–SiX_3$; X=Cl, OCH_3, OC_2H_5).

groups (such as carboxyl, amino, phosphoric, sulfonic, thiol) may be used as the terminal group of surfactants.

Some siliane coupling agents (alkylalkoxysilanes, alkylaminosilanes and alkylchlorosilanes) can form SAMs on metal and oxide surfaces under appropriate conditions.[28,30,31] In the case of agents that have three active groups for the surface, siloxane bonds are formed between adjacent molecules as well as between molecules and substrate surface, as shown in Fig. 6.10, SAMs of such organosilicon compounds exhibit excellent mechanical, chemical and thermal stabilities due to the siloxane network formed on the surface.[31]

Recently, SAMs have attracted great interest for improving interconnect (mechanical and electrical) properties of adhesive joints. For example, Wong and colleagues[32] have investigated the electrical properties of anisotropic conductive adhesive (ACA) joints when SAM compounds are introduced into the interface between the metal filler and the substrate bond pad. The ACA joints with SAM-treated conductive fillers and bond pads exhibited superior electrical properties to the non-treated joints.

6.4 Other influential factors determining bond strength of real adhesive joints

6.4.1 Internal stress generated in adhesives

Because adhesives always shrink during curing, residual stress (internal stress) is induced in the adhesive layer of joints.[1] Figure 6.11 shows schematically the generation of internal stress in a simple joint. Substrates 1 and 2 are assumed to be not fixed when the adhesive layer shrinks. Although the adhesive layer shrinks isotropically, compressive stress is not directly induced in the vertical direction by the shrinkage because the substrates can move simultaneously with shrinkage of the adhesive. However, since the adhesive layer adheres to the substrates, shear stress is generated in the horizontal direction, to warp the joint as shown in Fig. 6.11. Consequently,

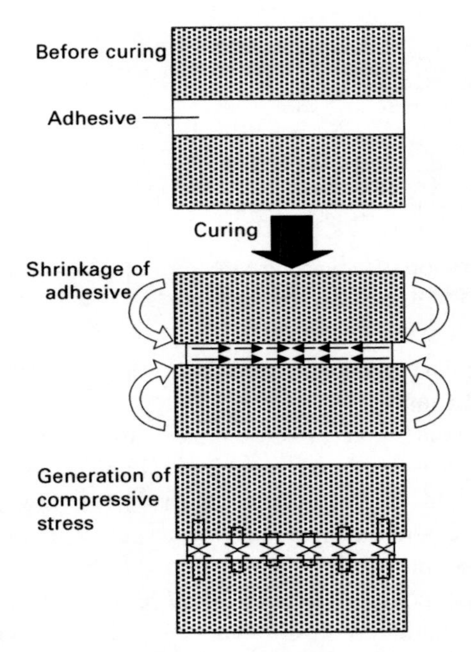

6.11 Generation mechanism of internal stress in adhesive joints due to shrinkage of adhesives.

compressive stress in the vertical direction is induced in the adhesive layer. The internal stress induced by the shrinkage of adhesives always decreases the bond strength of the joints.

The internal stress is considered to be generated through two steps during the adhesive curing process. In the case of thermosetting adhesives, adhesives shrink during the curing reaction (Fig. 6.12). In most cases, the curing process is performed at a higher temperature than the glass transition temperature (T_g) of fully cured adhesives. Thus, most of the internal stress generated in this step should be relaxed by micro-Brownian motion of polymer chains in the adhesive. The residual stresses after relaxation are accumulated as the internal stress in this step.

After curing the adhesive, a joint is cooled from the curing temperature to ambient temperature. During the cooling step, the adhesive layer of the joint undergoes significant shrinkage (cooling shrinkage), as shown in Fig. 6.12. In the temperature range below T_g of adhesives, there are no significant relaxation mechanisms to relax the stresses because the micro-Brownian motion of the polymer chains is frozen. Consequently, a large internal stress is induced during the cooling process from T_g to ambient temperature. The internal stress induced in this step is expressed by Equation 6.26:

$$P \propto \int_{T}^{T_g} E_a(\alpha_a - \alpha_s)dT \qquad [6.26]$$

where E_a and α_a are the elastic modulus and linear expansion coefficient of the adhesive, respectively, and α_s is the linear expansion coefficient of the substrate.

Equation 6.26 suggests a guideline for reducing the internal stress generated in the adhesive layer of the joints. Because the magnitude of internal stress is proportional to E_a and $\Delta\alpha = \alpha_a - \alpha_s$, the stress can be reduced by decreasing

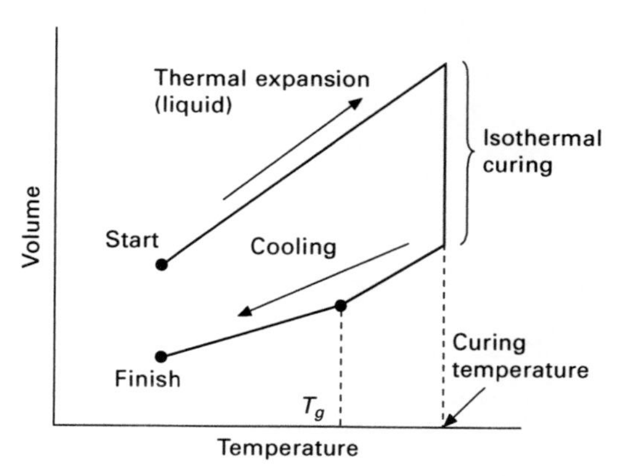

6.12 Shrinkage behavior of adhesives during curing and cooling processes.

at least one of these parameters. The magnitude of the internal stress is also reduced by decreasing T_g, but adhesives that have very low T_g are useless for high-temperature applications, for example, because a decrease in T_g results in decreased heat resistance of the adhesive (see in Section 6.5.1). Thus, to reduce internal stress, E_a and α_a should be controlled without significant decrease in T_g of the adhesive.

The magnitude of E_a can be controlled by the adhesive chemistry, and α_a can be decreased by mixing fillers with adhesives. However, introduction of fillers with high elastic modulus in adhesives simultaneously results in increased E_a, so that incorporation of fillers in adhesives is not always effective in decreasing internal stress. Furthermore, the interfacial adhesive strength significantly decreases when the adhesive contains an excessive amount of filler. Optimum material design is required for adhesives to control internal stress.

6.4.2 Mechanical behavior of adhesives and substrates

The fracture modes of adhesive joints are categorized into three modes, as shown schematically in Fig. 6.13. In the cohesive fracture mode, the bond strength of adhesive joints is governed by the fracture strength of the adhesives or the substrates.

When thermosetting adhesives (such as epoxy-based adhesives) are used for preparing joints, the degree of cure of the adhesives is a key factor for obtaining sufficient bonding strength.[33] The degree of cure is usually defined as the extent of conversion of the curing reaction of the adhesive. Of the methods for estimating the degree of conversion of adhesives, Fourier transform infrared (FTIR) spectroscopy and differential scanning calorimetry (DSC) are

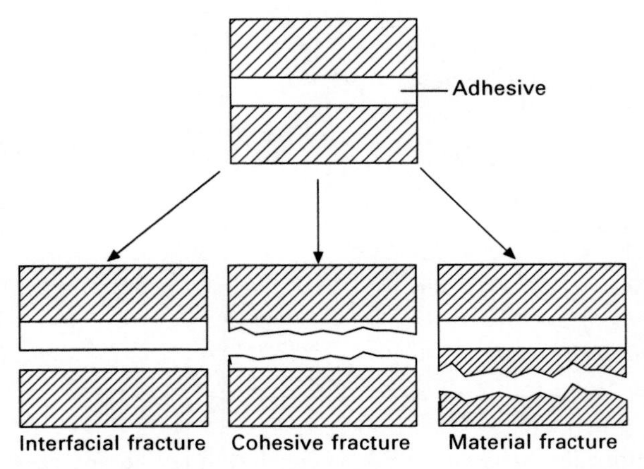

6.13 Schematic illustration of fracture modes of adhesive joints.

often utilized. Because the reactive functional groups in adhesive pastes can be measured quantitatively by FTIR spectroscopy, the degree of conversion is estimated by comparing the concentration of the reactive groups before and after curing. The heat of reaction that is released during curing is often detectable using DSC, and the degree of conversion can be estimated from the heat of reaction as an indicator for the curing reaction. Furthermore, kinetic analysis of the curing reaction of adhesives can be performed using DSC measurements to predict optimum curing conditions.

The values of T_g for adhesives often show a one-to-one relationship with the degree of conversion.[34,35] Figure 6.14 gives experimental T_g and degree of conversion data for an epoxy-based isotropic conductive adhesive (ICA), as an example.[36] In this case, the values of T_g are determined by the degree of conversion, regardless of curing temperature, hence the degree of conversion can be estimated by measurement of T_g.

An experimental result for the relationship between peel strength of adhesive joints and degree of conversion of adhesives[37] is shown next. Figure 6.15 shows the 90° peel strength of model joints composed of ITO glass substrate/epoxy-based anisotropic conductive film (ACF)/polyimide flex bonded at 180 °C. The peel test was performed at a test speed of 8.33 μm s^{-1}. The peel strength of the ACF joints increased significantly with increasing degree of conversion, and a drastic increase in peel strength observed at 40–60% conversion coincided with a significant increase in T_g. Hence, the drastic increase in peel strength is considered to be caused by the formation of a cross-linked polymer structure in the adhesive binder. The peel strength of the ACF joints increased further above 80% conversion.

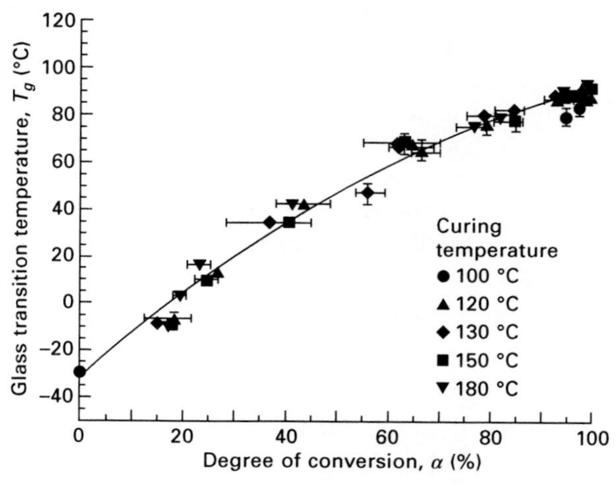

6.14 Relationship between degree of conversion (α) and glass transition temperature (T_g) in an epoxy-based isotropic conductive adhesive (ICA) cured at several temperatures.

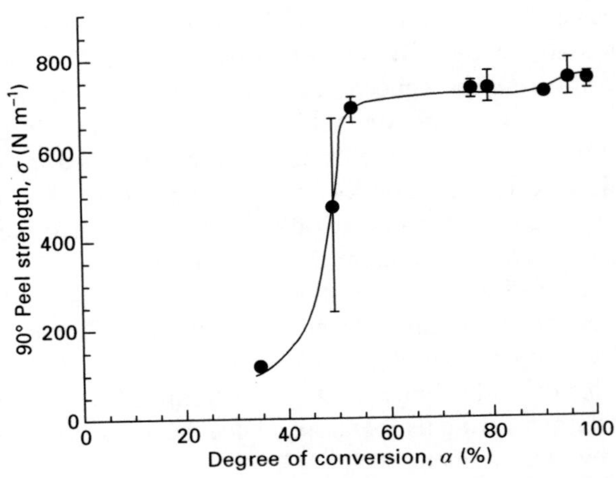

6.15 The 90° peel strength of model joints composed of ITO glass substrates and polyimide-based flex with an interdigitated Cu pattern, bonded using an epoxy-based anisotropic conductive film (ACF) at 180 °C for several bonding times. The peel test was conducted in the direction parallel to the Cu pattern at a test speed of 8.33 $\mu m\ s^{-1}$.

The mechanical behavior of adhesives usually gives a characteristic test speed dependence of the bonding strength of adhesive joints. As Andrews and colleagues[14,15] showed, adhesives exhibit significant viscoelastic deformation during the peel test of adhesive joints. The viscoelasticity of adhesives can affect the bonding strength of adhesive joints. Because the viscoelastic deformation of adhesives is a time-dependent phenomenon, adhesive joints usually exhibit higher bonding strength with increasing test speed of peel tests. The bonding strength is decreased by the stress relaxation due to viscoelastic deformation of adhesives when the peel test is conducted at slow test speed. Figure 6.16 shows the test speed dependence of a 90° peel strength of the model (epoxy-based) ACF joints between polyimide-based flex and ITO glass substrate bonded at 180 °C for 15 s (82.7% degree of conversion of the adhesive binder), as an example.

In addition to the effect of test speed on bonding strength of adhesives, there is also an effect of stressing-mode,[1] which is discussed in Section 6.5.1.

6.4.3 Relationship of physical factors at the interface to bonding strength

Physical effects (anchoring effects) that contribute to improving the bonding strength of adhesive joints can be realized when substrates have microscopic

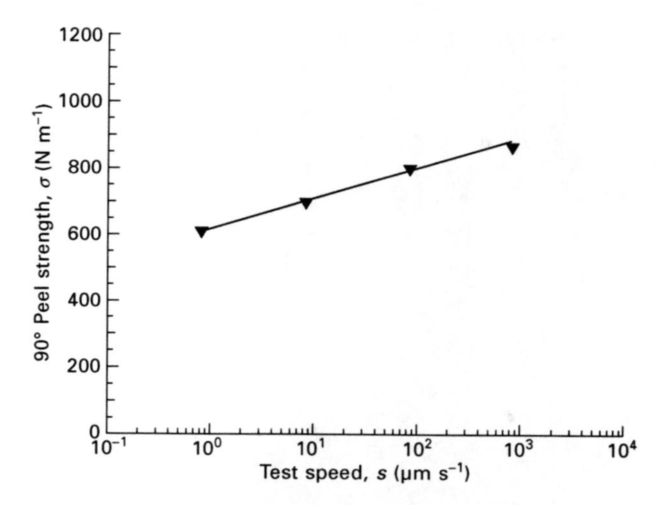

6.16 Test speed dependence of 90° peel strength of model ACF joints composed of ITO glass substrate and polyimide-based flex bonded at 180 °C for 15 s.

and macroscopic morphologies (anchor patterns) on their surface. The magnitude of the anchoring effects is known to depend on the geometrical factor of the anchor patterns.

Experimental peel test data for an epoxy-based ACF (anisotropic conductive film) joint[37] have been shown to explain the contribution of anchoring effects to the bonding strength. In these experiments, a polyimide-based double layer flex with an 18 μm thick interdigitated Cu pattern was used for preparing model joints with ITO glass substrates (Fig. 6.17a). A large number of dimples with dimensions of ~1 μm were also formed on the polyimide surface of the flex (Fig. 6.17b).

It was established by cross-sectional SEM observation of the model joints that the adhesive binder had completely infiltrated into the micro-dimples on the polyimide surface (see Fig. 6.18). As a consequence, the micro-dimples were likely to contribute to increased bonding strength of the joints due to an anchoring effect.

In addition, the macroscopic Cu circuit pattern appeared to exhibit an anchoring effect. Figure 6.19a shows peel strength profiles (measured at a test speed of 8.33 μm s^{-1}) of joints bonded at 180 °C for 15 s, in directions parallel and perpendicular to the interdigitated Cu pattern. The peel strength in the direction parallel to the Cu pattern was significantly higher than that in the perpendicular direction. During the peel test in the parallel direction, periodic oscillation of the peel strength was clearly observed (Fig. 6.19b), and was correlated with the interdigitated Cu pattern, since the period of the oscillation coincided with the Cu pattern. The peel strength profiles indicate

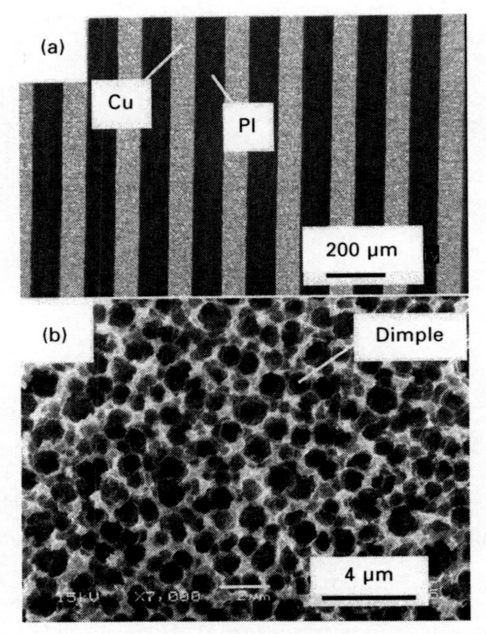

6.17 Polyimide-based flex that was used for preparing model ACF joints. (a) Interdigitated Cu pattern, (b) dimples formed on the surface of the polyimide-based film.

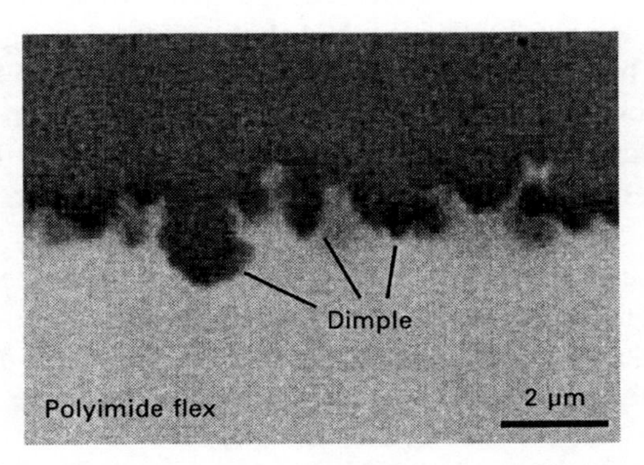

6.18 Cross-sectional SEM micrograph of the bonding interface between ACF and polyimide in the model joint.

that an anchoring effect due to the Cu pattern was effectively obtained in the parallel direction. Thus, anchoring effects due to the surface morphology of the flex can be induced cooperatively by the microscopic dimples on polyimide and the macroscopic Cu pattern.

Figure 6.20 shows peel strength profiles (measured at a test speed of 8.33 $\mu m\ s^{-1}$) in the parallel direction for ACF joints bonded at 140, 180 and 200 °C

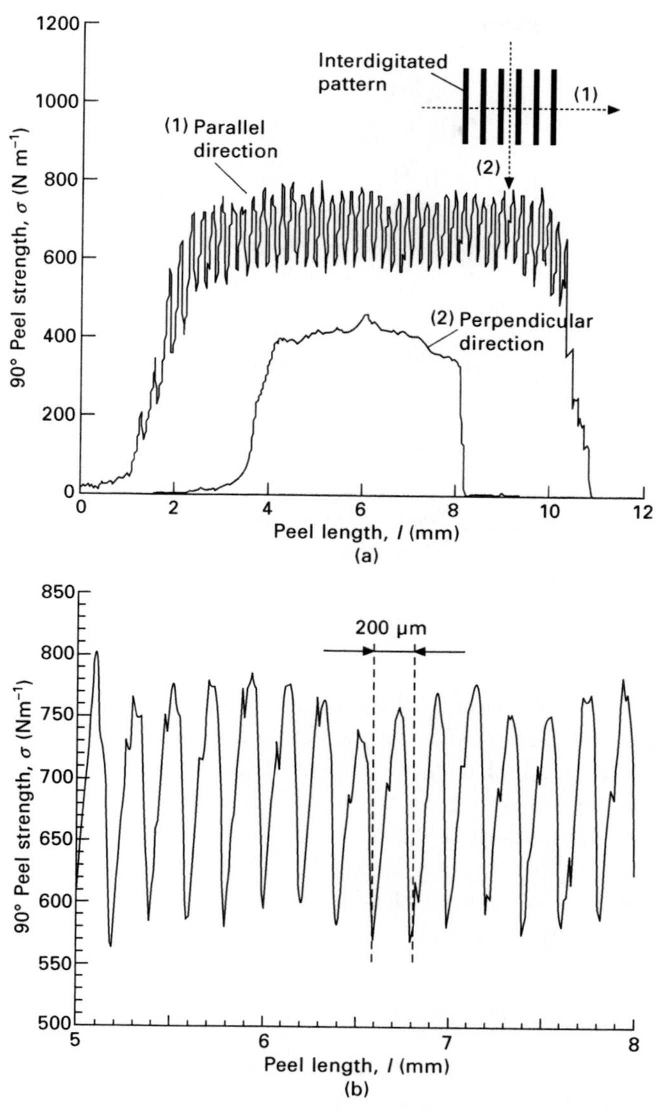

6.19 (a) The 90° peel strength profiles of the model ACF joints (bonded at 180 °C for 15 s) in the directions parallel and perpendicular to the interdigitated Cu pattern measured at test speed 8.33 μm s^{-1}. (b) Periodic oscillation of the peel strength observed during the peel test in the direction parallel to the interdigitated Cu pattern.

for 15 s. Because the degree of cure of the adhesive binder increased with increasing bonding temperature (9.3, 82.7 and 92.5% at 140, 180 and 200 °C, respectively), the joints bonded at higher temperatures provided higher peel

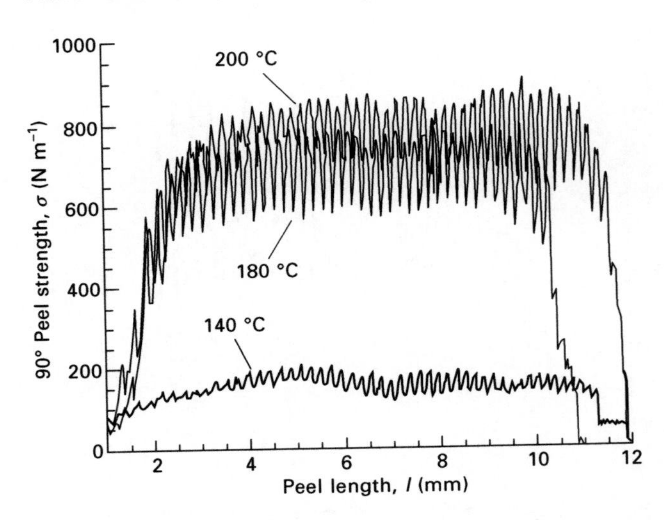

6.20 The 90° peel strength profiles of the model ACF joints bonded at 140, 180, and 200 °C for 15 s, in a direction parallel to the interdigitated Cu pattern. The peel test was conducted at a test speed of 8.33 μm s^{-1}.

strengths. The periodic oscillation due to the Cu pattern was observed even in the joints bonded at 140 °C, but the magnitude of the oscillations of peel strength was quite small in those joints. Hence, adhesives are required to have sufficient strength in order to obtain significant anchoring effects.

The surface morphology of the substrates is a necessary factor for realizing anchoring effects.[38] However, good wetting of adhesives at the interface and sufficient mechanical strength of adhesives and substrates are also needed to obtain significant magnitude of anchoring effects.

6.5 Effect of environmental factors

6.5.1 Temperature dependence of shear and peel strengths of adhesive joints

The bonding strength of adhesive joints is significantly different for different stressing modes. Figure 6.21 shows a schematic illustration of the relationship between elastic modulus of adhesives and bonding strengths (shear and peel strengths) of adhesive joints.[1] The shear strength of the joints tends to increase with increasing elastic modulus of the adhesives. By contrast, the peel strength of the joints exhibited a maximum value at a low elastic modulus. To obtain high peel strength of the joints, the stress relaxation ability of adhesives is the most important factor. The relationship between

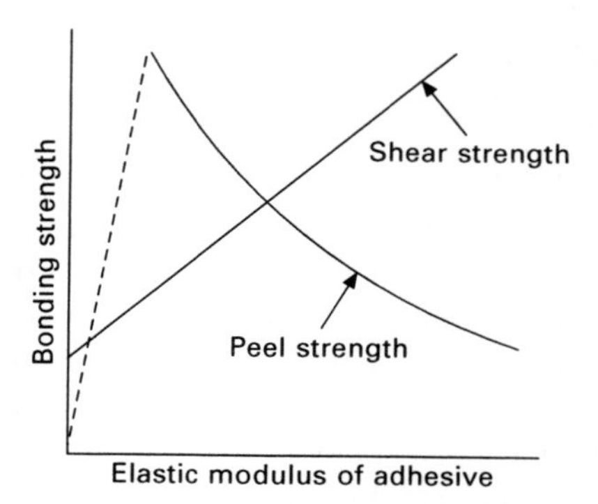

6.21 Schematic illustration of variations in shear and peel strength of adhesive joints with elastic modulus of adhesives.

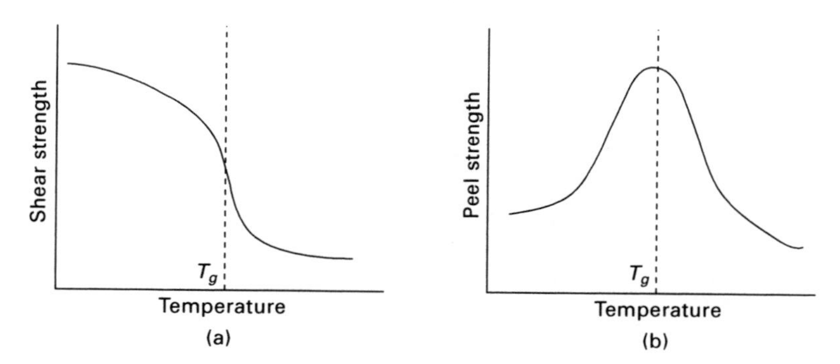

6.22 Schematic illustration of temperature dependence of (a) shear strength and (b) peel strength of adhesive joints.

elastic modulus and bonding strength illustrated in Fig. 6.21 should be taken into account in the design of adhesive joints.

To interpret the temperature dependence of the bonding strength, the elastic modulus dependence shown in Fig. 6.21 provides useful information. Shear strength of adhesive joints significantly decreases with decreasing elastic modulus (storage modulus) of adhesives in the temperature range around T_g of the adhesives, as illustrated in Fig. 6.22a. By contrast, the peel strength of joints and the loss modulus of adhesives exhibit a maximum value around T_g of adhesives since the stress relaxation effect of adhesives is significantly induced as shown in Fig. 6.22b.

If high shear strength is required to be realized together with high peel strength in the same adhesive joint, highly-elastic polymer-based composites containing rubber (elastomeric) fillers are good candidates. Fortunately, the rubber fillers also contribute to decreasing the internal stress generated during curing and cooling.

6.5.2 Effects of moisture absorption

General remarks on moisture absorption of adhesive joints

Moisture absorption in the adhesive layer is a serious problem in forming reliable adhesive joints. The amount of moisture absorption generally increases with increasing concentration of polar functional groups in adhesives. The polymer structure of adhesives contains free volume, as shown in Fig. 6.23a. The free volume spaces are considered to be divided into two categories namely micro-voids (which are different from large voids or pores with sizes greater than a few hundred nanometers) and more narrow spaces due to the density variation of polymer chains (Fig. 6.23b).[39] H_2O molecules can diffuse into the free volume spaces of adhesives from the surrounding environments. Since a large number of micro-voids are considered to exist at the interface between adhesive and substrate rather

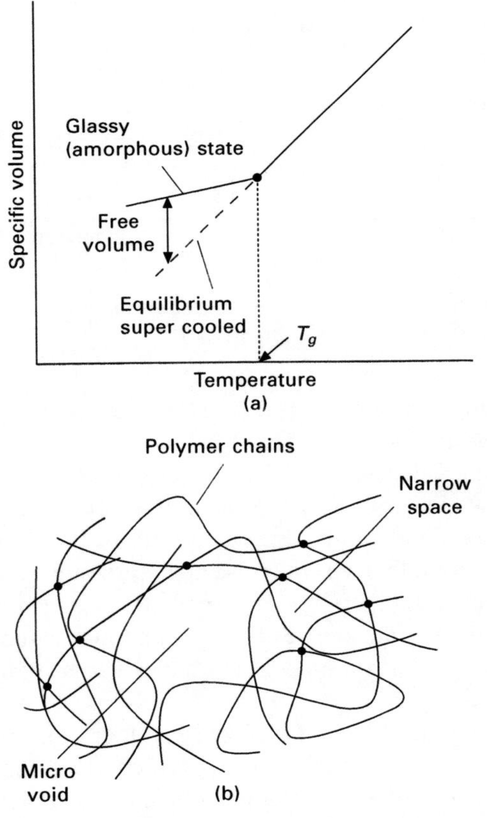

6.23 (a) Conceptual and (b) schematic models of free volume spaces in polymers.

than within the adhesive layer, the H_2O molecules predominantly diffuse along the interface.

Figure 6.24 shows the near-infrared (NIR, 7500–6000 cm^{-1}) absorption spectrum of an epoxy-based ACF exposed to an 85 °C/85%RH environment for 1000 h. From this spectrum, several types of H_2O molecules are suggested to exist in the free volume spaces. The molecules can exist both as free molecules (S_0, ~7075 cm^{-1}) and as molecules hydrogen-bonded to polymer chains in the adhesive.[40,41] The free molecules are likely to exist in the micro-voids. The bound molecules can be divided into S_1 (with one bond, ~6820 cm^{-1}) and S_2 (with two bonds, ~6535 cm^{-1}) species depending on the number of hydrogen bonds.

Various properties, including mechanical and electrical properties, of adhesive joints are influenced by the H_2O molecules that diffuse into the adhesives. For example, the bound H_2O molecules can affect mechanical properties of the joints since the intermolecular interactions are weakened at the interfaces as well as within the adhesive layer. In addition, the free H_2O molecules in micro-voids are considered to form a continuous channel within adhesives. Because chemical species such as ions and oxygen molecules can diffuse through the channel, chemical and electrochemical phenomena are induced in the adhesive joints due to exposure to humid environments. This section focuses on the effects of moisture absorption on the mechanical properties of adhesives.

Plasticization[42] of adhesives is always induced by moisture absorption via diffusion of H_2O molecules into the narrow free volume spaces as well as the micro-voids. The elastic modulus and T_g of adhesives are decreased, depending on the amount of moisture absorption. By contrast, the ability of

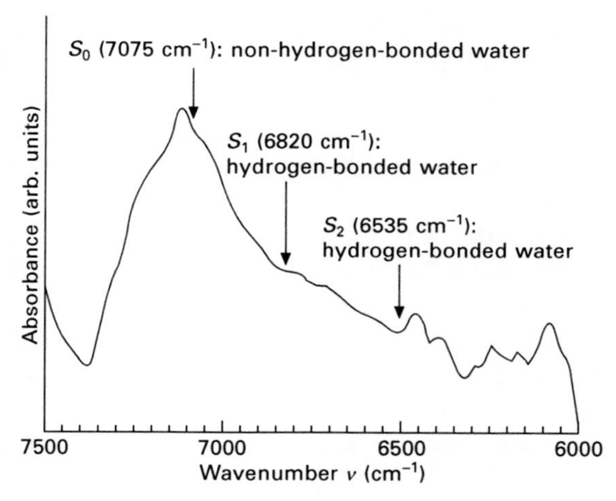

6.24 Near-infrared absorption spectrum of an epoxy-based ACF exposed to an 85 °C/85%RH environment for 1000 h.

adhesives to undergo stress relaxation is considered to be improved somewhat. The plasticization effect of adhesives usually disappears reversibly by drying the adhesives.

H_2O molecules that diffuse along the interfaces[43] can break the intermolecular interactions between adhesive molecules and functional groups on substrate surfaces, if the H_2O molecules create hydrogen bonds with the groups on the substrate surface. However, the adhesive molecules reversibly recover the intermolecular interactions with substrate surfaces by drying the adhesive joints, thereby removing H_2O molecules.[2] When covalent bonds are introduced at the interfaces by using silane coupling agents, the adhesive joints are often reported to exhibit high bonding strength even during and after exposure to humid conditions.

Moisture absorption sometimes promotes irreversible chemical and electrochemical reactions at interfaces. For example, metal surfaces such as Cu and Ni can be oxidized and hydroxylated during exposure to humid environments. If such irreversible reactions occur at a substrate surface, the bonding strength is not recovered, even after removing H_2O molecules.

When variations in the bonding strength of adhesive joints are discussed, the stressing modes should be borne in mind. The shear strength of joints is significantly decreased by plasticization and interfacial weakening due to moisture absorption. By contrast, the peel strength of joints is sometimes increased by enhancement of the ability of adhesives to undergo stress relaxation, accompanied by plasticization due to the moisture absorption.

Variation of mechanical properties of an Anisotropic conductive film (ACF) joint due to moisture absorption

Variation in the bonding strength of adhesive joints due to moisture absorption may be quite complex, depending on the situation. Experimental data for variation of the mechanical properties of ACF joints composed of ITO glass substrate and polyimide-based flex, that were referred to in Sections 6.4.2 and 6.4.3, due to exposure to an 85 °C/85%RH environment are shown as an example.

First, the moisture absorption behavior of the ACF specimens is explained. Figure 6.25 shows the increase in weight due to moisture absorption for non-cured and fully-cured ACF during exposure to the 85 °C/85%RH environment. The weight increment was almost saturated in the first few hours of exposure. Although the curing reaction of the adhesive binder was promoted in the non-cured specimen during exposure, the magnitude of moisture absorption increased with decreasing degree of conversion before exposure.

Variations in the values of T_g for the ACF specimens bonded at 140 and 180 °C for 15s during the exposure to the 85 °C and 85 °C/85%RH

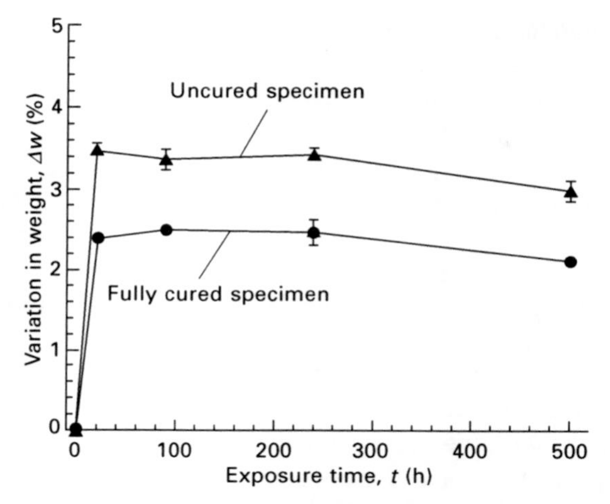

6.25 Variation in weight of epoxy-based ACF specimens during the exposure to the 85 °C/85%RH environment. The ACF specimens were uncured and fully cured before exposure.

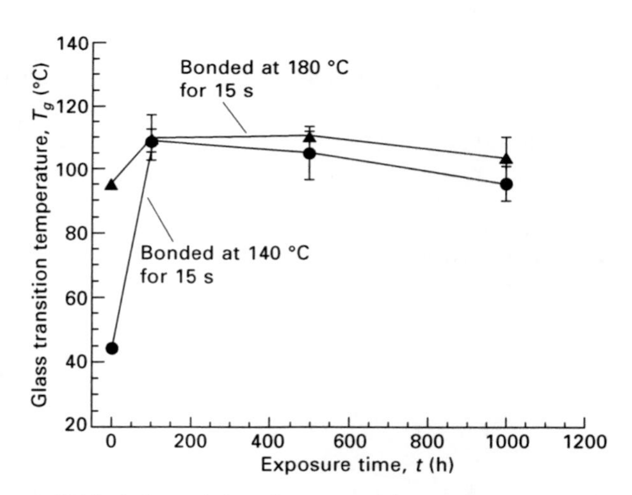

6.26 Variation of the glass transition temperature of epoxy-based ACF specimens bonded at 140 and 180 °C for 15 s, during exposure to an 85 °C/85%RH environment. The degree of conversion of the specimens bonded at 140 and 180 °C was 9.3 and 82.7%, respectively, before the exposure.

environments for 1000 h are shown in Fig. 6.26. The specimens were cured during the bonding process at 140 and 180 °C to 9.3 and 82.7% conversion, respectively. These specimens were subsequently post-cured for the first 100 h of exposure to these environments. It is apparent that the curing reaction (increase in T_g) was similarly promoted during exposure to these environments, although moisture absorption occurred significantly in the 85 °C/85%RH environment. The post-curing reaction of the adhesive binder occurred independently of moisture absorption during the early stages of exposure. The implication is that H_2O molecules predominantly diffused into the micro-voids during this period. Plasticization (decrease in T_g) of the specimens became significant in the later stages (more than ~200 h) of exposure.

Next, the effects of moisture absorption on the peel strength of the ACF joints are discussed. Figure 6.27 shows the 90° peel strength of ACF joints composed of ITO glass substrate and polyimide-based flex covered with Cu foil (with no interdigitated pattern), that were bonded at 180 °C for 15 s, as a function of exposure time to the 85 °C/85%RH environment. In the early stages (~100 h) of exposure, the peel strength of the ACF joints decreased although post-curing was promoted. By contrast, the peel strength increased during exposure for more than 200 h. The increase in the peel strength is thought to be caused by the increase in stress relaxation ability of the adhesive binder due to plasticization, because the peel strength significantly decreased after drying at ambient temperature in a vacuum chamber, as shown in Fig. 6.27.

Figure 6.28 shows the 90° peel strength of the ACF joints (bonded at 140 and 180 °C for 15 s) composed of ITO glass substrate and polyimide-based flex with a Cu interdigitated pattern in the direction parallel to the Cu

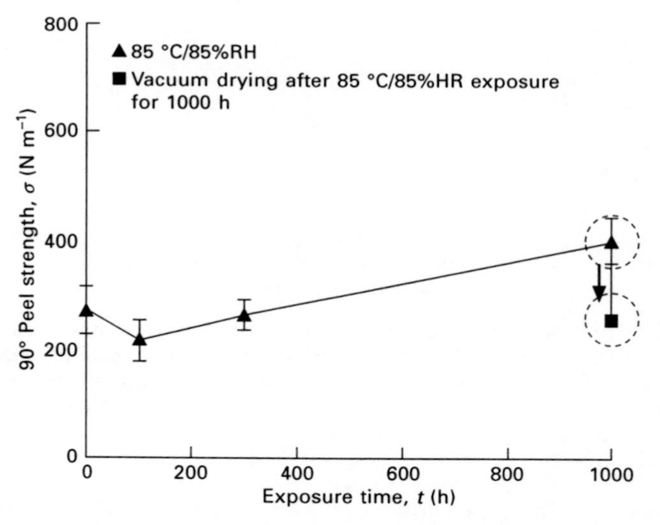

6.27 Variation in 90° peel strength of the ACF joints composed of ITO glass substrate and polyimide-based flex covered with a Cu foil (without any patterns) bonded at 180 °C for 15 s, during exposure to an 85 °C/85%RH environment. The peel strength of the joints after drying in a vacuum chamber is also shown.

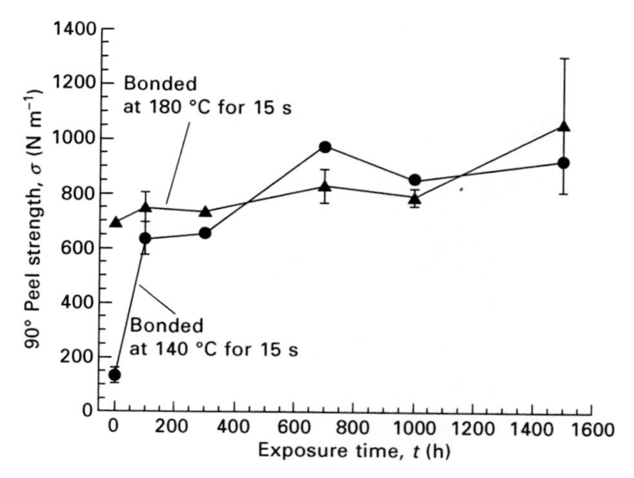

6.28 Variation in 90° peel strength of ACF joints composed of ITO glass substrate and polyimide-based flex with an interdigitated Cu pattern bonded at 140 and 180 °C for 15 s during the exposure to an 85 °C/85%RH environment. The peel test was conducted in a direction parallel to the Cu pattern.

pattern, as a function of exposure time to the 85 °C/85%RH environment. In the early stages of exposure, the peel strength of the joints significantly increased due to the post-curing of the adhesive binder. Subsequently, the peel strength exhibited an almost constant value regardless of bonding temperature and exposure environment. In this case, the anchoring effects due to the Cu pattern dominated the peel strength of the ACF joints, as discussed in Section 6.5.3. Consequently, the effects of the moisture absorption were not reflected clearly in the peel strength. The geometrical factor of adhesion interfaces is also important, in consideration of the mechanical reliability of adhesive joints.

6.6 Interconnections using electrically conductive adhesives

In the case of adhesives for interconnections, electrical conductivity is required to be ensured as well as mechanical bonding. This section discusses the relationship between mechanical bonding and electrical conductivity in adhesives for making interconnections.

The mechanism for electrical conductivity in electrically-conductive adhesives (ECAs) including ICAs and ACAs is discussed first. Figure 6.29 shows a schematic illustration of the cross-sectional microstructure of ICAs, which usually contain 40–50 vol% of metallic fillers (such as Ag) in order to form a percolation network that provides a conduction path for electrons.

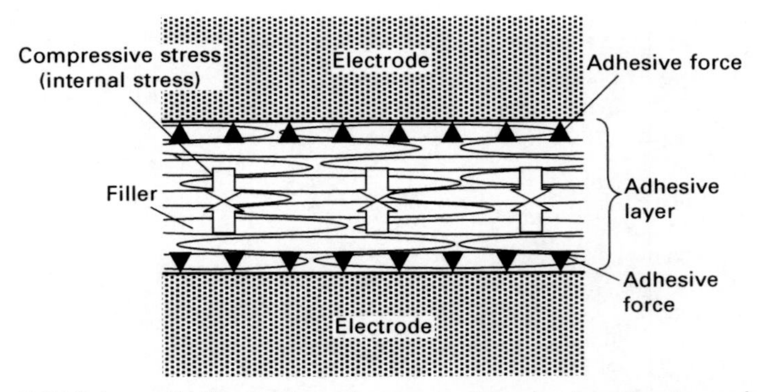

6.29 Schematic illustration of cross-sectional microstructure of ICAs.

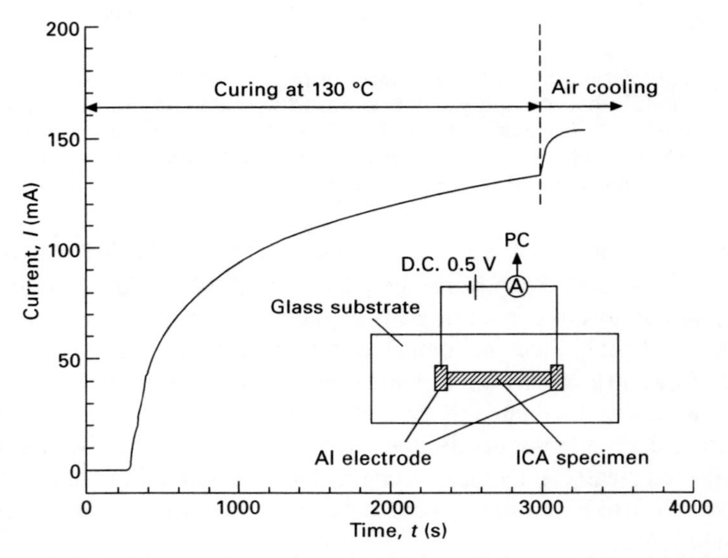

6.30 Result of *in situ* monitoring of electrical conduction in an epoxy-based ICA specimen during curing (at 130 °C for 3000 s) and cooling (air cooling) processes.

The electrical resistance of ICAs is divided into two components, namely intra-particle and interfacial (filler/filler and filler/electrode) resistance.[44,45] The interfacial resistance is further categorized into constriction and tunneling resistance.

During the curing process, the adhesive binder of the ICA adheres to the substrate surfaces. In addition, the curing and cooling shrinkages of the adhesive binder result in generation of internal stress. The mechanical bonding strength of the ICA joints is determined by these processes, and the conductive percolation network composed of metallic fillers is formed simultaneously.[46,47] Figure 6.30 shows the result of *in situ* monitoring of electrical conduction in an epoxy-based ICA specimen during curing (at 130 °C for 3000 s) and cooling (air cooling). Prior to curing, the electrical current was difficult to measure using this apparatus because the ICA had high electrical resistance. The specimen showed a significant decrease in electrical resistance (increase in electrical current in Fig. 6.30) during heating at the curing temperature. The onset of decrease in electrical resistance (for curing time ~300s) corresponds to 70–80% conversion from the viewpoint of the curing reaction of the adhesive binder. It is considered that the internal stress caused by the curing shrinkage began to be accumulated in the adhesive layer during this conversion range, because the stress-relaxation ability of the adhesive binder decreased due to the development of a cross-linked polymer structure in the binder. The internal stress (compression stress) can contribute to a decrease in the interfacial resistance, such as tunneling resistance, to generate electrical conductivity in the ICA specimen. Furthermore, decrease in the electrical resistance is also observed during the cooling process. The interfacial resistance of the ICA specimen is considered to be decreased by the internal stress generated by the cooling shrinkage of the adhesive.

Next, the generation of electrical conductivity during the bonding process of ACF is explained. Figure 6.31 shows a schematic illustration of the cross-sectional microstructure of ACF. The conductive particles (such as metal-coated polymer balls) are sandwiched between electrodes during the bonding process. Figure 6.32 shows the result of *in situ* monitoring of electrical conduction in a model (epoxy-based) ACF joint composed of ITO glass substrate and polyimide-based flex during the bonding (180 °C, 1.5 MPa, 15 s) and cooling (air cooling) processes. Significant plastic flow of adhesive binder and capture of conductive particles between the electrodes occurred concurrently just after starting the bonding process (~2 s). The conductive particles captured between the electrodes were significantly deformed to make conductive contacts with the electrodes during this period. Subsequently, the adhesive binder was cured to fix the conductive contacts between the particles and

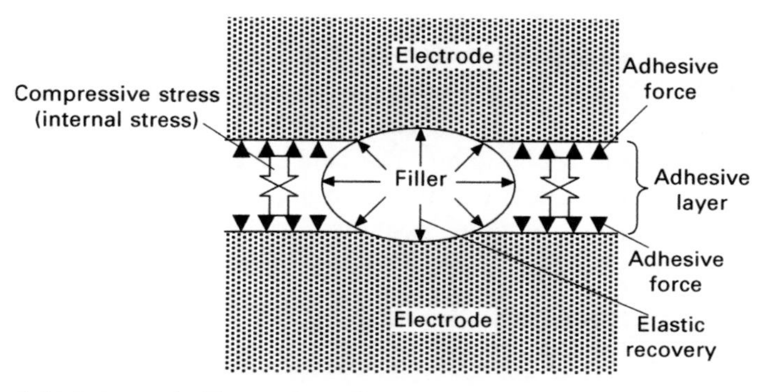

6.31 Schematic illustration of cross-sectional microstructure of ACAs.

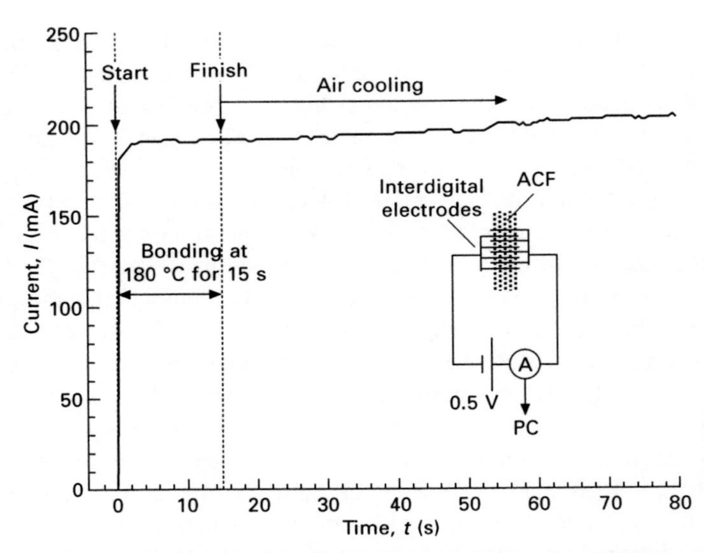

6.32 Result of *in situ* monitoring of electrical conduction in an epoxy-based ACF joint composed of ITO glass substrate and polyimide-based flex during the bonding (at 180 °C for 15 s with1.5 MPa applied pressure) and cooling (air cooling) processes.

electrodes. After removing the tool from the specimen, the electrical current gradually increased during the cooling process. This increase in electrical current indicates that the interfacial resistance decreased, which is attributed to generation of internal stress during the cooling shrinkage of the adhesive binder.

Based on the above discussions, the relationship between electrical conductivity and bonding strength of ICA and ACA joints is now considered. The electrical conductivity of ICAs is significantly influenced by the internal stress induced in the adhesive binder during the curing and cooling processes. In the case of ACAs, the elastic recovery stress of conductive particles captured between the electrodes also contributes to decreasing the contact resistance between the particles and electrodes, as well as the internal stress induced in the adhesive binder. Although these stresses are essential to obtain electrical interconnections, at the same time they have a negative effect on the bonding strength of the joints. The balance between these stresses and the adhesion force is the key to realizing high (mechanical and electrical) reliability of the adhesive joints, as shown in Figures 6.29 and 6.31.

6.7 Conclusions

The theoretical basis and influential factors for adhesion and bonding strength of adhesive joints have been explained in detail with the aim of formulating guidelines for development of advanced adhesives for electronics packaging applications.

The bonding strength of joints is determined not only by the thermodynamic work of adhesion, but also by various factors such as the viscoelastic deformation of adhesives. However, the role of interfacial adhesion strength is still important for realization of high bonding strength, as Andrews and colleagues established.

Interfacial adhesion usually occurs by intermolecular interactions, including van der Waals forces and hydrogen bonding. In such cases, the concepts of surface free energy and solubility parameter based on dispersive, polar and hydrogen bonding components are effectively applied for designing the chemical composition of adhesives. By contrast, these concepts are not valid when covalent bonds are introduced at interfaces using coupling agents. Since the common theories are not established for the introduction of covalent bonds at interfaces, individual considerations are required for the selection of reactive coupling agents from the viewpoint of reactivity of their functional groups. In any case, the chemistry of adhesive compounds is one of the most important subjects in developing advanced adhesives. The concepts of interfacial chemistry are also applicable to molecular adhesion in nanotechnology, such as with self-assembled monolayers (SAMs).

In addition to the chemistry of interfacial adhesion, modifying the geometry of interfaces to significantly induce the anchoring effect plays an important role in determining the bonding strength of adhesive joints.

Because the viscoelastic deformation of adhesives and the anchoring effect are effectively obtained when the adhesives are sufficiently cured, the curing process should be controlled in order to achieve higher bonding strength. Kinetic analysis for the curing reaction provides useful information for control of the process.

By contrast, internal stress due to the shrinkage of adhesives generated during the curing and cooling processes, has a negative effect on bonding strength, although the stress is essential to secure electrical conductivity in ECAs. Hence, a balance between the interfacial adhesive force and the internal stress should be sought to obtain highly reliable joints.

Microsystems will continue to be further diversified in the future. Consequently, the importance of advanced adhesives will increase in electronics packaging technology, and the design concepts for adhesive compounds based on the scientific fundamentals that are discussed in this chapter will be increasingly important for engineers and scientists in this field.

6.8 References

1. Japan Adhesion Society, *Setchaku Handbook*, 4th ed., *Nikkan Kougyo Shinbun-sha*, Tokyo (2007); *ibid.* 2nd ed. (1980).

2. R. D. Adams, *Adhesive Bonding: Science Technology and Applications*, Woodhead Publishing Ltd, Cambridge, UK (2005).
3. F. M. Fowkes, Determination of interfacial tensions, contact angles, and dispersion forces in surfaces by assuming additivity of intermolecular interactions in surfaces, *J. Phys. Chem.*, **66**, 382 (1962).
4. F. M. Fowkes, Attractive forces at interfaces, *J. Ind. Eng. Chem.*, **56**, 40–52 (1964).
5. D. K. Owens, R. C. Wendt, Estimation of the surface free energy of polymers, *J. Appl. Polm. Sci.*, **13**, 1741–1747 (1969).
6. D. K. Owens, Some thermodynamic aspects of polymer adhesion, *J. Appl. Polym. Sci.*, **14**, 1725–1730 (1970).
7. D. H. Kaelble, K. C. Uy, Reinterpretation of organic liquid-polytetrafluoroethylene surface interactions, *J. Adhesion*, **2**, 50–60 (1970).
8. Y. Kitazaki, T. Hata, Extension of Fowkes' equation and estimation of surface tension of polymer solids, *J. Adhesion Soc. Jpn.*, **8**, 131–142 (1972).
9. F. M. Fowkes, in *Acid–Base Interactions*, (ed. K. L. Mittal, H. R. Anderson Jr.), pp. 93–115, VSP, Utrecht (1991).
10. K. Kitaura, K. Morokuma, A new energy decomposition scheme for molecular interactions within the Hartree–Fock approximation, *Int. J. Quantum Chem.*, **10**, 325–340 (1976).
11. Y. Minamizaki, Y. Tanaka, K. Kobayashi, Molecular orbital calculations of interfacial energy in adhesion, *J. Adhesion Soc. Jpn.*, **40**, 282–288 (2004).
12. Y. Ikada, T. Matsunaga, Surface energy of polymers, *J. Adhesion Soc. Jpn.*, **14**, 427–434 (1978).
13. K. Takemoto, M. Mitou, *Setchaku no Kagaku*, Kodansha, Tokyo (1997).
14. E. H. Andrews, A. J. Kinloch, Mechanics of adhesive failure. I, *Proc. R. Soc. Lond. A*, **332**, 385–399 (1973).
15. I. Kamyab, E. H. Andrews, Interfacial and bulk contributions to peeling energy, *J. Adhesion*, **56**, 121–134 (1996).
16. P. Atkins, J. du Paula, *Atkins' Physical Chemistry*, 8th ed., Oxford University Press, Oxford (2006).
17. J. Kloubek, Development of methods for surface free energy determination using contact angles of liquids on solids, *Advances in Colloid and Interface Science*, **38**, 99–142 (1992).
18. K. Akamatsu, S. Ikeda, H. Nawafune, Site-selective direct silver metallization on surface-modified polyimide layers, *Langmuir*, **19**, 10366–10371 (2003).
19. A. Kawai, T. Kumagai, M. Takata, Dispersion and polar components of surface free energy of transition metal films measured by contact angle method in air, *J. Adhesion Soc. Jpn.*, **31**, 307–311 (1995).
20. A. Kawai, J., Surface free energy variation of sputtering-formed Au and Pt thin films after exposure to air, *Adhesion Soc. Jpn*, **32**, 39–43 (1996).
21. T. Suga, Y. Takahashi, H. Takagi, B. Gibbesch, G. Elssner, Structure of Al-Al and Al-Si_3N_4 interfaces bonded at room temperature by means of the surface activation method, *Acta Metall. Mater.*, **40**, Suppl., S133–S137 (1992).
22. H. Takagi, K. Kikuchi, R. Maeda, T. R. Chung, T. Suga, Surface activated bonding of silicon wafers at room temperature, *Appl. Phys. Lett.*, **68**, 2222–2224 (1996).
23. C. M. Hansen, *Hansen Solubility Parameters, A User's Handbook*, 2nd ed., CRC Press, Boca Raton (2007).
24. D. W. Van Kreveken, in *Properties of Polymers*, pp. 189–225, Elsevier, Amsterdam (1990).

25. A. F. M. Barton, *CRC Handbook of Solubility Parameters and Other Cohesion Parameters*, 2nd ed., CRC Press, Boca Raton (1991).
26. W. C. Bigelow, E. Glass, W. A. Zisman, Oleophobic monolayers; temperature effects and energy of adsorption, *J. Colloid Interface Sci.*, **2**, 563–591 (1947).
27. R. G. Nuzzo, D. L. Allara, Adsorption of bifunctional organic disulfides on gold surfaces, *J. Am. Chem. Soc.*, **105**, 4481–4483 (1983).
28. A. Ulman, Formation and structure of self-assembled monolayers, *Chem. Rev.*, **96**, 1533–1554 (1996).
29. R. K. Smith, P. A. Lewis, P. S. Weiss, Patterning self-assembled monolayers, *Prog. Surf. Sci.*, **75**, 1–68 (2004).
30. M. J. Wirth, R. W. P. Fairbank, H. O. Fatunmbi, Mixed self-assembled monolayers in chemical separations, *Science*, **275**, 44–47 (1997).
31. K. Hayashi, H. Sugimura, O. Takai, Frictional properties of organosilane self-assembled monolayer in vacuum, *Jpn. J. Appl. Phys.*, **40**, 4344–4348 (2001).
32. Y. Li, K.-S. Moon, C. P. Wong, Adherence of self-assembled monolayers on gold and their effects for high-performance anisotropic conductive adhesives, *J. Electron. Mater.*, **34**, 266–271 (2005).
33. M. A. Uddin, M. O. Alam, Y. C. Chan, H. P. Chan, Adhesion strength and contact resistance of flip chip on flex packages – Effect of curing degree of anisotropic conductive film, *Microelectron. Reliab.*, **44**, 505–514 (2004).
34. A. Hale, C. W. Macosko, H. E. Bair, Glass transition temperature as a function of conversion in thermosetting polymers, *Macromole.*, **24**, 2610–2621 (1991).
35. J. P. Pascault, R. J. J. Williams, Glass transition temperature versus conversion relationships for thermosetting polymers, *J. Polym. Sci.: Part B*, **28**, 85–95 (1990).
36. M. Inoue, K. Suganuma, The dependence on thermal history of the electrical properties of an epoxy-based isotropic conductive adhesive, *J. Electron. Mater.*, **36**, 669–675 (2007).
37. M. Inoue, K. Suganuma, Influential factors in determining the adhesive strength of ACF joints, *J. Mater. Sci.: Mater. Electron.* **20**, 1247–1254 (2009).
38. D. J. Arrowsmith, Adhesion of electroformed copper and nickel to plastic laminates, *Trans. Inst. Metal Finishing*, **48** (pt.2), 88–92 (1970).
39. Y. Tsujita, Gas sorption and permeation of glassy polymers with microvoids, *Progress in Polym. Sci.*, **28**, 1377–1401 (2003).
40. P. Musto, G. Ragosta, G. Scarinzi, L. Mascia, Probing the molecular interactions in the diffusion of water through epoxy and epoxy–bismaleimide networks, *J. Polym. Sci.: Part B*, **40**, 922–938 (2002).
41. P. Musto, G. Ragosta, L. Mascia, Vibrational spectroscopy evidence for the dual nature of water sorbed into epoxy resins, *Chem. Mater.*, **12**, 1331–1341 (2000).
42. P. Nogueira, C. Ramirez, A. Torres, M. J. Abad, J. Cano, J. Lopez-Bueno, L. Barral, Effect of water sorption on the structure and mechanical properties of an epoxy resin system, *J. Appl. Polym. Sci.*, **80**, 71–80 (2001).
43. M. P. Zanni-Deffarges, M. E. R. Shanahan, Diffusion of water into an epoxy adhesive: Comparison between bulk behaviour and adhesive joints, *Int. J. Adhesion and Adhesives*, **15**, 137–142 (1995).
44. L. Li, C. Lizzul, H. Kim, I. Sacolick and J. E. Morris, Electrical, structural and processing properties of electrically conductive adhesives, *IEEE Trans. Components, Hybrids, and Manufacturing Technol.*, **16**, 843–851 (1993).
45. J. E. Morris, in *Conductive Adhesives for Electronics Packaging* (ed. J. Liu), p. 36, Electrochemical Publications, Port Erin (1999).

46. D. Lu, C. P. Wong, Effects of shrinkage on conductivity of isotropic conductive adhesives, *Int. J. Adhesion and Adhesives*, **20**, 189–193 (2000).
47. M. Inoue, K. Suganuma, Effect of curing conditions on the electrical properties of isotropic conductive adhesives composed of an epoxy-based binder, *Soldering and Surface Mount Technology*, **18**, 40–46 (2006).

7

Modelling techniques used to assess conductive adhesive properties

C. BAILEY, University of Greenwich, UK

Abstract: Modelling techniques are now routinely used to help understand the manufacturability and performance of conductive adhesives when used in electronic, photonic and microsystems packaging. This chapter reviews the latest advances of the techniques for structural and conductive adhesives.

Key words: process modelling, finite element analysis, optimisation.

7.1 Introduction

Increasing global competition is a significant factor affecting the design of modern electronic and photonic products. While the product development time for electronic systems in the early 1980s was often years, portable computing and consumer products today have a time-to-market of only a few months. Such rapid times-to-market do not leave room for time-consuming trial and error approaches that have been the normal practice in the past. Adoption of modelling software technologies early in the design stage of product development saves money. It lets companies lower, if not eliminate, the number of costly physical prototypes, encourages change, and gives rise to organisations that deliver more products in less time than previously possible.

Technology roadmaps (ITRS, iNEMI) emphasise the requirement for improved design and modelling tools that permit integrated modelling and simulation of materials and processes to accommodate rapid advancements in technology. Modelling tools can help industry identify potential manufacturing defects very early in the design cycle and, more importantly, they can be used to provide optimal process conditions and material properties that will ensure success.

Virtual prototyping or computational modelling tools that predict thermal, electrical and mechanical phenomena are now playing a key part at the early design stage for product development and providing delivery of reliable products to market. Exploitation of these software technologies benefits companies by:

- minimising the amount of physical prototyping
- improving quality and performance
- identifying optimal properties and process conditions

- generating knowledge of the process
- getting products to market earlier
- reducing overall development cost.

7.2 Numerical modelling technologies

At the heart of computational modelling are the mathematical equations that describe the physical processes taking place during manufacture and influence the product when it is in service. For example, one of these processes could be the transfer of heat from a chip to its surroundings. These equations relate the physics of the process to external influences taking place. These include the governing classical equations of physics for fluid flow, heat transfer, stress and electromagnetics.

Numerical techniques, such as the finite element method, are used to discretise these mathematical equations that are usually represented by partial differential equations representing the governing physics taking place, and the behaviour of the materials that make up the electronic or photonic device. Continuum mechanics modelling tools can be classified as:

- *computational fluid dynamics (CFD)*, solving phenomena such as fluid flow, heat transfer, combustion and solidification
- *computational solid mechanics (CSM)*, solving deformation, dynamics, stress, heat transfer, and failures in solid structures
- *computational electromagnetics (CEM)*, used to solve electromagnetics, electro-statics and magneto-statics.

Until recently the majority of continuum mechanics codes focused on the prediction of distinct physics, but now there has been a strong push by software vendors to develop multi-physics or co-disciplinary tools that capture the complex interactions between the governing physics such as fluidics, thermal, mechanical and electrical.

Tools based on numerical optimisation techniques bring enormous advantages by offering an automated, logical and time-efficient approach in identifying the best model and/or process parameters. The interested reader is referred to Vanderplaats (1999) for more details on the most common numerical optimisation techniques. Reliability-based design optimisation formulations offer convenient and automated virtual exploration of the design space and identification of the best configuration of input parameters from the viewpoint of the objective function and uncertainty (probabilistic) influenced requirements. Although different design problems have been solved using optimisation procedure, fully integrated or coupled simulation–optimisation software modules are only just appearing and much more is required to fully capture process variation and uncertainty into optimisation calculations.

Design for manufacturing and reliability of microsystems packaging that include adhesives is an extremely complex engineering task. The complexity of the advanced adhesives often makes real prototyping and testing difficult or expensive. Therefore it is essential to incorporate, at the early design stage, methods that exploit not just numerical simulation for the physical behaviour but also techniques that allow for quantification and optimisation of the risk and reliability of the systems. Deterministic and stochastic simulation models are now emerging as valuable tools in modern design, and they provide help managing and mitigating the associated failure risks.

Several commercially available software environments are available for predicting thermal, mechanical and electrical performance for electronic packages. Table 7.1 details a number of well-known codes. It should be noted that codes, and the results that they can produce, do not replace experimentation and physical testing. They compliment experimentation and allow design engineers and researchers to identify key physical phenomena leading to product failure and trends early within the product design cycle.

Figure 7.1 illustrates the interaction between physics-based modelling tools and optimisation in predicting the behaviour and reliability of microsystems devices from fabrication, packaging, testing and qualification, and finally in-service performance.

7.3 Modelling applied to packaging processes

Modelling of typical assembly or packaging processes is discussed in this section. Those discussed are (i) deposition of conductive adhesives and (ii) bonding of components using conductive adhesives.

7.3.1 Deposition of conductive adhesive materials

Stencil printing is one of the processes that can be used to deposit adhesive paste at precise locations on the printed circuit board (PCB) pads in order to prepare for the placement and bonding of electronic components. In this process, a squeegee blade moves the conductive adhesive over the surface

Table 7.1 Example of some commercial software codes and capabilities

Software	Capabilities
ANSYS	Multiphysics
FLUENT	Fluid flow, heat transfer
ABACUS	Stress, heat transfer
FLOTHERM	Thermal
PHYSICA	Multiphysics
COMSOL	Multiphysics
ROMARA	Optimisation
Optimus	Optimisation

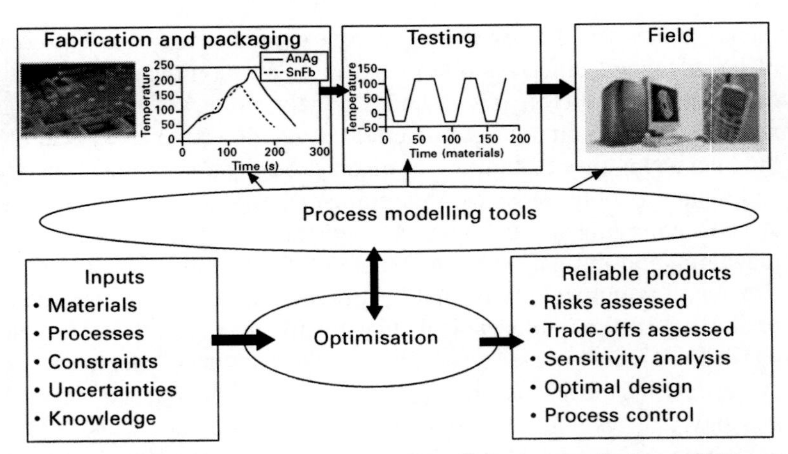

7.1 Optimisation-driven numerical modelling for predicting reliable packaging of microsystems.

of a stencil with a particular pattern of apertures. As a result of the high pressure in the conductive adhesive induced by the squeegee blade, the adhesive is forced to fill the stencil apertures.

The overall conductive adhesive composition exhibits non-Newtonian rheological properties. This behaviour enables the paste to flow into the apertures with a low viscosity when the shear rate is high due to the action of the moving squeegee blade. After the removal of the stencil, the viscosity increases again in the absence of shearing, a phenomenon that helps the adhesive to remain in place. Computational fluid dynamics can predict the movement of conductive adhesive across a stencil surface. For example, simulation of the motion of the material can be undertaken using the classical Navier–Stokes equations with the following viscosity model (Nguty and Ekere, 1999):

$$\frac{\eta - \eta_\infty}{\eta_0 - \eta_\infty} = \frac{1}{1 + K\lambda^m} \quad [7.1]$$

where η is the apparent viscosity, η_0 and η_∞ are the viscosity at zero and infinite shear rate respectively, λ is the strain rate and K and m are experimentally obtained constants. Figure 7.2 shows a schematic of the printing process and associated CFD predictions for adhesive (ECA) and solder using the classical continuum approach (Glinski *et al.*, 2001).

Other adhesive dispensing processes, such as dispensing from a jetting or dispensing from a syringe can be used to deposit paste. In these processes, fluid properties such as surface tension and contact angle have a significant impact on flow behaviour and quality of the deposited adhesive in the printed circuit board. Again, a non-Newtonian flow model is required to predict the flow behaviour of the adhesive. In addition to this, free-surface algorithms

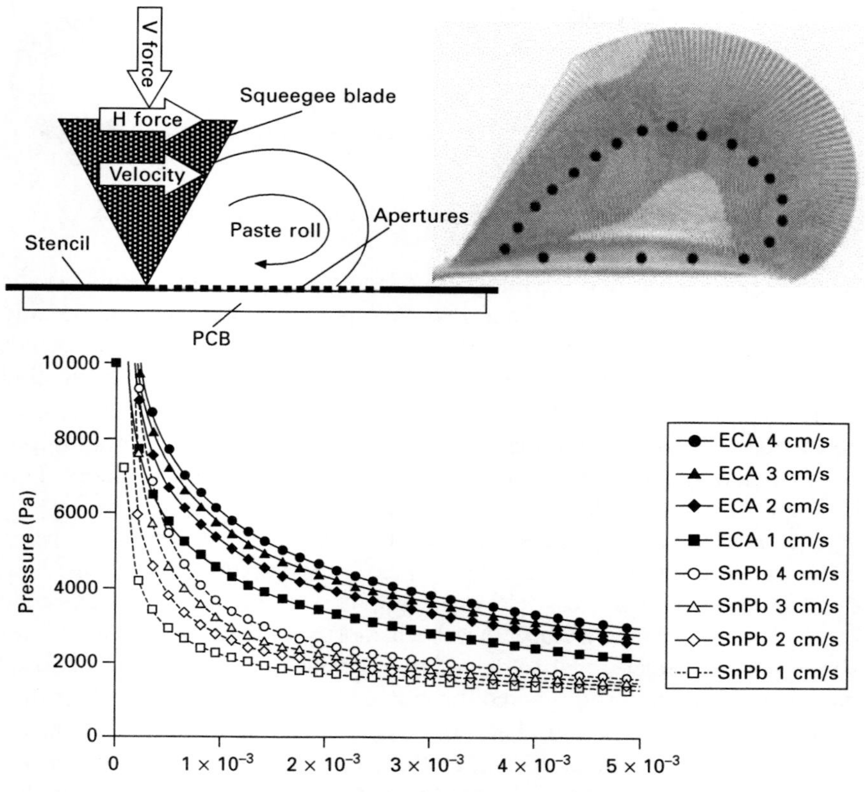

7.2 Modelling predictions for the flow of solder material in stencil printing.

are required to accurately predict the movement of the adhesive interfaces with the air and PCB. Identification of these relevant adhesive properties for use in computational fluid dynamics models is important. Chen and Ke (2006) have provided a methodology that identifies these properties using a few dispensing experiments.

7.3.2 Bonding of components using conductive adhesives

Anisotropic conductive films (ACFs), with many distinct advantages such as extremely fine pitch capability, freedom from lead, and environmental friendliness, are being widely used in fine-pitch flip chip technologies (Liu, 1999). A typical ACF flip chip is shown in Fig. 7.3. The conductive particle is a nickel/gold-coated polymer ball with a diameter here of 3.5 micrometres. The diameter of the conductive particles in ACF materials is generally

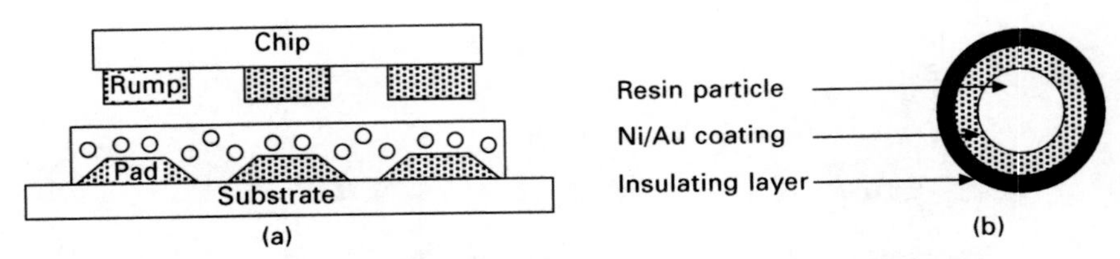

7.3 ACF flip chip (a) and structure of a conductive particle (b).

several micrometres and the thickness of the particle metallisation is in the nano scale, which is about 50nm. If the die is 11mm in its length, the ratio of the two is approximately 1: 200,000! In addition, there are thousands of conducting particles in a typical ACF material used to bond a flip chip component to a substrate. This means that an 'exact' model, which includes all the particles and interconnections, would require millions if not billions of computational mesh elements to be used in a finite element model. This is simply not achievable with today's computer technology.

Computer modelling analysis, in particular finite element analysis, is being used as a powerful tool to predict the behaviour and responses of ACF particles during the bonding process. However, previous modelling work (Mercodo *et al.*, 2003; Wei *et al.*, 2002) has been mostly limited to the analysis of simplified two-dimensional models. Three-dimensional models have focused on the micro-domain and ignored the global effects at package level, or they have modelled the whole package and used gross assumptions at the micro interconnect level (Kim and Jung, 2006; Rizvi *et al.*, 2005). The recognised difficulty here is due to the vast range of length-scales in an ACF flip chip assembly, and the large number of conductive particles. Therefore, a three-dimensional macro-micro modelling technique is required in order to provide the ability to accurately model the behaviour of the conductive particles. Two models, one macro and one micro, with very different mesh densities, have been used to model the bonding process. The macro model is used to predict the overall behaviour of the whole assembly. The displacements obtained from this macro model are then used as the boundary conditions for the micro model so that the detailed stress analysis in the region of interest can be carried out. This macro–micro modelling technique enables more detailed three-dimensional modelling analysis of an ACF flip chip than previously. Figure 7.4 details this approach.

During these simulations, it is important to accurately represent the behaviour of all of the materials in the simulation. For the die and substrate, the usual assumption is that these materials will have elastic properties. For adhesives, the visco-elastic behaviour needs to be represented. This visco-elastic behaviour also needs to be cure-dependent to accurately represent the curing process taking place during packaging. An excellent review on this is provided by Yang (2007), who details a number of constitutive models

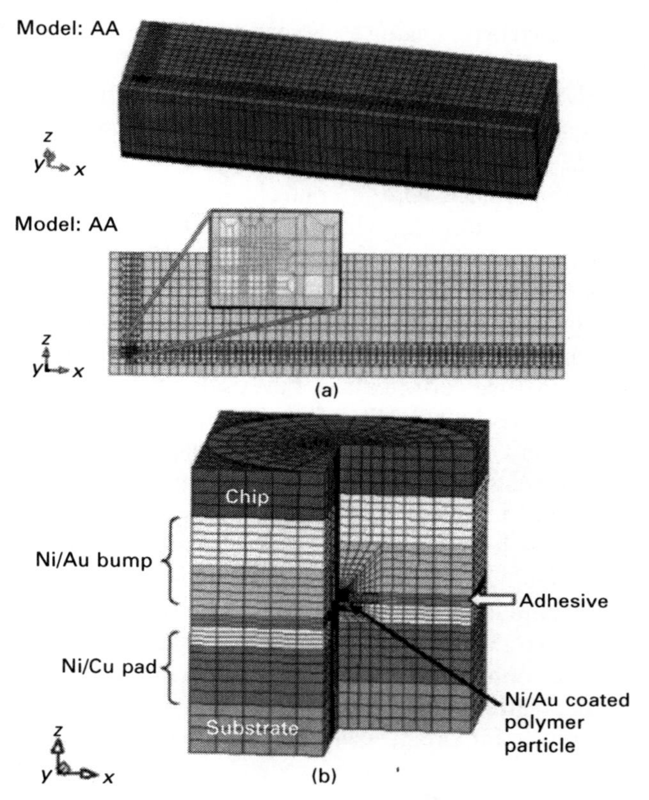

7.4 Micro–macro modelling approach for bonding of ACF particles. (a) Macro model, (b) micro model.

for adhesive materials used in electronic packaging. The cure behaviour is obtained from experimental techniques such as DMA and DSC analysis and these visco-elastic cure-dependent models are used to predict the behaviour of adhesives for Quad Flat no Leads (QFN) and flip chip packages.

7.4 Modelling the thermal, electrical and mechanical performance of adhesives

7.4.1 Modelling the thermal and electrical performance of adhesives

Owing to the increase in functionality and miniaturisation of electronic products, the issue of heat dissipation has become a critical design criterion. Past package generations dissipated very low power, so thermal management issues could easily be resolved. This is not the case with many of the packages and systems used today. To help passively remove heat from the package, some adhesives are used in packaging as thermal interface materials. Modelling has been utilised extensively by the electronic packaging community to predict the sensitivity of package design and materials, such as die adhesive, mould compound, and board level underfill, to optimise thermal performance. These simulations aim to compliment the JEDEC standards for thermal management. Of importance here is controlling the junction temperature of the die, which, if it is too hot, will affect the reliability of the package. Computational fluid dynamics and conduction-only solvers have been used for this task. These provide the ability to simulate both active and passive cooling technologies. Thermal models for staked die systems in which the properties of die attach and BGA underfill adhesives have been investigated to optimise overall thermal performance of the package. In this analysis CFD models were used to predict heat dissipation due to thermal conduction and natural convection, and to validate a package model for a staked die BGA on JEDEC environments. Such an approach has significantly reduced the thermal design cycle and time-to-market.

Characterising the electrical contact resistance of conductive adhesives is also very important. This requires models that realistically predict the contact resistance for multiple particles in contact between conductive tracks. An example of this is the anisotropic conductive adhesive. In assessing the overall electrical performance of such adhesives, parameters such as number of particles, spatial distribution of the particles, and magnitude of the initial bonding force, will be important. Chin and Hu (2007) have detailed both theoretical and finite element models for predicting the electrical performance of anisotropic conductive adhesives. Unlike previous work in this area, they demonstrated a multiple particle model which did not assume that the electrical contact resistance for each particle was the same. In this work they look into account (i) non-identical contact areas, (ii) elastic recovery, (iii) electrical interaction between the particles and (iv) electrical interaction between particles and surrounding contact surfaces. A multi-coupled finite element modelling approach was investigated to predict the final electrical contact resistance. Conclusions from this detailed study demonstrate that an increase in the number of particles leads to a decrease in electrical contact resistance, but this also depends strongly on the bonding force used to assemble the die. To minimise contact resistance, a uniform spatial distribution of the particles is preferred and the quality of the contact interface between the particles and the conductive tracks is very important.

7.4.2 Modelling the mechanical performance and reliability of adhesives

The reliability of adhesives can be affected by environmental conditions such as temperature, vibration and moisture. During qualification, an electronic package will be subjected to thermal cycling, vibration and humidity tests to assess its suitability in the field. These environments impose stresses on the package and the adhesive materials that can lead to reliability problems.

When a package is exposed to temperature changes, individual materials' coefficients of thermal expansion (CTE) result in each material expanding or contracting by a different amount. This leads to stresses throughout the package, which can cause damage. When the package is exposed to moisture, differences in the coefficient of moisture expansion of each material means that the package will behave in a similar manner to when it is exposed to temperature changes and again this causes stress throughout the package.

The key aim of modelling here is to accurately predict the stress and damage imposed on the packaging materials due to these environments. Figure 7.5 summarises the modelling steps to undertake such an analysis. As can be seen, finite element analysis is used to predict the damage parameters in the materials when subjected to temperature, vibration or humidity loading. These damage parameters, such as stress, strain or creep strain energy, are used in a damage model to assess the impact of damage on the reliability of the material. Again, to undertake this type of modelling, it is important to have high quality input data to the models. This include constitutive laws for the materials, suitable boundary conditions for the model, and finally a suitable damage model and failure criteria.

Standard finite element codes can be used to predict the stress–strain behaviour for the materials in the package and, from these results, the relevant damage parameters can be extracted for use in the damage model. This is the physics-of-failure approach to reliability assessments of electronic packages.

Modelling the behaviour of electronic packages containing conductive

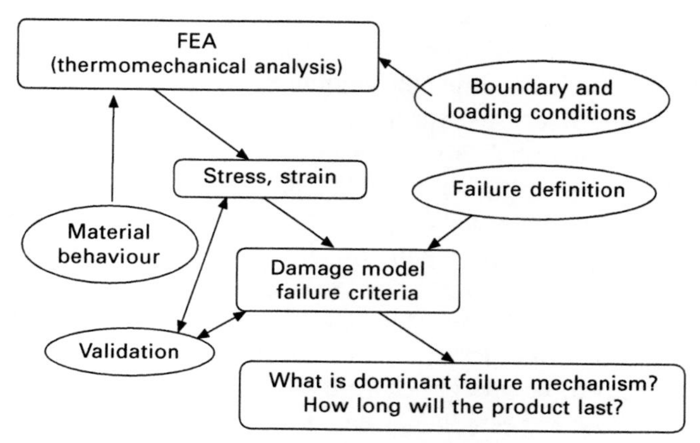

7.5 Modelling predictions for the flow of solder material in stencil printing. FEA, finite element analysis.

adhesives when subjected to temperature and humidity changes has been undertaken by a number of groups. Of importance are the adhesion properties between the adhesive and the substrate, where four primary mechanisms have been proposed (Ferguson and Qu, 2006). In their paper, it is assumed that the dominant mechanism can be explained by adsorption theory, and a thermodynamic engineering model is proposed to predict the loss of adhesion at the interface due to humidity changes

Finite element methods have also been used extensively with damage models, such as fracture mechanics to model the delamination of interfaces. Harries and Sitaraman (2001) used such an approach to predict the delamination propogation of peripheral array packages when subjected to thermal loads. Tsai and co-workers (2007) used finite element analysis to investigate the behaviour of ACF materials for chip-on-glass applications. In their analysis they found that warpage of the package was influenced by both temperature and humidity changes, and that to minimise this effect, the mismatch between the CTE of the materials in the package must be controlled. Also, that larger ACF fillet thicknesses can give rise to greater levels of warpage and hence damage.

Yin *et al.* (2005, 2006) investigated the stresses in flip chip packages and ACF materials when the package was subjected to an autoclave test environment. Multi-scale models were used to represent the package and the ACF conductive particles. In this study the location of cracks was identified and this was supported by experimental findings. The cohesive zone model has also been used to model cracks in anisotropic conductive adhesives (Zhang *et al.*, 2009)

Underfills are widely used in the packaging industry to offset the damaging effects of CTE mismatch on interconnections such as solder joints between a chip and the substrate in a flip chip assembly. The effectiveness of an underfill in reducing the impact of CTE mismatch mainly depends on its thermal–mechanical properties, such as the Young's modulus, CTE, and CME. These properties can be modified by adding filler particles such as silica into the polymer matrix. Because of the low filler content, no-flow underfills have higher CTE than traditional capillary underfills. To investigate the effect of this on the lifetime of a flip chip solder joint, Lu *et al.* (2002) modelled a flip chip's fatigue lifetime under cyclic thermal–mechanical loading for a range of underfill properties. Figure 7.6 shows the 3D FEA model used in the modelling, and Fig. 7.7 shows the predicted lifetimes. The results show that flip chips using no-flow underfills have significantly lower lifetimes than traditional underfills. In order to achieve the highest reliability, the CTE of the underfill needs to be brought down to about 20 ppm/°C.

Recently there has been increased interest in the use of prognostic algorithms to predict the degradation of electronic packaging materials in the field (Pecht, 2008). The existing prognostic approaches for electronics

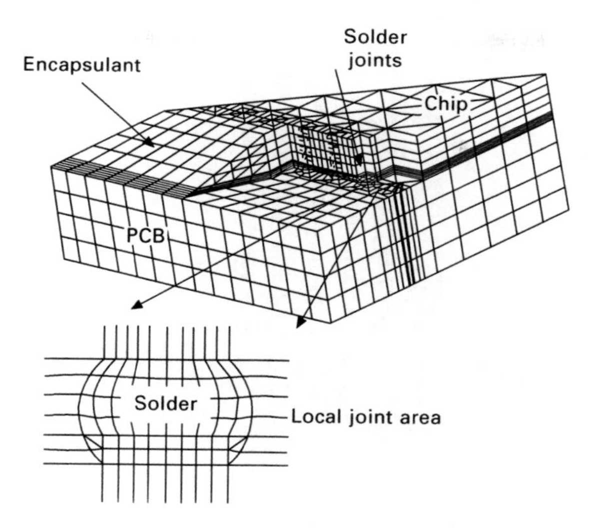

7.6 Flip chip computer model used to study the effects of the no-flow underfill material.

can be classified into the following categories: (i) data-driven methods, (ii) model-driven methods, and (iii) fusion techniques combining the previous methods. Data-driven techniques use statistical methods to assess when damage is occurring at an increasing rate in the materials. Model-driven methods use reduced order models to predict the damage parameters in the materials and then suitable damage models to estimate remaining useful life. To the author's knowledge, these techniques have not yet been used to predict the damage to adhesive materials when subjected to field conditions. However, this will be an area of interest to the reliability community in the future, to accurately estimate the accumulating damage in adhesive materials used in electronic packages when subjected to real field conditions.

7.5 Future trends

Although numerical modelling tools are now routinely used in the design of electronic packages, there are a number of key capability challenges for these tools to address in the future. For example:

(i) *Multi-physics modelling*. Electronic packaging processes are generally governed by close coupling between different physical processes. For example, in the curing process of adhesives there is close coupling between fluid dynamics, temperature, phase change, and stress. Numerical modelling tools are now addressing the need for multi-physics calculations, but more work is required to capture the physics

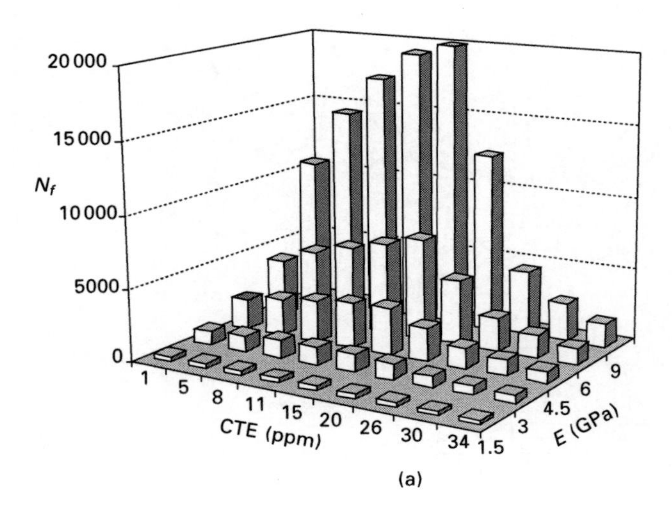

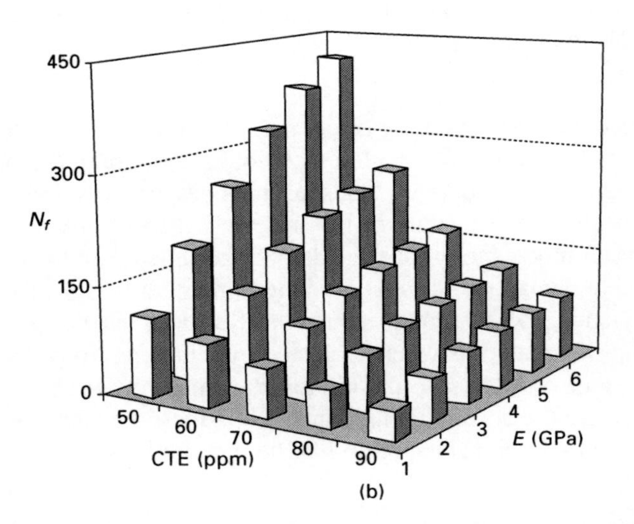

7.7 Predicted lifetime of flip chip solder joint. (a) Traditional underfill, (b) no-flow underfill.

accurately in these calculations and to identify relevant failure models for electronic packages.

(ii) *Multi-discipline analysis*. Thermal, electrical, mechanical, environmental, plus other factors are important in the design and manufacture of electronic packages. Numerical modelling tools that allow design engineers from different disciplines to trade-off their requirements early in the design process will dramatically reduce lead times.

(iii) *Multi-scale modelling.* Electronic packaging quality and reliability is increasing, governed by phenomena taking place across the length scales (nano – micro – meso – macro). Modelling techniques that provide seamless coupling between simulation tools across the length scales are required.

(iv) *Materials data.* The modelling community must work closely with the metrology community to ensure that suitable materials data and failure criteria are documented for adhesive materials. It should be noted that model predictions are very dependent on the materials data inputs used in the model.

(v) *Fast calculations.* Numerical modelling software that solves highly non-linear partial differential equations, using the finite element method for example, are computer intensive and slow. There is a need for reduced-order models (or compact models) for many electronic packaging simulations. Although not as accurate as finite element analysis, they provide the design engineer with the ability to quickly eliminate many unattractive designs early in the design process.

(vi) *Life-cycle considerations.* Major life-cycle factors, such as reliability, maintenance and end-of-life disposition, receive limited visibility in numerical modelling tools. Future models will include all life-cycle considerations, such as product greenness, reliability, recycling, disassembly and disposal.

(vii) *Variation risk mitigation.* Electronic packaging simulations usually ignore process variation, manufacturing tolerances and uncertainty. Future models will include these types of parameters to help provide a prediction of manufacturing risk. This can then be used by the design engineer to enable them to implement a mitigation strategy.

(viii) *Integration with optimisation tools.* Numerical optimisation techniques bring enormous advantages by offering an automated, logical and time-efficient approach to identify the best process/design parameters for reliable electronic packaging. Although different design problems have been solved using optimisation procedures, fully integrated or coupled simulation-optimisation software modules are only just appearing and much more is required to fully capture process variation and uncertainty into optimisation calculations.

(ix) *Modelling through the supply chain.* Numerical modelling tools require high-quality input data in terms of materials data and failure models. Many companies are now using these modelling tools and there is an increasing requirement for companies within each other's supply chain to gather and provide relevant modelling data. This is now starting to take place but much more effort is required.

7.6 Conclusions

The use of modelling tools and techniques in assessing the performance of adhesives in electronic packaging has come a long way over the last 15 years. Models are now being used to predict the deposition of adhesive materials, their behaviour during the bonding (packaging) process and finally their thermal, electrical and mechanical performance. In the future, the predictive ability of these models will increase as more materials data becomes available and as computational resources increase.

7.7 References

Chen X, Ke H (2006), Effects of Fluid Properties on Dispensing Processes for Electronics Packaging. *IEEE Trans. on Electronics Packaging Manufacturing*, **29** (2): 75–82.

Chin M, Hu S (2007), A Multi Particle Model for the Prediction of Electrical Contact Resistance in Anisotropic Conductive Adhesive Assemblies, *IEEE Trans. on Components and Packaging Technologies*, **30**(4): 745–753.

Ferguson T, Qu J (2006), Predictive Model for Loss of Adhesion of Molding Compounds from Exposure to Humid Environments, *Electronics Components and Technology Conference*, pp. 1408–1411, Pub IEEE.

Glinski G, Bailey C, Pericleous K (2001), A Non-Newtonian Computational Fluid Dynamics Study of the Stencil Printing Process, *Journal of Mechanical Engineering Science, Proc IMechE*, **215**(4): 437–446.

Harris R, Sitaraman S (2001), Numerical Modeling of Interfacial Delamination Propogation in a Novel Peripheral Array Package, *IEEE Trans. on Components and Packaging Technologies*, **24**(2): 256–264.

Kim JW, Jung SB (2006), Effects of Bonding Pressure on the Thermo-mechanical Reliability of ACF Interconnection, *Journal of Microelectronic Engineering*, **83**(11–12): 2335–2340.

Liu J (1999), *Conductive Adhesives for Electronics Packaging*, Electrochemical Publications Ltd, pp. 234–248.

Lu H, Hung KC, Stoyanov S, Bailey C and Chan YC (2002), No-flow underfill flip chip assembly – an experimental and modelling analysis, *Microelectronics Reliability*, **42**: 1205–1212.

Mercodo LL, White J, Sarihan V and Lee TYT (2003), Failure Mechanism Study of Anisotropic Conductive Film (ACF) Packages, *IEEE Transactions on Components and Packaging Technologies*, **26**(3): 509–516.

Nguty TA and Ekere NN (1999), The rheological properties of solder and solder pastes and the effect on stencil printing, *Rheologica Acta* **39**: 607–612.

Pecht M (2008), *Prognostics and Health Management of Electronics*, Wiley, ISBN 0470278021.

Rizvi MJ, Chan YC, Bailey C and Lu H (2005), Study of Anisotropic Conductive Adhesive Joint Behaviour Under 3-point Bending, *Journal of Microelectronics Reliability*, **45**(3–4): 589–596.

Tsai M, Huang C, Chiang C, Chen W, Yang S (2007), Experimental and Numerical Studies of Warpages of ACF-bonded COG Packages Induced from Manufacturing and Thermal Cycling, *IEEE Trans on Advanced Packaging*, **30**(4): pp. 665–673.

Vanderplaats GN (1999), *Numerical Optimisation Techniques for Engineering Design: With Applications*, VR&D, Colorado Springs.

Wei Z, Waf LS, Loo NY, Koon EM and Huang M (2002), Studies on Moisture-induced Failures in ACF Interconnection, *7th Electronics Packaging Technology Conference (EPTC)*, Singapore, pp. 133–138.

Yang D (2007), *Cure-dependent Viscoelastic Behavior of Electronic Packaging Polymers*, Delft University of Technology, ISBN: 90-9022180-9.

Yin CY, Lu H, Bailey C and Chan YC (2005), Moisture Effects on the Reliability of Anisotropic Conductive Films, *6th International Conference on Thermal, Mechanical and Multiphysics Simulation and Experiments in Micro-electronics and Micro-systems*, Berlin, pp. 162–167.

Yin CY, Lu H, Bailey C and Chan YC (2006), Macro–Micro Modeling Analysis for an ACF Flip Chip, *Journal of Soldering and Surface Mounting Technology*, **18**(2): 27–32.

Zhang Y, Fan J, Liu J (2009), Multiscale Delamination Modelling of an Anisotropic Conductive Adhesive Interconnect Based on Micropolar Theory and Cohesive Zone Model, *International Conference on Electronic Packaging and High Density Packaging (ICEPT-HDP)*, pp. 160–163.

8

Adhesive technology for photonics

M. A. UDDIN and H. P. CHAN, City University of Hong Kong, Hong Kong

Abstract: Optical fiber communication systems involve generation, guiding and control of light. In such systems, optical devices can be made using different materials, and they are generally bonded with optical fibers using various types of adhesive. Among the myriad types of adhesive, polymeric optical adhesives offer a great potential for low cost and mass production in both device fabrication and component packaging. The performance of photonic components depends mainly on manufacturing process limitations and selection of compatible adhesive materials for the process. This chapter provides an overview of both fiber packaging and device fabrication issues. First, packaging processes of photonic devices are briefly introduced, together with major failure issues. Then the materials and fabrication issues of polymeric photonic devices are discussed. These critical issues need to be addressed in order to reliably manufacture these adhesive-based photonic devices and components.

Key words: adhesive, photonic packaging, polymer photonic device, material optimization, process optimization, spin coating, interfacial adhesion, reliability.

8.1 Introduction

Owing to the increasing and high demand for photonic components for handling our increasingly networked society, tremendous efforts have been made to open the door for technologies that meet the economic criteria, technical specifications and rapid manufacturing rate requirements without sacrificing high performance.[1] Polymeric adhesives are being increasingly used for a variety of applications in photonics because of their structural flexibility, easy processing and fabrication capabilities at low cost and high yields.[2] However, as a relatively new technology, polymeric adhesives have some limitations and drawbacks in this application. Therefore, it is very important to understand their limitations and how to use them optimally. The development of an adhesive with advantageous optical, mechanical and thermal properties, which will facilitate the use of adhesive bonding in the joining of various components with different characteristics, is essential. Nevertheless, no single adhesive can be used for a wide variety of applications, and thus one of the essential tasks is finding the most suitable adhesive for a specific application. The initial desirable properties include (i) high bond strength, (ii) fast curing rate, (iii) high stability, (iv) uniform film formation, and (v) low stress generation.[3] Therefore, the main objective of this chapter is to outline the issues related to adhesive applications in photonics with a general emphasis on their process optimization in fabricating planar photonic devices and their packaging with optical fiber links.

There are some fundamental differences in the use of adhesives for photonic and for electronic applications. For electrical interconnection, these concerns are related to the presence of stresses and the issue of structural integrity, as well as the need to maintain high electrical conductivity; in optical applications, they are mainly characterized by the refractive index of the component material, and aiming at minimizing any light transmission losses.[4]

8.2 The major characteristics of adhesives for photonic applications

Polymeric adhesives are versatile, and very importantly, they are able to spread and interact on the surface of the substrate material. For this highly technical application, adhesives are generally designed for both performance and processability. As a result, they can dramatically differ in their characteristics, even if their chemistries are relatively similar. Such types of adhesives should have an appropriate molecular structure to provide controllable optical properties that they need to develop for various optical devices and assemblies. The major characteristics of adhesives for photonic applications are briefly explained in the following:[5]

8.2.1 Viscosity

Viscosity is defined as the resistance of a fluid to flow. It is very important in the case of thin film deposition, and in dispensing through capillary processes. The viscosities of adhesives range from very liquid (low viscosity) to viscous (high viscosity) and are very sensitive to temperature. It is important to control the fluid flow during the packaging and to control the film thickness in device fabrication.

8.2.2 Curing profile

The adhesive is generally supplied in the form of a reactive, cross-linkable monomer blend. The process of converting it from a liquid to a solid state is termed ‘curing’. The adhesive is generally cured by heat, or by light, or even by a combination of both (known as dual cure). The cure schedule is a combination of the temperature or power and the amount of time to which an adhesive must be exposed in the curing environment for ‘complete’ curing. Different cure schedules result in different properties of the cured adhesive. Different cure schedules as well as different adhesives are necessary for specific applications. The optimal curing condition depends on many variables and is often optimized through testing and experience.

8.2.3 Refractive index

Refractive index (RI or *n*) is of critical importance for photonics applications such as optical waveguide devices and their packages. It is an important property in determining the light harvesting efficiency of fibers and other devices. However, substituents and backbone atoms that control the refractive index also effect Raleigh scattering losses, and mechanical, thermal and surface properties of the system.[6] The optical design of the components and packages is totally based on their unique refractive index characteristics and good optical clarity.

8.2.4 Transparency

Specific applications require transparency windows in specific regions of the spectrum, from the UV/visible to the near infrared. Intrinsic losses result from the physical and chemical structure of the polymer, due to absorption and scattering. Absorption of light in the ultraviolet, visible and near infrared regions of the electromagnetic spectra is directly related to polymer chemistry. Extrinsic losses in optical fibers and waveguides arise from impurities and additives that absorb light, as well as from inclusions and core-cladding interface imperfections that scatter light.

8.2.5 Optic coefficient

The thermo-optic (TO) and electro-optic (EO) coefficients of polymers play a vital role in determining a device's performance. A large thermo-optic coefficient can favor the reduction of power consumption for both switching and attenuating devices because it corresponds to a small temperature change and thus a small power input for causing the necessary change in the refractive index of polymer optical device.[7] Devices based on a large electro-optic response also offer greatly increased rates of information transmission by enhancing optical network speed, capacity and bandwidth for data networking and telecommunications.[8]

8.2.6 Adhesive strength

High adhesive strength is a critical parameter for multi-layer interconnections that are sensitive to shocks encountered during fabrication, handling and lifetime.[9] Therefore, the most important test for any adhesive is that it should give joints that are strong and durable. Although ways do exist of assessing the quality of joints by ultrasonic non-destructive testing, the ultimate test is to measure the force or energy needed to break the joint.

8.2.7 Moisture resistance

Usually, polymeric adhesive materials are sensitive to the surrounding environment, such as high humidity. Small amount of absorbed moisture can change in the refractive index, optical transmission, glass transition temperature and mechanical integrity of this class of materials.[10] Thus, device performance may easily be affected by changes in those main critical parameters. Moisture resistance is determined by exposing the cured adhesive to humid conditions or submerging it in a water bath for a defined period of time and temperature. It is expressed as the percent gain when the initial weight and the weight after the moisture exposure are compared.

8.2.8 Operating temperature

Polymeric adhesive materials can easily degrade or vaporize under high temperatures. Therefore, some important borderline temperatures should be distinguished for each type of polymeric adhesive. These are: (i) the maximum continuous operating temperature at which the adhesive can be exposed for an unlimited period of time without any damage; (ii) the maximum intermittent temperature at which it is possible to exceed the maximum continuous operating temperature for a short period (up to several hours) without noticeable damage to the adhesive; (iii) the degradation temperature which is a good indicator of the thermal stability of the adhesive material; (iv) the glass transition temperature (T_g), at which the cured adhesive transitions from a 'glass-like hard' to a 'rubber-like soft' state. Below the T_g, the adhesive is hard and rigid, and above the T_g it becomes rubber-like (comparable to a pencil eraser).[5]

8.3 Types of adhesive used in photonics

There is a wide selection of polymeric adhesives available for various applications and no simple way to classify the optical adhesives. They can be categorized according to their chemical family, according to their cure method, or according to their function. In all cases, there is some overlap among the adhesives types. In the fiber-optics industry, the adhesives are mostly based on epoxy or acrylate types. The term 'epoxy' refers to a chemical group consisting of an oxygen atom bonded to two carbon atoms forming a ring structure. The simplest epoxy is a three-member ring structure known by the term 'alpha-epoxy' or '1,2-epoxy'. The term acrylate refers to a chemical group consisting of a carbon–carbon double bond bonded to an ester functional group –COOR. Other types of adhesive are urethanes, silicones, cyanoacrylates, etc.

On the basis of their curing method, they can be classified as thermally curable, light curable or moisture curable.

Based on the adhesive usage, they can be classified as fixing adhesive, bonding adhesive, mounting adhesive, temporary adhesive, structural adhesive, locking adjustable adhesive, coating adhesive, film adhesive or sealant adhesive.

8.4 Major applications of adhesives in photonics

In the photonics industry, adhesives are mainly used for two different purposes: (i) for packaging of photonic components, and (ii) for fabricating planar photonic devices. Other applications include the encapsulation of LEDs, providing conductive media, hosting other active materials, and as anti-reflective coatings for solar cells, displays and contact lenses. This chapter will discuss the first two (main) applications of adhesives in photonic applications.

8.5 Adhesives for photonic packaging

Packaging, and in particular pigtailing, is becoming an increasingly important issue as optical networks move from the wide area network (WAN) domain to the local area network (LAN) domain, with a resultant pressure to increase the production and reduce the cost.[11] In general, photonic packaging fulfils two functions:

(i) To provide robustness for the optical component in order to enable it to survive in different application environments such as temperature, humidity and mechanical stress.
(ii) To provide connections between photonic chips and optical fibers so that the optical signals can be transmitted between them.

8.5.1 Typical example of photonic packaging

The passive optical power splitter is one of the key elements in a passive optical network (PON), which equally splits the signal power from the optical line terminal (OLT) in the central office (CO) to each optical network unit (ONU). It can branche and couple waves without converting optical transmissions into electric signals, for connecting households to telecommunication carriers in optical communication networks.[12] Planar lightwave circuits (PLC) can provide various key practical devices for such optical networks because of their suitability for large-scale integration, long-term stability, and mass-production capability.[13] In order to utilize the integration capability of PLC devices, the input and output fibers have to be connected to the PLCs. The fiber connected splitters are required to exhibit not only high optical performance (such as low loss, wavelength flatness and low polarization dependence) but also long-term reliability. However, to attain a reliable low loss splitter, connection methods must be precise and meticulous.[14]

Figures 8.1a and b show typical schematic configurations of an unpackaged and a packaged PLC-type optical splitter, respectively. The unpackaged 1×8 PLC optical splitter includes a PLC (splitter) chip, and single-channel (input) and eight-channels (output) fiber arrays. The three parts are mounted by an adhesive, as shown in Fig. 8.1a. However, in a packaged device, the mounted optical splitter is secured in its housing by applying fixing adhesive between the Al fixing blocks and the fibers. Moreover, an adhesive is used to bond the rubber boots and the end of fiber arrays in the housing, as shown in Fig. 8.1b. It is very clear that different types of adhesive are needed for bonding different parts of the PLC package. In fact, a large number of different materials are ultimately required for assembling the splitter packages. Table 8.1 shows the physical properties of materials used in an adhesive-based optical splitter package. Figure 8.2 shows the appearance of a bonded PLC optical splitter package in an aluminum package.

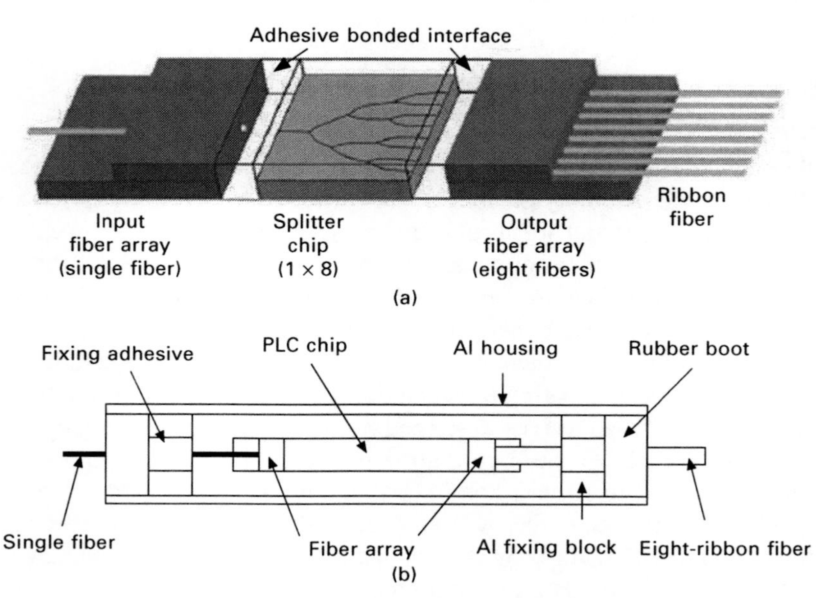

8.1 (a) Schematic configuration of unpackaged PLC optical splitter. (b) Schematic configuration of packaged PLC optical splitter.

Table 8.1 Physical properties of materials used in adhesive-based packaging of an optical splitter

Description	Material	CTE (ppm/°C)	Modulus (PSI × 10^6)
Fiber	Fused silica	0.55	10.500
Fixing adhesive	UV-cure epoxy	68	0.100
Bonding adhesive	UV-cure epoxy	55	0.100
V-groove adhesive	UV-cure epoxy	55	0.100
Rubber boot	Silicone	150	0.005
Housing/fixing block	Aluminum	24	10.007
PLC chip	Quartz	0.55	10.500
Fiber array substrate	Pyrex	2.60	21.750

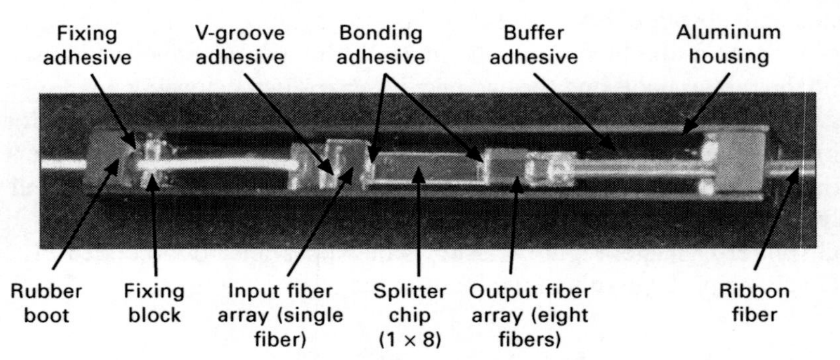

8.2 Appearance of a bonded PLC optical splitter package in an aluminum housing.

8.5.2 Advantages of adhesive bonding in photonic packaging

Until now, soldering, laser welding and adhesive bonding methods have been employed for the assembly of optical fiber arrays. The soldering process's need for metallization and high heat loads can affect reliability. Laser welding is regarded as having long-term stability, but it also requires metallization, high power lasers, and careful control of a variety of parameters. Adhesives in photonic packaging have many advantages in both cost and convenience as compared to the above-mentioned alternative conventional methods.

Currently, adhesives offer advantages in terms of mass-productivity and low-cost.[15] The adhesives described not only perform the function of bonding, but also have the high degree of light transmittance and other properties required to form a bond most suitable from an optics point of view. They can also be cured by both heat and by light, without affecting the fiber alignment. Light curing provides a number of economic advantages: rapid through-cure, low energy requirements, room temperature treatment, and non-polluting and solvent-free formulations. In this way, heat-sensitive materials in the assembly are not damaged during cure.[11]

8.5.3 Packaging process of photonic devices

The low-cost packaging process of photonics devices is a challenging task, because the design rules for the packaging are significantly more complex than those found in the semiconductor industry. For semiconductors, advances in wafer processing technology have resulted in a packaging process that is automated and planar. However, for optical components, the front-end process is significant, and the controlled assembly processes and critical assembly tolerances create a challenge during the package design. The selection of a particular method depends on a series of criteria, which include reliability, temperature excursions of subsequent packaging processes, package materials used, and constraints imposed by active alignment and automated assembly. Therefore, the selection of appropriate adhesive materials and methods of attachment are very important in determining the stability and reliability of the packaged device.[11] A schematic of the typical adhesive-based bonding process of an eight channel fiber array is shown in Fig. 8.3. The packaging process consists mainly of alignment, adhesive dispensing and adhesive curing.

Alignment methods

There are two main alignment methods in device packaging. They are: (i) *Active alignment technique*: In this method during the assembly process, coupled power from a photonic component is monitored by an optical power meter. The alignment accuracy is optimized by maximizing the measured coupled optical power. The technique applies iterative movements of the components while monitoring the coupled output power to acquire precise positioning. This technique is typically slow and therefore increases the production cost in both semi and fully automated packaging.[16] (ii) *Passive alignment technique*: In this technique, all components are assembled without electronically activating the dies and measuring the light output. Passive alignment determines the aligned position by fiducial marking, device features or by mechanical precision fixture. As a result, the techniques decrease the assembly steps and reduce the manufacturing complexity. Thus, the technique is suitable for mass production due to the fact that it is intrinsically simple and fast.[17]

Adhesive dispensing

Nowadays, various types of adhesive dispensing tools are available for use in photonic applications. They are hand syringes, micropens, ink jets, dispense jets, and auger valves. Manual adhesive dispensing is not suitable for fast assembling processes. Therefore, the dispensing technique must be

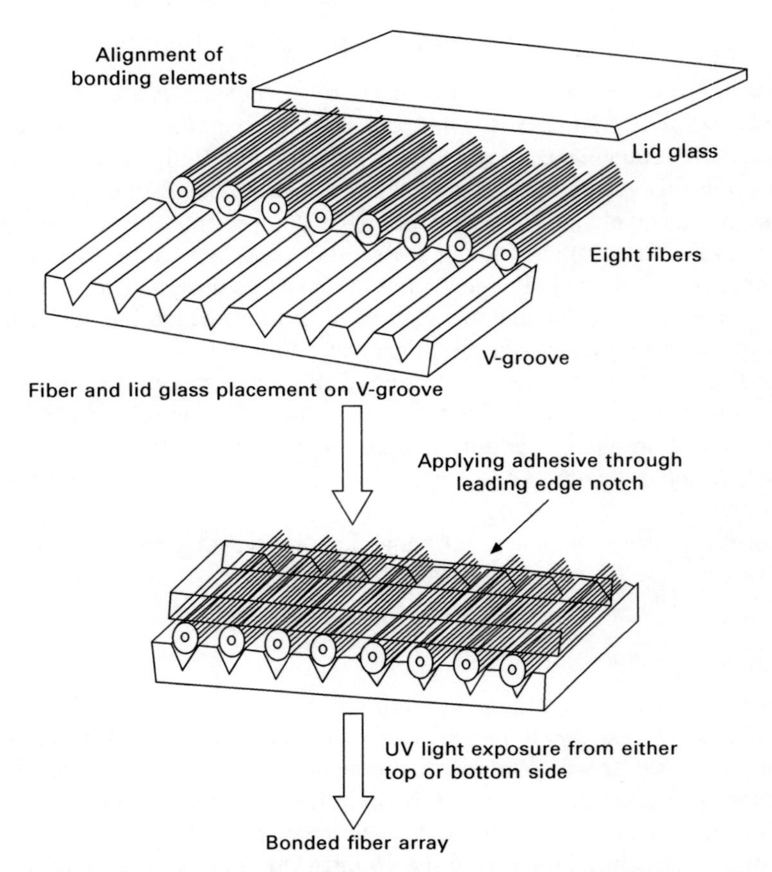

8.3 Schematic of the bonding process of eight-channel fiber array.

compatible with the components assembly process and rates. In addition, the continuous miniaturization of photonic components and devices requires new micro and nano-dispensing techniques. One of the promising techniques for such micro- and nano-dispensing is the valve technique, which provides a dispensing system with accurate control of the dispensing needle tip, as well as improved repeatability of the deposited materials.[18]

Adhesive curing

The curing process of the adhesive is an important factor in determining the performance of the component. Traditional thermal curing may not be suitable for adhesive curing in photonic applications. Most optical adhesives are photosensitive and require a certain type of light-source for radiation curing. Others are moisture-sensitive and may require a particular level of humidity for curing. Some adhesives can be cured in few seconds and others may require several days to achieve a full cure. They may be curable at room temperature or may require an elevated temperature. Sometimes there is a need to compromise between the cure schedules of cure time and temperature. However, the final properties of the cured adhesive often depend upon the details of the curing process and schedule.[19]

8.5.4 Major failure issues for photonic packaging

Failure is normally identified by adhesion failure, an increase in delta core pitch and insertion loss. The major failure issues are discussed below:

Surface contamination

As the integrated optical components are getting smaller with the use of advanced materials, contaminant-free active surfaces are crucial to obtain high-yield reliable products. Therefore, an important part for product reliability achievement is the control of contamination to ensure good bondability between various mating surfaces.[20] Figure 8.4 shows contamination-induced delamination between the V-groove fiber block shoulder and the cover lid. Contaminants may be introduced in the packages during the fabrication and also from the environment. Contamination is caused mainly by poor

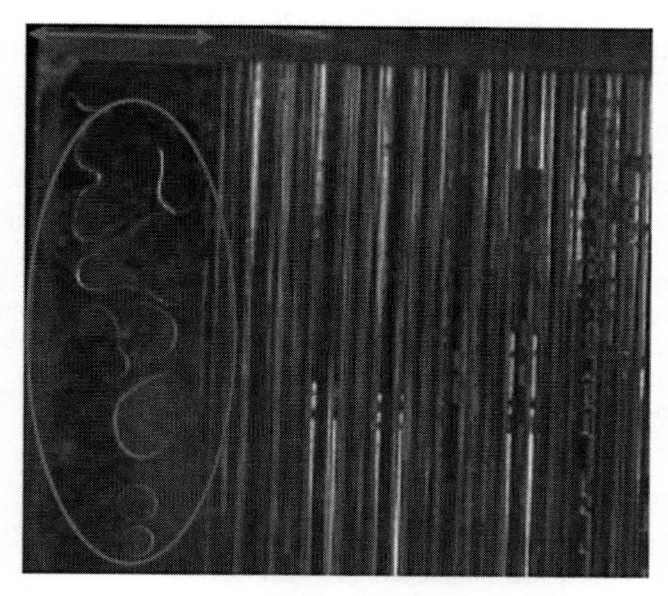

8.4 Optical microscopic photograph of contamination-induced delamination between V-groove fiber block shoulder and cover lid.

process control and is present in the form of residues, mold release agents, anti-oxidants, carbon residues or other organic compounds on the bonding surface. The contaminant may case serous defects, such as:

- Greatly degraded bonding surface of the photonic package.
- Weakened adhesion force resulting in poor performance. (When such a phenomenon occurs, an optical fiber may readily slip off from the substrate due to adhesion failure.)
- Osmotic pressure built up at the contaminant surface, initiating delamination.
- Less hydrogen bonding between adherents causing a decrease in the adhesion strength. (The adhesion depends strongly on hydrogen bonds between the oxygen atoms on the quartz layer and the hydrogen atoms of the bonding adhesive). Hence, the total adhesion depends strongly on the cleanliness of the bonded surface and the availability of oxygen atoms to form hydrogen bridges.[21]

Therefore, developing suitable photonic packages to minimize deterioration, delamination, cracking or peeling is essential. Because impurities must be thoroughly removed before the bonding process in order to eliminate the delamination problem and enhance the adhesion of photonic packages, plasma treatment-based surface modification is used for surface cleaning and increased surface roughness before bonding. Plasma's physical and chemical energy can be used to remove micron-level contamination. Also, roughening of the surface will increase the total contact area at interfaces, which significantly increases adhesion between the adhesive and the substrate.[22]

Entrapped air bubbles

Air bubbles entrapped in the adhesive are also a concern for reliable adhesion. Air bubbles may be entrapped during the adhesive flow process. Figure 8.5 shows optical microscopic photos of entrapped air bubbles in an adhesive-based eight channel fiber array package. Trapped air bubbles are best avoided because they can cause adhesive delamination when the array is exposed to temperature cycling. Such defects also provide a propagation path for stress cracking. The voids can nucleate at the interface and propagate through the interconnection, resulting in a loss of adhesion and maybe failure under low force.[23]

Misalignment

The position of the optical device in passive alignment is defined by the geometry of the device. The impact of this misalignment can be described by the position of the packaged fiber in the V-groove as shown in Fig. 8.6. If the fiber makes contact with the V-groove side walls, the center of the fiber

8.5 Optical microscopic photograph of entrapped air bubbles in an adhesive-based eight-channel fiber array package.

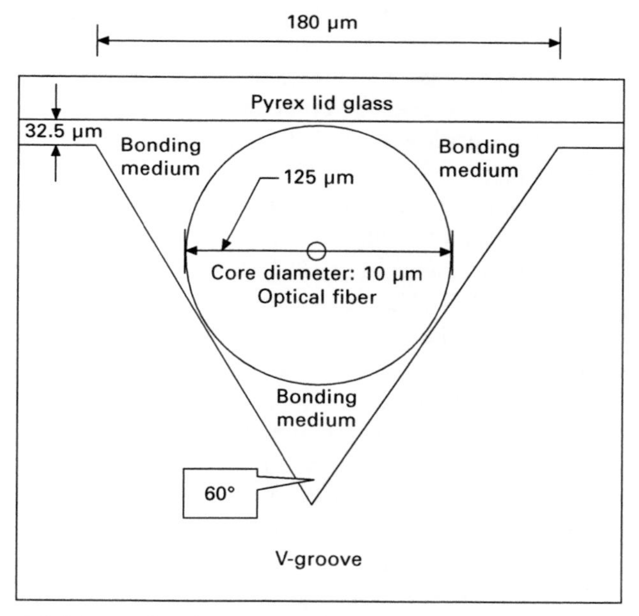

8.6 Schematic of a packaged fiber in V-groove.

can be located, as the radius is known.[11] Therefore, the surface feature of the V-groove side wall is another significant factor affecting the fiber alignment. If the center of the fiber has an offset less than 2 μm to the theoretical center position, the fiber is declared as an aligned fiber. The buoyancy of the adhesive under the fiber can cause it to float upwards. Usually the cover plate is used to press the fiber against the side-wall of the V-groove. However, the optical properties are affected if the pressing process is not well-controlled. The stress applied by the pressing cover may deform and even damage the optical fiber. Moreover, the pressing process may introduce voids and bubbles in the adhesive and this leads to reliability problems.[23]

Uneven curing of the adhesive

Uniform adhesive curing and uniform adhesive bondline are essential for minimizing the stress within the fiber package as well as for preventing actual failure. Uneven curing of adhesive in the package can generate high interfacial stresses upon heating or cooling of the structure during fabrication, assembly, or in field use. Propagation of the resulting delamination along an interface can degrade or destroy the functionality of the system. Therefore, interfacial delamination due to the uneven curing of adhesive, is one of the primary concerns in photonic package designs. Alignment and shrinkage of the fiber array also depend on effective light-ray penetration during the adhesive curing process. Due to the complexity of interconnects, it is also interesting to consider how illuminated light propagates through the uneven interfaces. High light-reflectance from any interface of the assembly reduces the light intensity for the next layers and induces uneven curing of adhesive. Shadowing due to light bending or optical element shape can also cause incomplete or uneven curing.[24]

The uneven-curing-induced delamination effect has been extensively studied.[10] The shaded area and delamination were much greater when light was exposed from the bottom side of the V-groove rather than from the top side, as compared in Fig. 8.7. This is due to the geometrical shape of the bonding element. Minimum shaded area and delamination were found at the middle fiber when light was exposed from the top side. These were greater when the light was exposed from the bottom side, and were at a maximum at the outermost fiber. These effects are very severe for large value of Δn, the refractive index difference between the adhesive and cladding materials. It is concluded that that the delamination problem can be minimized by using a UV-curable adhesive having the same or slightly higher reflective index than that of the cladding material. It is also recommended to light expose from the top side, and the lower pitch V-groove is preferred for fiber packaging. This type of fiber array results in a more reliable assembly and also increases productivity.

Curing shrinkage

Shrinkage is the volume prior to curing compared with the volume attained when fully cured. This volume shrinkage will cause a micro shift between the

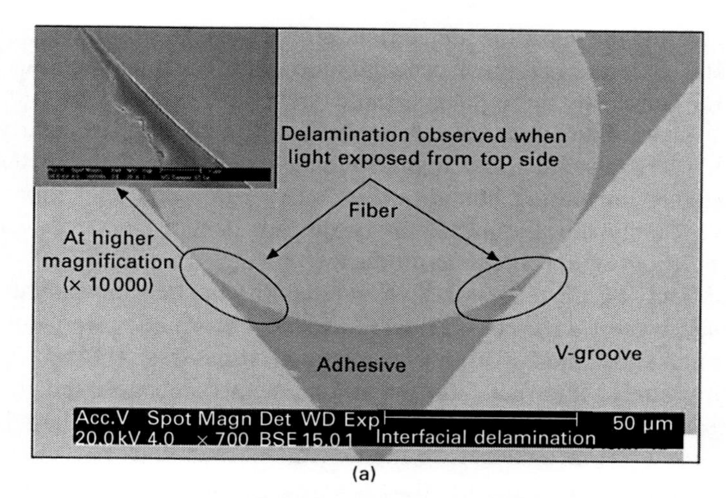

(a)

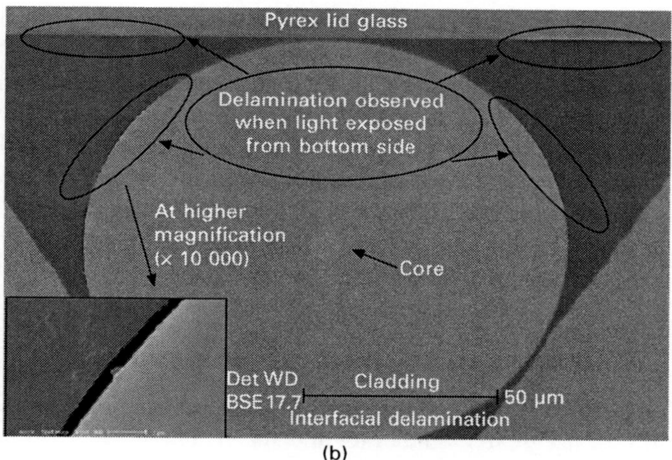

(b)

8.7 Typical end face images of the delaminated fiber in a V-groove: (a) when exposed from the top side and (b) when illuminated from the bottom side.

adhesive surface and the device. Movement or relaxation of the fiber bond joint during the cure process is a typical causes of this failure mode. During the long post-cure cycle, most adhesives that have been gelled only, will relax as they begin to heat, and transform into a fully cross-linked polymer only after passing through a period of lower viscosity and relaxation.[25]

Coefficient of thermal expansion mismatch

When packaged, the coefficient of thermal expansion (CTE) of the adherents, fiber and bonding materials have to be well matched. If there is a difference in CTE among the constituent materials, stresses and strains in the packages are bound to occur. The stress concentration of the bonding materials in the V-groove caused by these phenomena cannot sufficiently adapt to the thin bonding layer. The stress caused by the bond increases particularly with a larger CTE mismatch and higher Young's modulus of the adhesive.[26] Alternatively, increasing humidity causes expansion and relaxation of the adhesive.[27] Any misalignment among the optics will cause optical loss, resulting in out-of-specification product.

The effect of CTE of the adhesive on reliability has been investigated.[21,28] Two adhesives of different CTE were used.[21] The samples were subjected to thermal shock and a highly-accelerated stress test (HAST) for the reliability study. Interfacial delamination, delta core pitch and insertion loss measurements were used to characterize the packages. The adhesive having less CTE mismatch with the constituent materials has showed the best performance. In order to reduce such degradation in the performance of the fiber array assembly, it is recommended to select bonding adhesives having a close CTE match with the bonding substrate.

In an unpackaged optical device, the CTE mismatch-induced stress can relax without increasing insertion loss during the environmental test. However, in the packaged device, the stress cannot release and increases the insertion loss. The CTE mismatch effect also induces fiber bending inside packages, which ultimately initiates fiber cracks. Numerical simulation has been used to confirm the experimental results and interpretation. These numerical calculations were in good agreement with the experimental measurements.[28]

8.6 Adhesives used in photonic devices

The use of photons instead of electrons to transmit communication signals offers significantly greater bandwidth and speed. Therefore, particularly for applications involving high-speed processing or communications, photonic devices provide a significant performance edge. Optical waveguide is an alternative to copper wire, which is used to carry signals for voice and data. Waveguide technology shares a number of advantages for optical interconnections. It provides well-controlled and accurately directed connections for optical transmission. They do not require impedance matching, as do-high speed electrical interconnections, and thus support extremely high bandwidths. It can also provide additional functions to enhance the coupling efficiency or to incorporate component redundancy.[29,30]

8.6.1 Example of a photonic device

Most optical devices including lasers, modulators, switches, power splitters, directional couplers and filters are in the form of optical waveguides. Optical waveguides are dielectric structures where the central material, called the core, is surrounded by another material, called the cladding, of a lower refractive index. Figure 8.8 schematically shows a typical structure of a fully buried or embedded optical waveguide. It shows that the fabrication process of the optical waveguide includes the deposition of a multi-layered thin-film structure. This multi-layered structure supports electromagnetic waves, which are guided in the core region as the wave propagates along the z-direction. These waves are stable and their energy generally stays mostly within the core region.[31]

8.6.2 Materials issues for photonic devices

Several light-guiding inorganic materials capable of multiple functions (such as modulation, switching/attenuation, wavelength conversion, and amplification) are under intensive investigation, including lithium niobate ($LiNbO_3$), silicon dioxide (SiO_2) on silicon, and III-V compound semiconductors. Such materials

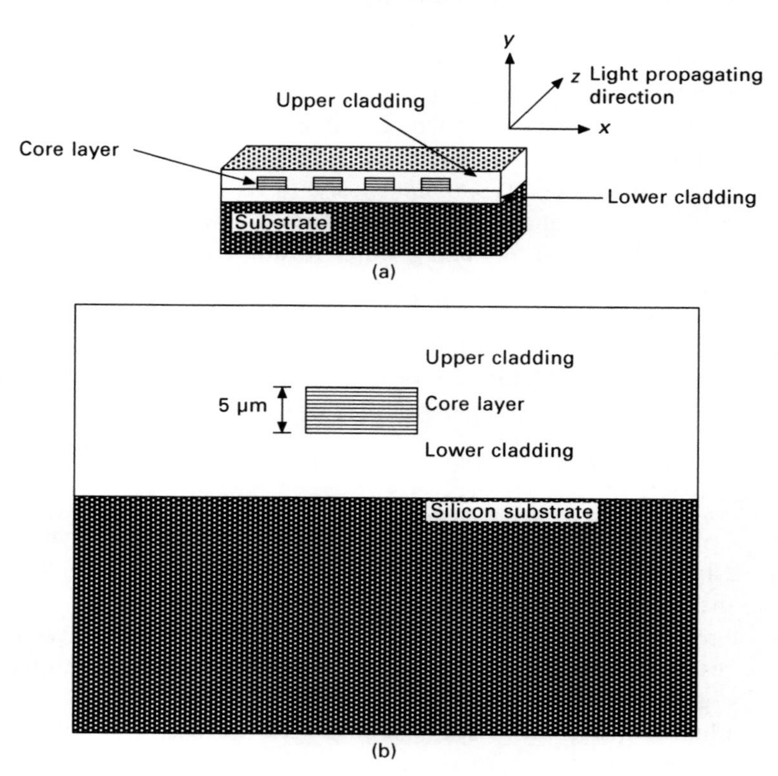

8.8 (a) Schematic 3D diagram of fully buried or embedded optical waveguide. (b) Schematic cross-sectional view of single fully buried or embedded optical waveguide.

exhibit polarization-dependent loss for many active and passive devices. Sharp and polarized absorption and emission lines result in an amplification bandwidth too narrow to be useful for multiple wavelength systems such as wavelength division multiplexers and bit-parallel multi-wavelength interconnects. Such characteristics limit their use for optical gain circuits in specific fiber systems unless polarization converters are used. These materials are strictly substrate selective due to the lattice matching required for single-crystal thin-film growth. Finally, the fabrication costs associated with these materials are very high, which seriously jeopardizes the commercialization of the end products.[32] As an alternative, organic adhesives, typically composed of polymers, offer many advantages over the aforementioned materials for the fabrication of optical devices.[33] These include:

- Polymers are usually in an amorphous state that can provide a wider bandwidth of amplification if an appropriate gain mechanism is identified.
- The microstructure can be easily engineered to provide desired optical parameters such as bandwidth of transparency, high electro-optic (EO) coefficient values, and temperature stability for specific applications.
- The thermo-optic (TO) coefficient ($\Delta n/\Delta T$) of polymeric material is an order of magnitude more than that of SiO_2; as a result, a polymer-based thermal optical switch can potentially perform both switching and variable attenuation functions simultaneously. In the case of metropolitan-area-network applications that require arrays of optical switching devices with millisecond switching times, a combined switch and variable optical attenuator is very attractive from the viewpoint of cost and power consumption.
- Unlike any of the inorganic materials that cannot be transferred to other substrates, the polymeric passive and active devices proposed herein can be easily integrated on any surface of interest.

Of the different kinds of polymer used in this application, epoxy is the best suited due to:[34]

- Low cost compared with acrylic and polyimide based polymers.
- High thermal stability. The resins have a high T_g (ca 200 °C) and are therefore expected to have good thermal stability.
- Precise controllability of refractive index. By mixing several epoxy resins together, the refractive index can be precisely controlled in the region of 1.48–1.60 with a 0.001 order of accuracy.
- Photo-curable property. The resin can be easily patterned with a conventional mask process using a UV light source.
- Acceptable optical loss. This issue can be minimized when epoxy is used as the cladding layer.

8.6.3 Processing issues in device fabrication

Recent advances in the development of organic molecules and polymeric epoxy adhesives with high optical quality and performance have led to a maturing of the PLC device field and have brought commercialization closer. However, the physical, optical and mechanical properties of the epoxy adhesive are very critical for reliable fabrication and mainly depend on the processing. They are all interrelated and it is not possible to look at one property alone to determine suitability. A balance of these properties is obviously needed for reliable fabrication of PLC devices. Only a clearer fundamental understanding of the deposition, fabrication and degradation mechanisms can allow manufacturers to develop highly reliable, low-cost and better performance polymer photonic devices.[35] The performance of polymeric epoxy adhesive-based PLC devices depends mainly on the following issues:

Thin film deposition

Polymeric thin film is the fundamental building block of polymer-based optical devices. One of the simplest and most common techniques of applying thin polymers films onto wafers is spin coating. The process solely involves the dispensing of an excessive amount of fluid onto a stationary or slowly spinning substrate and then spinning it at high speed. Therefore, it is useful to understand the behavior of complex mixed solutions under conditions of rapid fluid flow and convectively-driven evaporation that occur during spin coating.[36]

Curing conditions

The curing of epoxy adhesive is of a great importance in applications. It is a process of conversion from a liquid to a solid state, accomplished by a chemical reaction in the epoxy resin. Therefore, curing is believed to be critical to develop the ultimate mechanical and optical performance of polymer devices. A minimum degree of curing is needed to provide a certain level of that performance. An under-cured adhesive is not optimized to acquire those performances, especially in regards to environmental resistance and dimensional properties. However, over-cured adhesives also become brittle, resulting in greater stress on the adhesive bond and interfaces.[37,38]

Stability

In this application, after the deposition of the initial adhesive layer, additional heating and chemical etching is required for the fabrication of the subsequent step of photonic device manufacture. Therefore, the polymeric adhesive material should have sufficient thermal and chemical stability to withstand typical fabrication processing and operating conditions with good performance.[39] Knowledge of stability, degradation and mode of decomposition under the influence of heat and chemical solution is very important in process optimization. The threshold gives an indication of the ultimate processing conditions that can be used during the subsequent fabricating and operating processes. A proper understanding of potential degradation mechanisms can greatly aid the appropriate selection of material and process parameters in the fabrication and extend the outdoor longevity of the product.[40,41]

Mechanical strength

The curing conditions and stability of the polymeric adhesive film have an influence on the strength of the adhesion of the coated film to the substrate. High adhesion strength is a critical parameter of multi-layer interconnections, which are fragile to the shocks encountered during fabrication, handling and lifetime. Concerning this technology, surface finish or surface roughness is another important parameter that controls the state of adhesion. The interfacial strength also depends on the environmental and processing conditions of subsequent fabrication and operation processes.[42] Thermo-mechanical failures are caused by stresses and strains generated within an optical device due to mismatch in the coefficient of thermal expansion (CTE) among different materials during thermal loading from the environment or internal heating in service operation.[25]

Surface condition of adherend

Continuous effort is being made to improve interfacial adhesion in polymeric adhesive-based PLC devices. Interfacial strength is fundamentally related to surface attachment, and the properties and condition of the adherend surface are of paramount importance. The silicon wafer is commonly used as a substrate for the fabrication of photonic devices. The main advantages of silicon wafers is that it is easy to cut into pieces as the samples need to be sectioned to the area of interest, or for easier handling, and therefore there is no need to polish the end faces of the devices. However, the adhesion of polymeric adhesives to the silicon surface is comparatively poor.[43] An oxidized silica-containing (thin layer of silica on silicon) silicon wafer can be use as a substrate and is becoming very attractive due to its matching refractive index with optical polymers. Thermal oxidation of silicon is easily achieved by heating the substrate to temperatures typically in the range of 900–1200 °C. Another potential alternative is to use a thin metal layer on a silicon wafer to improve the adhesion of the polymeric adhesive to the substrate. A thin metal layer can be deposited rapidly on silicon wafers by a typical deposition process such as sputtering, or thermal or E-beam evaporation. Thus, a careful treatment and modification of the surface before the adhesive deposition is essential for realizing a strong interfacial bond.[27]

8.6.4 Fabrication techniques of planar polymer photonic devices

In this application, the critical issue is the ability to pattern the micro or nanostructure in a high-throughput and cost-effective manner. The patterning of polymer photonic films can be done by many techniques. Typical techniques that have attracted great attention in recent years are given in Fig. 8.9 and the limitations are briefly discussed as follows:

Photolithography technique

The very common photolithography process can be applied to the simple polymeric material but still faces many problems regarding the material and process. The technique includes many steps such as metal deposition as a hard mask, photoresist deposition and patterning, reactive ion etching, and, finally, metal and photoresist etching. The problems associated with the process include chemical attack during the metal and photoresist etching, roughness, and stress-induced scattering loss.[44]

Embossing technique

The embossing technique is appropriate for mass production and fabrication of high-precision polymer microstructures for optical components. The method additionally needs application of optimum pressure and temperature/UV to the polymer. Therefore, it is essential to consider how mixed solutions behave under the conditions of applied rapid pressure and temperature/UV through the embossing master. Inevitably, the replication process can introduce an additional roughness to the waveguides, which adds significant scattering loss contribution to the overall waveguide loss. The method, however, has also some limitations regarding the achievable aspect ratio and is not suitable for constructing vertical waveguide interconnects among multiple layers. Until now, the problem of fine alignment has not yet been solved. The fabrication of an ultra-high resolution mold is also a difficult task. Fabrication of reliable and cheap imprint devices is therefore clearly another challenge.[45]

Direct laser writing technique

The direct UV or E-Beam writing technique has also recently been used since it is much simpler for the fabrication of complex devices. The technique

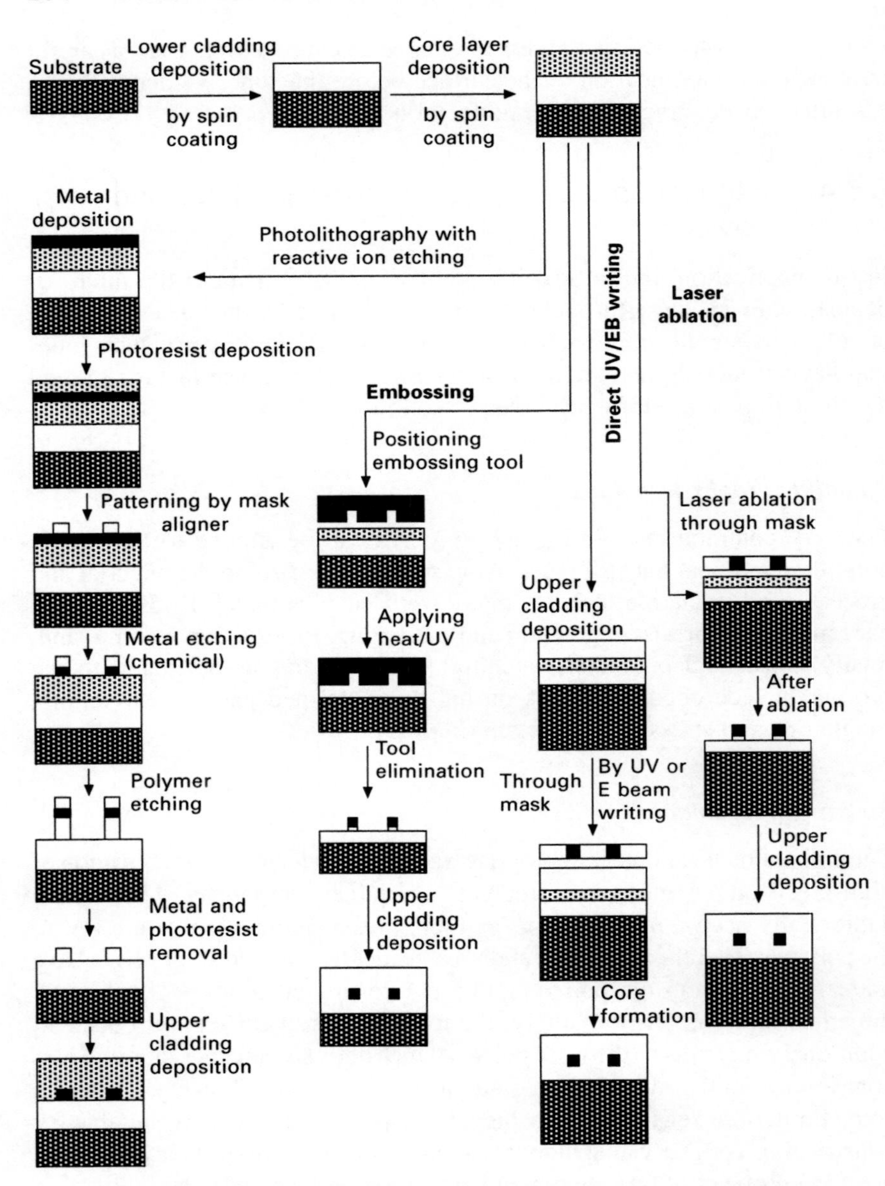

8.9 Overview of the fabrication processes of a polymer channel waveguide using different techniques.

involves beam irradiation on photosensitive material for waveguide fabrication, which locally changes the refractive index of polymer film. However, the overall performance of the optical device mainly depends on beam intensity, wavelength, spot size, scan speed and energy fluencies. Parameters such as post-exposure baking time, vertical and horizontal resolution, hardness and sensitivity are also all very crucial. Sophisticated equipment and numerous beams scanning a large area are required.[46–49]

Laser ablation technique

Laser ablation is a low-cost manufacturing technique and is compatible with current PCB production for optical interconnections. However, depending on the laser wavelength and the material, this can have the characteristics of ablative photodecomposition, or rapid heating and vaporization. These processes may introduce material degradation. Sidewall roughness induced scattering loss is also an issue with this technique.[50]

There are also some other ways of fabricating polymer optical waveguides, such as injection molding,[51] diffusion and ion exchange,[52] etc. All the techniques have some sort of process induced limitations. Therefore, there is a major need for practical as well as scientific information about the proper handling and properties of polymeric materials for the fabrication of reliable photonics products. Recent studies[38,43,53] have focused mainly on proper polymer processing and the fabrication of reliable photonics devices. The challenges are to achieve compatibility with other electro-optic systems and processes, end-use reliability, reproducibility, stability, acceptable performance, and cost, commensurate with added value.

8.7 Typical challenges for reliable fabrication of photonic devices

The most typical challenges for the production of reliable polymer photonic devices are uniform adhesive curing, higher stability, higher adhesion strength, and environmental reliability. They can all cause device failure at different stages. Figure 8.10 is an example of the interfacial failure of a channel waveguide. The sample was prepared by a photolithography technique using benzocyclobutane (BCB) as core, and silica and epoxy as the lower and upper cladding material respectively. The figure shows that there is considerable delamination between the polymer films at the interfaces. To find out the root cause of the delamination or failure, extensive research work has already been done, which included substrate surface analysis and optimum polymer curing. It was found that the bulk polymer on the silicon substrate showed good behavior and no interfacial delamination. However, the spin-coated epoxy adhesive showed degraded properties for the fabrication of waveguide device where most failures often occur. Figure 8.11 shows a typical failure and peeling out of the thin film from the substrate. Therefore, the spin-coated polymer films have been extensively studied. Some of the recent findings are summarized as follows:

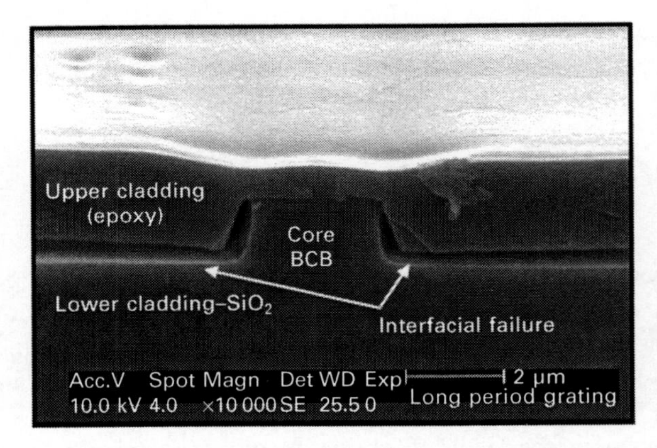

8.10 Typical SEM picture of the interfacial failure (larger delamination) in a channel waveguide.

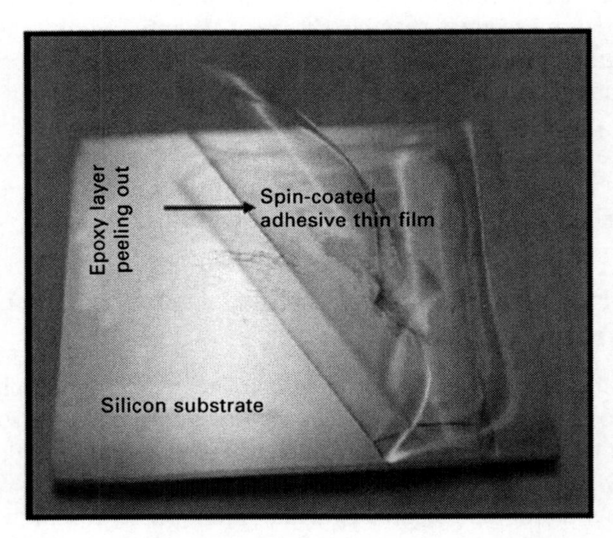

8.11 Typical interfacial failure of thin polymer film from the silicon substrate.

8.7.1 Degree of curing of spin-coated polymeric films

Thin polymer adhesive films can be deposited by a variety of techniques with different complexity and applicability. The choice of deposition technique depends upon the physicochemical properties of the material, the film quality requirements and the substrate being coated. The final properties of these films also depend on their morphologies, which are largely affected by the polymer chain orientation and the state of aggregates. One of the simplest and most common techniques of applying thin films onto wafers is spin coating.

The degree of curing of a spin-coated polymer film (over three locations of the deposited sample indicated in Fig. 8.12) was measured using the same FTIR method. Figure 8.13 shows the effect of spin coating on the

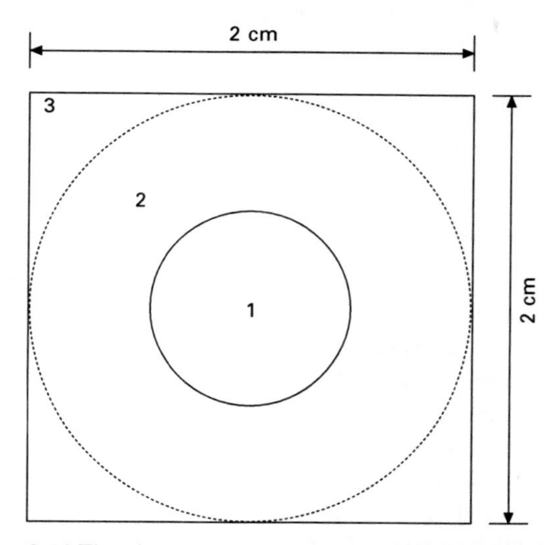

8.12 The three measurement locations used for the spin-coated adhesive film.

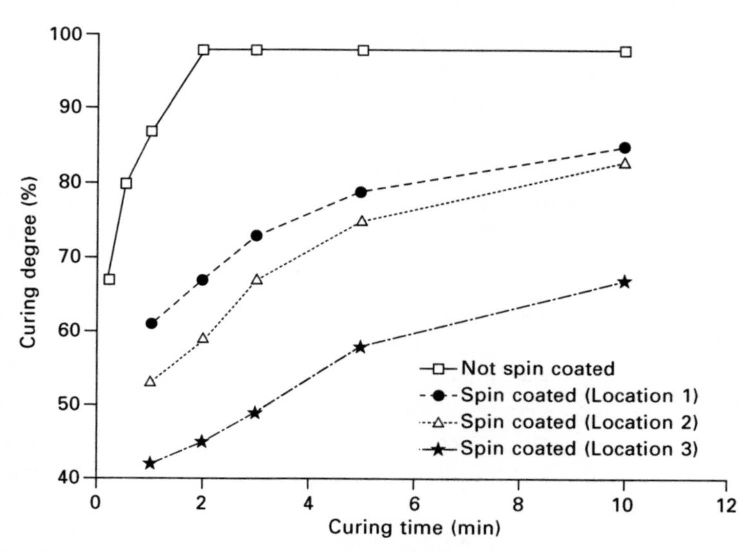

8.13 Effect of spin coating on the curing behavior of epoxy adhesive at different locations of the substrate for different UV exposure time (min).

curing rate of the epoxy adhesive at the different locations for different UV exposure times. The figure shows that the curing rate decreases abruptly for the spin-coated adhesive film compared with the adhesive that was not spin-coated. The degree of curing also varies at different locations of the spin-coated substrate. At the center of the substrate it is higher than at the other locations. The slower reaction rate is due mainly to changing the material properties during spinning. Spin-coating involves transient fluid flow and mass transfer in a medium that experiences drastic changes in properties. Both fluid viscosity and solvent–solute diffusivity can change by several orders of magnitude from the initial stages.[54]

Factors affecting the curing rate of spin-coated polymer film

The spin-coated liquid usually contains a volatile solvent that evaporates during spinning, leaving behind a thin solid film. The solvent during spin coating is most influential for obtaining good spin-coated films. Its evaporation has a tremendous influence on the control of the final film thickness after mass transfer and spreading of the solvent.[55] Evaporation rates are influenced by such factors as the temperature of the fluid, ambient temperature, heat conductivity, molecular association, molecular weight, vapor pressure, surface tension, humidity, latent heat of evaporation and vapor density. A high degree of uniformity in the evaporation rates leads to a similar uniformity in the final polymer film.[56]

Solvent evaporation couples liquid-phase diffusive transport of solvent toward the interface to gas-phase convective mass transport of solvent away from the interface. The solvent vapor is carried away in a stream of air that is drawn down axially toward the surface and driven radially across it by a centrifugal pumping action created by the spinning substrate.[55] Solvent is continually evaporating from the film, altering the material properties (e.g. viscosity, diffusivity, vapor pressure, surface tension) and eventually producing a film of finite thickness. As a result, the reaction rate of the spin-coated adhesive is decreased due to the following two factors.

(i) Solvent evaporation. Owing to evaporation during spinning, the viscosity of the adhesive increases. Increased viscosity reduces the mobility of the polymer molecules and slows the reaction rate. The number of solvent molecules also decreases and the free volume increases.[57] Owing to the lack of solvent molecules, the reaction rate and the total amount of reacted species is reduced, hence the lower curing degree of spin-coated adhesive films.
(ii) Composition gradient. Evaporation also establishes a composition gradient at the surface as volatile species leave and less volatile components are left behind.[58] The top surface becomes a high volatile species-rich

layer and the bottom is of low volatile species. Constituents need to further inter-diffuse for chemical cross-linking reaction during the UV exposure. This takes additional time which decreases the curing reaction rate substantially compared to the adhesive that has not been subjected to the spinning process.

The curing degree of a spin-coated adhesive is lower than that without spinning, even after a long (10 min) UV exposure. At the center of the substrate, the rate of evaporation is low due to the low angular momentum of the fluid. On the other hand, higher evaporation rates are caused by the combination of both radial and axial flow at the edge of the substrate. Therefore, the evaporation, as well as the composition gradient, varies from the center to the edge of the substrate. They are lower at the center and higher at the edge. The cross-linking reaction rate is higher in the center (Location 1) than at other locations of the substrate. Owing to the variation in curing degree and surface tension, the topography also varies at different locations of the spin coated adhesive film. Figure 8.14 shows the surface topography at the corner of the spin-coated adhesive film after curing for 10 min. The figure clearly shows how the topography differs from the center to the corner due to the spinning process.

Before curing, the fluid properties dictate a constant angle at the solid–liquid–gas interface. The result is a thick edge bead confined at the wafer's edge. Another reason for such a film pattern is the increased friction with air at the periphery, resulting in an increased evaporation rate that causes a dry skin to form at the edge and impedes fluid flow. As a result, the fluid in the

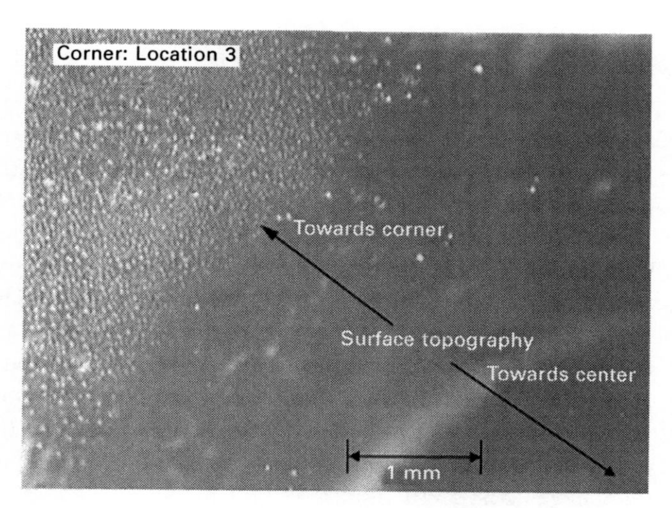

8.14 Surface topography near the corner of the spin-coated adhesive film after curing for 10 min.

center of the substrate still being driven out by centrifugal force begins to flow over the dry film and dries, resulting in a build-up of the edge bead.[59] The thickness and the viscosity of the edge bead is very high compared to other locations, so the degree of curing at Location 3 of the spin-coated adhesive film is very low.[60] As the free volume shrinks during the curing of the film, the film size is reduced and a few millimeters of the substrate becomes unoccupied by the adhesive.

The uneven curing induces the internal stresses, shrinkage and interfacial delaminations in the devices. This type of adhesive should not be used in spin-coating for the fabrication of polymeric thin films of optical devices. Therefore, when designing coating solutions, solvents should be selected with relatively low vapor pressures to reduce the solvent evaporation; also to reduce the composition gradient during spinning and the variation of curing speed observed after spin coating. This is one of the major criteria for selecting a spin-coating solution for the fabrication of thin films for PLC devices.

8.7.2 Stability of spin-coated polymeric adhesive films

The curing reaction rate of a spin-coated epoxy adhesive is much slower than that of one without spinning. The reaction rate at the center of the substrate is also higher than at other locations. The slower reaction rate is mainly due to changing material properties during spinning.[38] These changes, as well as the curing degree or cross-linking density of the cured material, greatly influence the stability of the adhesive.

Thermal stability

Thermo-gravimetric analysis (TGA) can be used to explore the thermal stability of spin-coated epoxy adhesive. Figure 8.15 shows a typical TGA diagram (weight loss during the temperature rise) of a cured epoxy adhesive (Samples A and B) that was not spin coated. The weight-loss profile for those two samples was similar throughout the weightloss process and differed only in initial degradation temperature. A lower initial degradation temperature was observed for Sample A (70 °C) than for Sample B (100 °C). Because Sample A was not post-cured by heat after UV curing, Sample B was more stable than Sample A. Sample A may have contained absorbed moisture in the adhesive existing in a state of free or loosely-bound water which started to evaporate at low temperature. It also indicates that UV light energy only is not sufficient for optimum curing of UV-curable epoxy adhesive, which needs post thermal exposure for moisture evaporation. The major concern involving the presence of moisture at elevated temperatures is hydrolytic degradation. Hydrolysis is a chemical change, which occurs

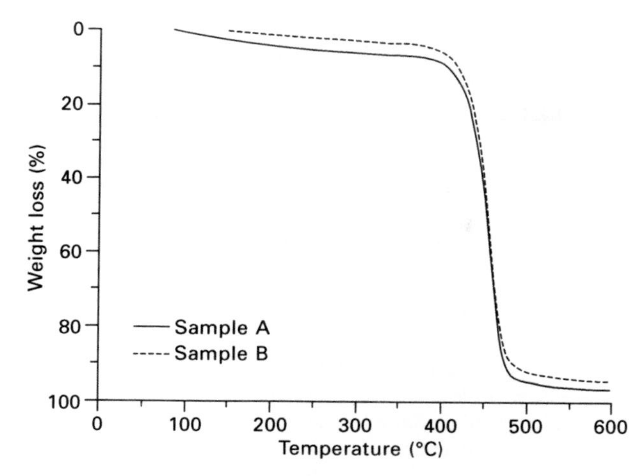

8.15 Thermo-gravimetric results of the epoxy adhesive that was not spin coated.

when moisture is present above or near the glass transition temperature (T_g) of the polymer. Hydrolytic degradation causes random chain scissions to occur, which brings about a reduction in molecular weight and, in turn, a reduction in the mechanical integrity of the cured adhesive. During post thermal exposure of an anhydride-cured epoxy, the cross-linking density also increases.[61] The anhydride group and hydroxyl group of the resin react to form ester cross-links. These ester networks are more thermo-stable than other types of linkage.[62]

The thermal stability of a *spin-coated* adhesive film (Sample C, over three locations of Fig. 8.12) was also measured using the same TGA method. Figure 8.16 shows the corresponding TGA curve of Sample C at three different locations alongside the TGA curve of Sample B. A comparison of the stability change due to the spin-induced degradation of the epoxy adhesive determined here with other previously found characteristic temperatures is shown in Table 8.2. The results show that the thermal stability decreased for the spin-coated adhesive film compared with the adhesive that was not spin coated.

Thermal stability also varied at different locations of the spin-coated substrate. At the center of the substrate it was higher than at the other locations. The lower thermal stability is mainly due to changes in material properties to various degrees at various location of the spin-coated epoxy adhesive during the spinning process. As previously described, the spin-coated liquid usually contains a volatile component that evaporates during spinning, leaving behind a thin solid film. Solvent evaporation from the film alters the material properties (e.g. viscosity, diffusivity, vapor pressure, surface

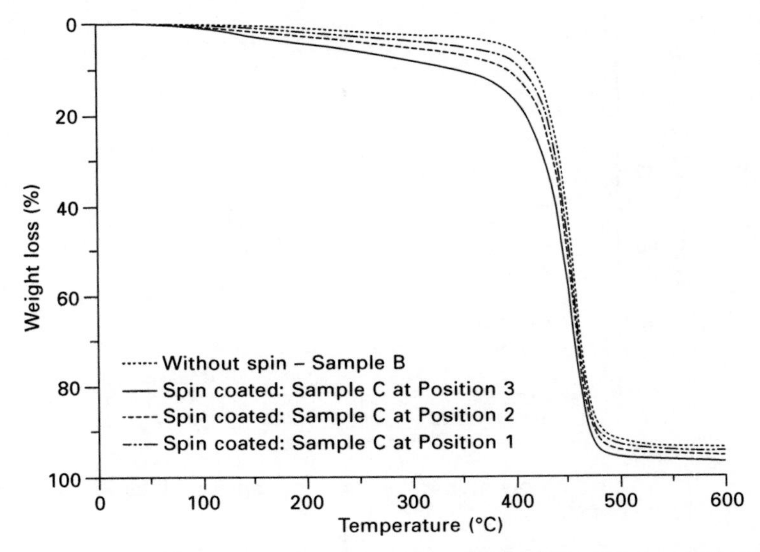

8.16 Effect of spin coating on the thermal stability of epoxy adhesive.

Table 8.2 A comparison of thermal stability changes due to degradation of epoxy adhesive

	Characteristic temperature (°C)		
Sample type	Initial degradation	Degradation	Decomposition
Sample A	70	408	448
Sample B	100	416	450
Sample C			
Position-1	90	407	445
Position-2	70	385	438
Position-3	55	338	432

tension). The high degree of uniformity in the evaporation rates leads to a similar uniformity in the final polymer film properties.[57]

Chemical stability

Chemical stability is the material's ability to withstand change from chemical contact. This issue, involving corrosive fluid exposure, should be evaluated to ensure chemically stable polymeric thin films for optical waveguides. An immersion test shows that the spin-coated epoxy surfaces deteriorate to various degrees at various location of the coated epoxy adhesive. Figure 8.17a shows an optical micrograph of Sample B after immersion in a nickel etchant. It is very clear that the without-spin sample surface was chemically stable in that solution with almost no change in surface morphology. However,

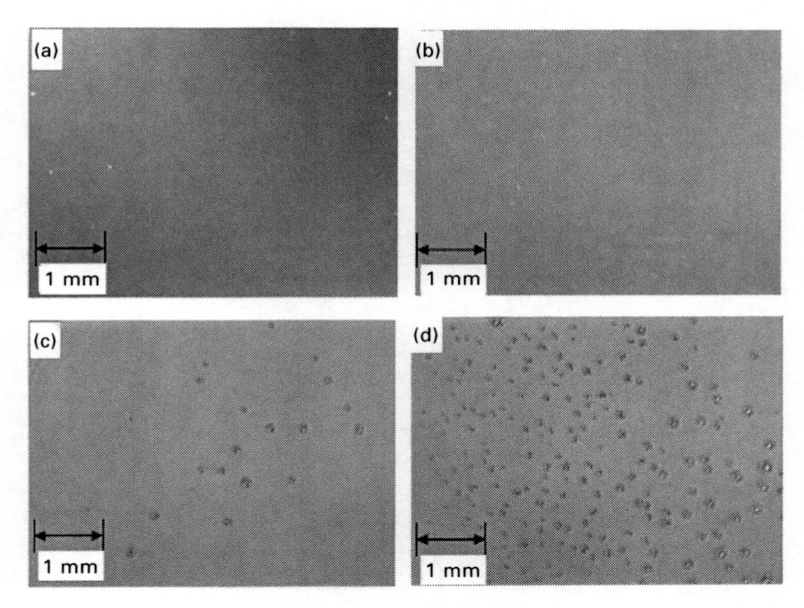

8.17 Optical micrograph of the cured epoxy after immersion in nickel (Ni) metal etchant: (a) without spin, (b) spin-coated Position 1, (c) spin-coated Position 2, (d) spin-coated Position 3.

the spin-coated sample (Sample C) showed different results. Figure 8.17b–d displays the optical micrographs of different locations on the spin-coated cured adhesive after immersion in the etchant. The chemical attack led to porosity at the polymer interface, which resulted in an excessive increase of optical loss. However, it was more stable in a chromium etchant than in the nickel. Figure 8.18 shows the optical micrograph of Sample C after the immersion in chromium etchant. The spinning also affected the refractive index of the cured adhesive. Figure 8.19 shows the refractive index of the spin-coated adhesive (Sample C) before and after immersion in chemical solutions. Before immersion, a higher refractive index was found at the corner and after immersion the higher drop also indicated lower stability at that portion. Moreover, the change of refractive index in the chromium etchant was lower than that in the nickel.

Factors affecting the stability of spin-coated polymer film

Earlier studies have found that the curing reaction rate of a spin-coated epoxy adhesive is much slower than that of the adhesive without spinning. The reaction rate at the center (Location 1) is also higher than at other locations on the substrate. The slower reaction rate is due mainly to changing the material properties during spinning. It clearly indicates the spin-induced

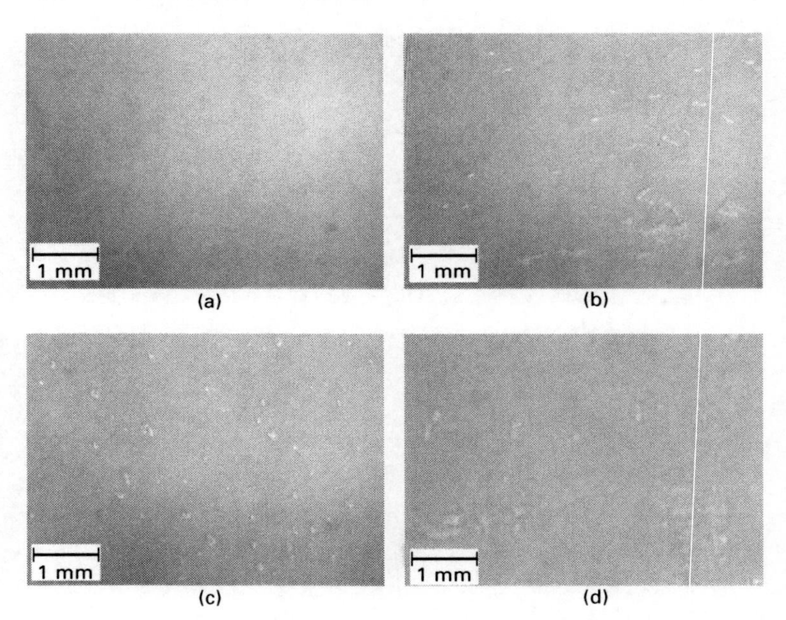

8.18 Optical micrograph of the cured epoxy after the immersion in chromium (Cr) metal etchant: (a) before immersion, (b) spin-coated Position 1, (c) spin-coated Position 2, (d) spin-coated Position 3.

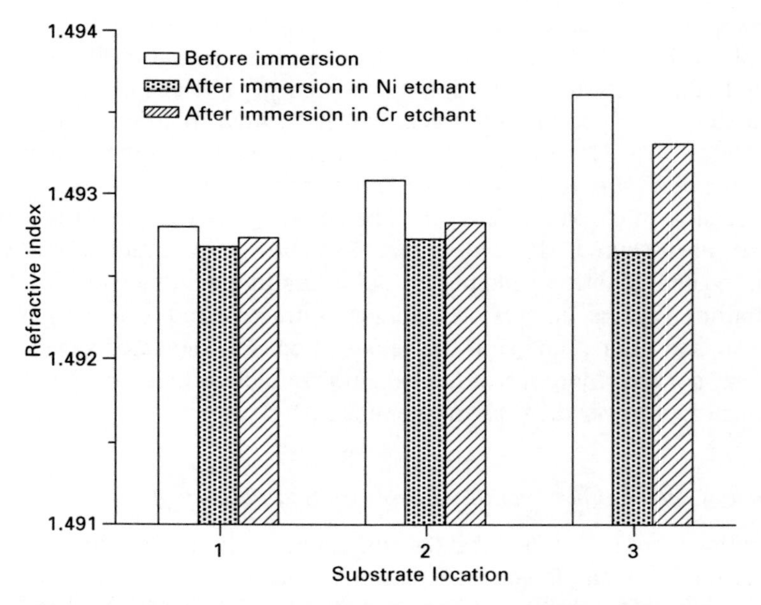

8.19 Refractive index of spin-coated cured adhesive (Sample C) before and after immersion in metal (Ni and Cr) etchant chemical solution.

degradation of spin-coated epoxy adhesive during the spinning. Compared with the no-spin material, the thermal and chemical stability decrease is due mainly to:

(i) *Mechanical degradation during spinning.* Under the influence of spinning forces, intermolecular interactions between certain molecules at certain sites of the polymer are disrupted. Spin-induced main chain rupture might also initiate the de-polymerization of linear polymers.[63] This type of chain scission plays a significant role in the thermal and chemical stability of polymeric materials.[64] During the heating or chemical etching, less energy is needed to fragmentize the polymer, hence there is a lower stability in spin-coated polymeric film.

(ii) *Changes in the ratio of resin and hardener.* Epoxy resin is more volatile than the hardener (amine). Therefore, the evaporated amount of epoxy resin is larger than that of the amine during the spinning. As a result, the correct mix ratio of amine and epoxy, to ensure the complete reaction of that reactive component is disrupted. Since the amine molecules 'co-react' with the epoxy molecules in a fixed ratio, unreacted amine remains within the matrix after the cure reaction; this alters and affects the final properties of the adhesive.

(iii) *Void formation.* Owing to the variation in spinning force, surface tension and the topography also vary at different locations in the spin-coated adhesive film. Figure 8.14 clearly shows how the topography differs from the center to the corner, due to the spinning process. As shown, there are a large number of voids at the corner and almost no voids at the center of the substrate. The void formed during the spinning and curing of the epoxy adhesive released from the coated adhesive layer at the heating or etching step and induced the excessive weight loss or porosity.

(iv) *Cross-linking density.* Lower curing degree, higher mechanical degradation and more void formation decreases the cross-link density. Decreasing cross-link density – that is, increasing the distance between reactive sites, usually has the effect of reducing thermal and chemical resistance by decreasing the compressive and tensile modulus as well as the impact strength. Less thermal or chemical energy is required to degrade a given mass into its volatile products.[65]

At the periphery of the substrate, the polymer experiences the highest spinning force, more voids and lower cross-link density. Therefore, the lowest thermal and chemical stability is observed at the periphery of spin-coated adhesive due to the rapid fluid flow and convectively-driven evaporation that occur during spin coating. On the other hand, at the center of the substrate, the highest stability is observed due to low stress, no voids and higher cross-linking density.

To overcome the problems, it is recommended that lower spin speeds be used during spin coating. Also less volatile reactive components of the spin coating solutions are preferred, having higher intermolecular forces that allow a greater part of the thinning behavior to occur without significant degradation of materials during the spinning process. Adhesives were also found physically and chemically more stable in chromium etchant solution than in nickel etchant. Therefore, chromium is also proposed to be used as the hard mask in photolithography processes during the fabrication of PLC devices.

8.7.3 Interfacial adhesion of spin-coated polymeric films

As the curing reaction rate of spin-coated epoxy adhesive is much slower than that of the adhesive without spinning, the reaction rate at the center (Location 1) is also higher than at other locations on the substrate. The slower reaction rate is due mainly to the material's changing properties during spinning. Lower thermal and chemical stability for the spin-coated adhesive was also found due to rapid fluid flow and convectively-driven evaporation that occur during spin coating. This was also the case at the periphery of the spin-coated polymer substrate due to the highest spinning force experienced, more voids and low cross-link density. On the other hand, at the center of the substrate, higher stability was observed due to the lowest stress, no voids and higher cross-linking density. These results clearly indicate the spin-induced degradation of the adhesive during the spinning.[53]

Different adhesion strength at different location

The properties of an adhesive depend upon the degree of cross-linking or completion of the curing reaction. During the curing, the polymer chains become locked together and their movement consequently becomes restricted. Cross-linked polymer chains are chemically bound together to give a three-dimensional 'chicken wire' molecular structure or chemical network. The higher the curing degree, the stronger the chemical bonding and the better adhesion strength at the adhesive interface. Shear strength was measured for investigating the interfacial adhesion of the spin-coated epoxy layer on a silicon substrate. The result shows that the adhesion strength is higher at the center and lower at the boundary of the substrate. Figure 8.20 shows the average shear strength of the deposited and cured adhesive layer at three different positions; the figure shows results for the heat-exposed adhesive and also the non-heat-exposed adhesive. Different adhesion strengths were found at different positions of the same sample. As the curing degree and stability of the adhesive are higher at the center and lower at the border side, consequently, the adhesion strength is also higher at the center and lower at the border side.

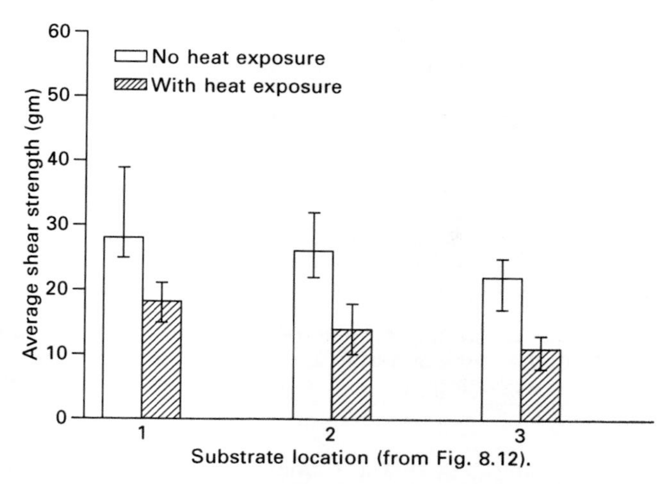

8.20 Average shear strength at three different locations for substrates with and without heat exposure.

Interfacial adhesion after heat exposure

After heat exposure, interfacial adhesion decreases substantially at all locations of the substrate. The heat exposure was performed at a high temperature, 250 °C, which is 130 °C above the T_g (120 °C) of the adhesive material. The epoxy adhesive exhibits a lower coefficient of thermal expansion (CTE) (64 ppm/°C) below its T_g than that above its T_g (143 ppm/°C). The difference between the CTE of the epoxy adhesive and silicon is also very high (4 ppm/°C for silicon vs. 143 ppm/°C at higher temperature). Therefore, the adhesion strength of the heat-exposed sample substantially decreased due to the following reasons:

(i) Above T_g, the amorphous or semi-crystalline polymer is transformed to a rubbery viscous state which lowers the mechanical integrity of the adhesive.
(ii) When the temperature rises in the solid, it expands, and this thermal expansion is directly proportional to the CTE of the material, its length and the temperature change. In this bi-material 'sandwich' structure, the layers are rigidly connected to each other. Therefore, when they attempt to expand in accordance with their CTE, each layer imposes a force along the interface to expand in an identical manner, and causes stresses to appear. The generation of these stresses may be understood from the sketch presented in Fig. 8.21. The generated stresses at the interface decrease the adhesion strength.
(iii) Figure 8.21 also shows that all the stresses are ultimately concentrated at the border side. Therefore, the highest stresses are experienced at the border side and the lowest at the center. Due to the highest developed

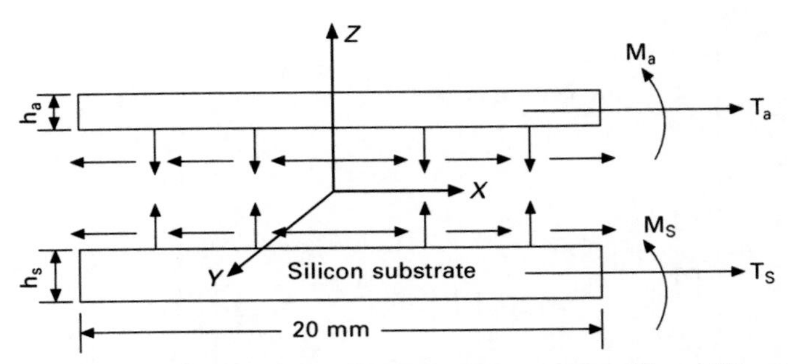

8.21 Free-body diagram of bi-material structure. T_a and T_S are the tensile strengths of the adhesive and the substrate respectively; M_a and M_S are bending moments of the adhesive and substrate respectively.

stresses at the interfaces of the border side, the loss of mechanical strength is also higher in that portion. As a result, the highest drop of interfacial adhesion is observed at that position.

It follows that the adhesive should not be exposed at above T_g during the subsequent fabrication process or in its operating life. In other words, an adhesive material with a T_g lower than the curing temperature of the waveguide core material should not be used as the lower cladding of an optical waveguide. (Otherwise the adhesive has to be processed at or above its T_g thus weakening its adhesive strength.)

Interfacial adhesion on plasma-treated substrates

Considering the mechanical interlocking theory of adhesion, the substrate was plasma-treated to increase the surface roughness and adhesion strength. Figure 8.22 shows a 3-D AFM image of untreated and plasma-treated substrate surfaces for different plasma conditions. Table 8.3 shows the surface roughness data for these different pre-treatment conditions.

It is well known that plasma etching increases surface roughness. However, it has been found that surface roughness decreases after etching in the highly reactive process gas, sulfur hexafluoride (SF_6, 100%), but increases again with the addition of oxygen to the gas. The following is a discussion of the mechanism of plasma etching of the silicon substrate in order to increase the surface roughness using SF_6 or mixture of $SF_6 + O_2$. Etching in SF_6 (100%) causes the gas phase to consist of F and SF_x ($1 \leq x \leq 5$) formed by electron impact dissociation. Its interaction with the silicon surface causes the formation of a non-volatile, thin fluorosilane SiF_x layer ($1 \leq x \leq 4$) with a thickness of 1 to 3nm. After SF_6 etching, the thin fluorosilane (SiF_x) layer covers the silicon surface and decreases the roughness of the substrate. However, if O_2 is added to the feed gas, besides F and SF_x, sulfur oxyfluorides (SO_2F_2, SOF_2,

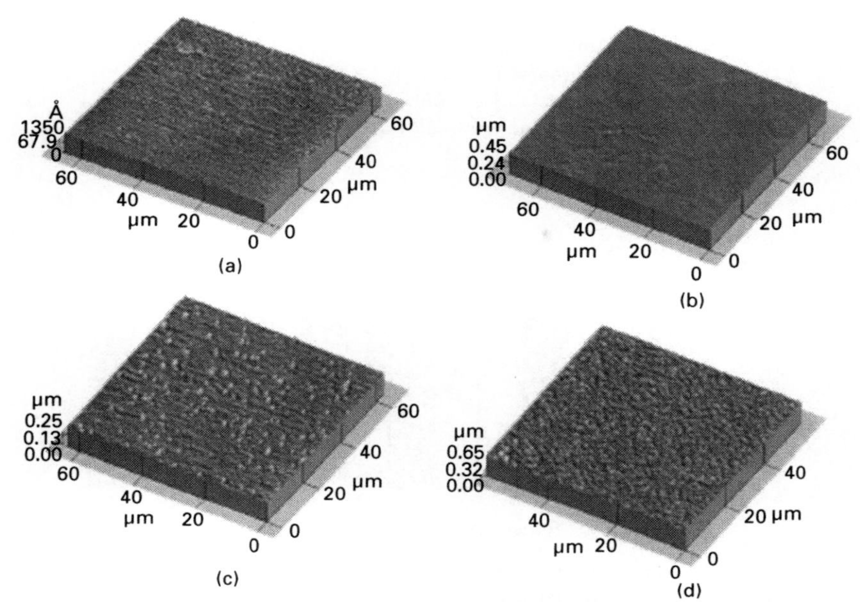

8.22 3-D AFM image of untreated and plasma-treated substrate surface for different plasma conditions: (a) without plasma treatment, (b) 100% SF_6 plasma treatment condition, (c) 90%-SF_6 + 10% O_2 plasma treatment condition, and (d) 80 % SF_6 + 20% O_2 plasma treatment condition.

Table 8.3 Surface roughness of silicon substrate for different surface conditions

Surface condition	Rp-v	rms roughness	Average roughness	Mean height
1	331 Å	20.2 Å	15.2 Å	106 Å
2	223 Å	12.1 Å	8.1 Å	92.9 Å
3	637 Å	52.2 Å	31.6 Å	131 Å
4	0.161 μm	126 Å	97.3 Å	398 Å

Rp-v, height difference between the highest and lowest point in the measured region.

SOF_4) are also produced in the discharge. The formation of oxyfluorides is due to the reaction of oxygen with SF_x radicals. The oxyfluorides are powerful etching agents. Thus they inhibit the formation of fluorosilane on top of the substrate and increase the surface roughness by etching. More O_2 added to the feed gas results in higher surface roughness.[66]

It is also well known that increasing surface roughness by plasma etching usually improves adhesion. However, the adhesion strength after plasma treatment has been shown to be lower than that of samples with no plasma treatment. Figure 8.23 shows the average shear strength at different positions, with and without plasma surface treatment. With such deterioration in the adhesion strength, it is necessary to investigate the mechanism behind it.

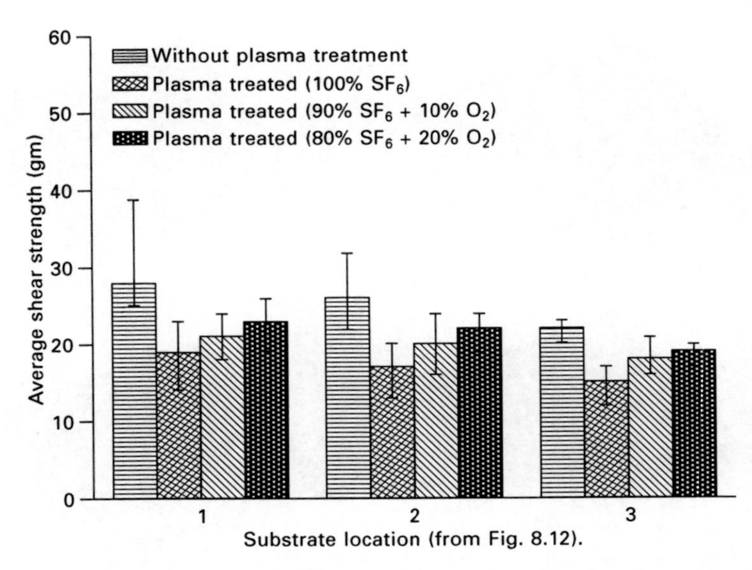

8.23 The average shear strength at different positions for untreated samples and samples with plasma surface treatment.

The mechanisms for adhesion include physical adsorption (van der Waals force), chemical bonding (covalent, ionic or hydrogen bonds), diffusion (inter-diffusion of polymer chains), and mechanical interlocking of irregular surface. With regards to adhesion mechanisms in the absence of surface treatment, mechanical interlocking has little role because the substrate surface was found to be very smooth (R_a = 20.2 Å). The role of inter-diffusion of polymer chains or other strong bonds, such as covalent or ionic bonding, are not considered here, as there was no surface deformation and smearing in the fracture surfaces. Therefore, only the Van der Waals force and hydrogen bonds are considered to be responsible for the adhesive bonding in the untreated condition. Although after plasma etching, surface roughness increased, the adhesion strength decreased. The probable reason may be.

(i) The roughness of substrates is generally a contributing factor only if the coating penetrates completely into all the irregularities of the surface and wets the surface. Failure to completely penetrate can lead the less coating-to-interface contact than the corresponding geometric area and will leave voids between the coating and substrate. The increased roughness can then lead to decreased adhesion, since trapped air bubbles in these voids allow an accumulation of moisture.
(ii) During the plasma etching, the surface structure may changes chemically, which may suppress the mechanical interlocking effect and reduce the adhesion strength.[43]

It is recommended that adhesives which exhibit different adhesion strengths at different parts of the substrate should not be used. The internal stresses developed in the devices may damage the functionality of these systems. An adhesive material with T_g lower than the curing temperature of the waveguide core material should also not be used for the lower cladding of an optical waveguide. An adhesive material with higher T_g is recommended. Silicon substrates that were plasma-treated to improve adhesion were found to be inefficient in increasing the adhesion strength. Lower adhesion strength was unexpectedly observed after plasma treatment, even for greater surface roughness. The changes caused by plasma etching of the silicon wafer surface are not yet clearly understood.

8.7.4 Interfacial adhesion on different substrates

Different adhesion strengths were found for different type of sample and process conditions. Even for the same substrate and process condition, different interfacial adhesions were found on different portions of the substrate.[43] However, the differences within the same substrate were not large and therefore the distribution did not show up well. Here, the results present the averages (solid bar) and the full distribution of the data values (36 samples for each bar) of the interfacial adhesion for each corresponding substrate structure with respect to a specific process condition. Figure 8.24 shows a comparison of shear strength for three different types of substrate structure with different types of process condition. The results are summarized below:

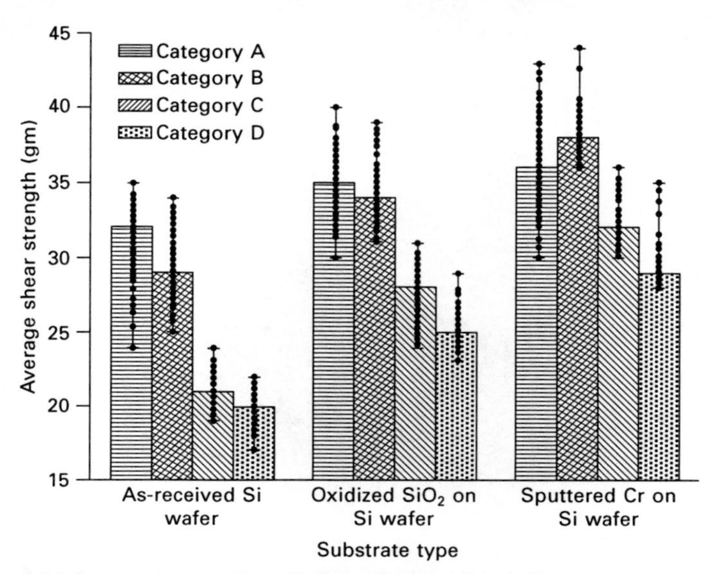

8.24 Comparison of interfacial adhesion for different surface structure and processing.

(i) Among the surfaces evaluated, the thin sputtered Cr-containing surface had the highest strength, followed by oxidized SiO_2 surfaces. Lower adhesion strength was observed for as-received silicon wafers.
(ii) The adhesion strength slightly decreased with silicon and silica surfaces but increased with the Cr surface due to applying heat treatment on the spin-coated thin adhesive layer before fabrication of the shear button.
(iii) However, the interfacial adhesion decreased substantially due to exposure to a damp heat condition (75 °C/95% RH/168 hours) after shear button preparation, with all substrates. The degradation was much higher with the silicon substrate than with the silica and Cr containing substrate.

To understand the variations in the adhesion strength on different substrates, it is necessary to undertake a morphology study of the substrate surfaces. In these experiments, the morphological study was carried out under an atomic force microscope (AFM). Among the morphology, one of the known variables affecting the adhesion is surface roughness, because of its effect on wetting at the bonding surfaces and mechanical interlocking. Therefore it was measured to quantify the surface morphology. Three-dimensional AFM images of three substrate surfaces are also shown in Fig. 8.25.

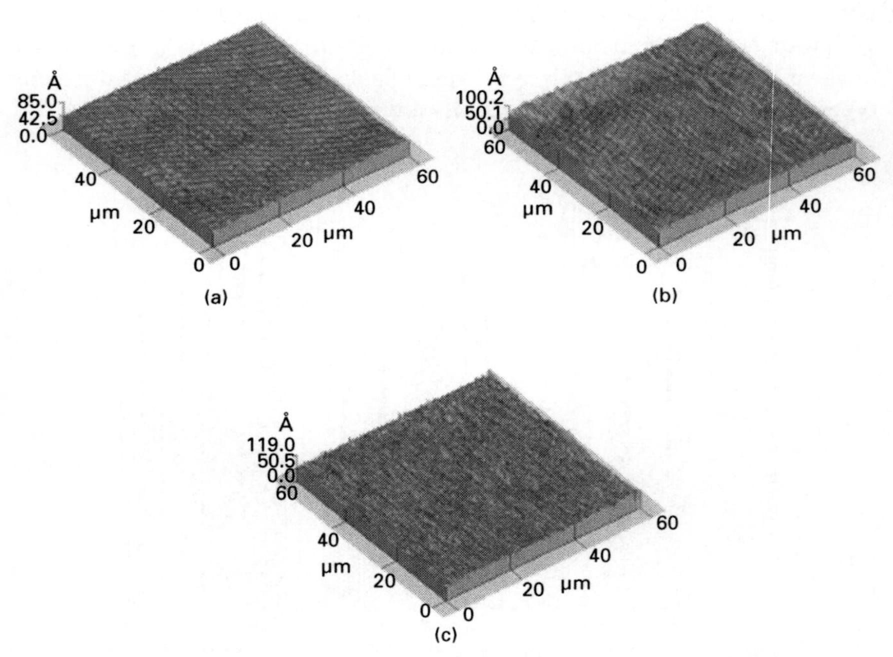

8.25 3-D AFM image of substrate surface for different surface treatments: (a) as-received silicon wafer, (b) oxidized SiO_2 on silicon wafer and (c) sputtered chromium on silicon wafer.

Interfacial adhesion on silicon surface

With regards to mechanisms of adhesion, mechanical interlocking has little role for adhesion on the silicon substrate because the substrate surface was found to be very smooth (average surface roughness, R_a = 1.81 Å). The role of inter-diffusion of polymer chains or other strong bonds such as covalent or ionic bonding are not considered here, as the silicon surface is very inactive to the polymeric adhesive and there was no deformation or smearing on the fracture surfaces. Only weak bonds such as Van der Waals force or hydrogen bonds are considered to be responsible for the bonding on the as-received silicon wafer.[43]

Interfacial adhesion on silica surface

The average surface roughness (R_a = 3.32 Å) was higher than on the silicon surface and therefore the role of mechanical interlocking for adhesion was higher. Beside this, covalent bonds are formed during the polymerization, since functionalities such as hydroxyl groups are generated during the epoxy curing reaction.[67] The hydroxyl produces polymeric chains of–$Si(OH)_2$-O-$Si(OH)_2$-OH groups with the silica surface which can link up in many different ways to form a three dimensional network and increase the adhesion strength.[68]

Interfacial adhesion on Cr surface

Among the surfaces evaluated, the thin sputtered Cr-containing surface had the highest average surface roughness (R_a = 4.71 Å) and this contributed to its higher adhesion strength. Also, for polymer–metal interfaces, other major adhesion mechanisms such as diffusion (or inter-diffusion), Lifshitz–van der Waals interaction, molecular interaction (acid–base interaction) and chemical adhesion (covalent bond) should have contributions for higher adhesion strength.[69]

8.7.5 Effect of heat treatment on the interfacial adhesion

Heat exposure was performed at a high temperature (275 °C). Above its T_g, the amorphous or semi-crystalline polymer is transformed to a rubbery, viscous state and this reduces the mechanical integrity of the adhesive. Therefore, the adhesion strength of a heat-exposed sample normally decreases. In the case of silicon and silica surfaces, it was also found that the interfacial adhesion decreased due to the heat treatment of the spin-coated adhesive film. But the decreased adhesion strength on silicon surface is not high, because, the adhesive used here had a higher T_g. Also, we used a lower spin speed to reduce the spin-induced degradation. Therefore, it is confirmed that an adhesive with a high glass transition temperature and use of a low spin speed is appropriate to reduce the degradation of the polymer photonic devices.

However, for the sputtered Cr surfaces, the adhesion increased after heat treatment. The possible reason for this is that chemicals bonds are formed at the interface, usually as a result of a charge transfer from the metal to the polymer. The metals (Ti, Cr, Zr, Al) can strongly bond to oxidized (due to heat treatment at 275 °C) polymer surfaces; when the hydrophilic groups on the polymer surface make contact with the metal layer, electrons are transferred from the metal to the hydrophilic groups, resulting in the formation of a charge transfer complex, which enhances the adhesion between the metal and the polymer.[69]

$$\mathrm{M} + \mathrm{O{=}C} < \rightarrow \mathrm{M{-}O{=}C}< \qquad [8.1]$$

Formation of M-O-C complex (M=Cr, Ti etc.)

$$2\mathrm{Cr} + (\mathrm{O})\text{-polymer} \rightarrow \mathrm{Cr_2O_3} + \text{organic fragments} \qquad [8.2]$$

Formation of Cr(III) oxide by a redox reaction between Cr and oxygen-containing functional groups of the polymer will increase the adhesion strength. Also there is a higher possibility of typical inter-diffusion between the metal–polymer interfaces increasing the adhesion strength.[12,16]

Effect of damp heat on the interfacial adhesion

Epoxy-based adhesives absorb moisture and experience hygroscopic swelling in humid environments, hence degrading the adhesion strength and elasticity. Any uncured adhesive can be severely attacked by moisture during the reliability test. Such hydrolytic attack breaks the ester linkages ('R–(C=O)–OR') of the polymer chains and creates two new end groups, a hydroxyl and a carbonyl. Hydrolyzation of the adhesive would appear to weaken its mechanical strength and adhesion to the substrate. The reduced adhesive strength also induces delamination on that interface. However, the reduction in adhesion strength is very much larger on silicon surfaces, followed by oxidized SiO_2 surfaces. A smaller reduction in adhesion strength was observed for the Cr surface during the temperature humidity test, because the formation of chemical bonds at the metal–adhesive interface by charge transfer is stronger than both the covalent bond at silica–adhesive interfaces and other weak bonds such as Van der Waals force or hydrogen bonds in the silicon–adhesive interface. Therefore, the Cr-containing interface is more reliable in a humid environment compared with the other two interfaces.

It is recommended that a thin metal layer (such as Cr) on the silicon wafer be used to increase the adhesion and reliability of polymer photonic devices. Oxidized silica on the silicon wafer is an alternative choice at the expense of reducing adhesion. However, using a silica layer has the advantage over using a Cr layer in that one fabrication step can be reduced, since the silica layer itself can effectively act as the lower cladding of the device.

8.8 Conclusions

Polymeric adhesives have gained much attention, and significant technological progress has been made recently in the photonic industry in order to meet the requirements of high-speed and large-capacity transmission of information at low cost. Thus the manufacturing infrastructure and packaging process for polymer-based *photonic devices* is fully deployed and is capable of meeting the demands of current optical modules. It is, however, essential to obtain a better understanding of the materials and process optimization in manufacturing adhesive-based *photonic components*. Within each area, major research findings and recommendations have been given in this chapter. These insights should be very useful for adhesive manufacturers in formulating better polymers with favorable performance for this application. In contrast, the industry is still learning about polymer materials and the process technologies have yet to be fully developed. It is believed that only after adhesive materials and processes have been broadly understood, incorporated and the manufacturing infrastructure built, will polymeric adhesives be widely used and eventually replace other conventional materials and techniques. Therefore, there is still a long way to go in terms of polymer advancement; the corresponding process development and infrastructure must be built before polymeric adhesives become a vital part of the PLC device. However, from polymer evolution history and the current explosive technological breakthroughs, it is certain that polymeric adhesives will be more widely utilized and will play an important role in the photonic industry.

8.9 References

1. K. S. Kim, 'On the evolution of PON-based FTTH solutions,' *Information Sciences*, Vol **149**(1–3), 21–30, 2003.
2. M. A. Uddin and H. P. Chan, 'Materials and process optimization in the reliable fabrication of polymer photonic devices, *Journal of Optoelectronics and Advanced Materials*, **10**(1), 1–17, 2008.
3. B. G. Yacobi, and M. Hubert, Introduction, *Adhesive Bonding in Photonics Assembly and Packaging*, American Scientific Publishers, Ch. 2, pp. 7–24, 2003.
4. E. Suhir, The Future of Microelectronics and Photonics and the Role of Mechanics and Materials, *IEEE/CPMT Electronic Packaging Technology Conference*, pp. 18–28, 1997.
5. *The Major Characteristics of Adhesives; Epoxy Technology*; http://www.epotek.com/techtips/The-Major-Characteristics-Of-Adhesives.pdf.

6. J. P. Harmon, *Polymers for optical fibers and waveguides: an overview*, ACS Symposium Series, **795**, Chapter 1, pp 1–23.
7. Zhiyi Zhang, Ping Zhao, Peng Lin and Fengguo Sun, Thermo-optic coefficients of polymers for optical waveguide applications, *Polymer*, **47**(14), 4893–4896, 2006.
8. P. N. Prasad and D. J. Williams, *Introduction to Nonlinear Optical Effects in Molecules and Polymers*, Wiley, New York, 1991.
9. M. A. Uddin, W. F. Ho, C. K. Chow and H. P. Chan, Interfacial adhesion of spin coated polymeric thin film for the fabrication of polymer optical devices, *IEEE/TMS Journal of Electronic Materials*, **35**(7), 1558–1565, 2006.
10. W. F. Ho, M. A. Uddin and H. P. Chan, Minimising the humidity effect of epoxy-based polymer optical waveguide devices through proper design, *J. Adv. Polym. Sci.*, submitted for publications.
11. M. A. Uddin, H. P. Chan, T. O. Tsun, and Y. C. Chan, Uneven curing induced interfacial delamination of UV adhesive bonded fiber array in V-groove. *Journal of Lightwave Technology*, **24**(3), 1342–1349, 2006,
12. Z. Huang, H. P. Chan and M. A. Uddin, Low-loss ultra-compact optical power splitter using a multi-step structure, *Applied Optics*, **49**(10), 1900–1907, 2010.
13. M. Kudo, T. Terashima, and T. Nakamura, *Fifth Asia–Pacific Conference on Communications and Fourth Optoelectronic and Communications Conference*: October 18–22, 1999 Friendship Hotel, Beijing, China, 1687, 1999.
14. N. Momotsu, Y. Noguchi, K. H. Seng, T. Omori, and H. Hosoya, *Fifth Asia–Pacific Conference on Communications and Fourth Optoelectronic and Communications Conference*: October 18–22, 1999 Friendship Hotel, Beijing, China, 1636, 1999.
15. M. A. Uddin, H. P. Chan, K. W. Lam, Y. C. Chan, P. L. Chu, K. C. Hung and T. O. Tsun, *IEEE Photonics Technology Letter*, 16(4), 1113, 2004.
16. Heilala, K. Keränen, J.-T. Mäkinen, O. Väätäinen, K. Kautio, P. Voho, P. Karioja, LTCC technology for cost-effective packaging of photonic modules, *Assembly Automation*, **25**(1), 30–37, 2005.
17. R. Boudreau, Passive optical alignment methods, *1997 International Symposium on Advanced Packaging Materials*, Braselton, GA, USA, pp. 180–181.
18. C. Perabo, *Critical Adhesive Requirements for Optoelectronic Packaging*, Loctite Corporation, 2001.
19. N. Murata, Adhesives for optical devices, *48th IEEE Electronic Components and Technology Conference*, pp. 1178–1185, 1998.
20. B. S. Mitchell, Formation and characterization of highly interfacial hybrid nanocomposites, *Reviews on Advanced Materials Science*, **10**(3), 239–242, 2005.
21. K. W. Lam, M. A. Uddin, H. P. Chan, Reliability of adhesive bonded optical fiber array for photonic packaging, *Journal of Optoelectronics and Advanced Materials*, **10**(10), pp. 2539–2546, 2008.
22. Shi D L, He P Surface modifications of nanoparticles and nanotubes by plasma polymerization, *Reviews on Advanced Materials Science*, **7**(2), 97–107, 2004.
23. M. A. Uddin, M. Y. Ali and H. P. Chan, Materials and fabrication issues of optical fiber array, *Reviews on Advanced Materials Science*, **21**(2), 155–164, 2009.
24. H. F. Woods, Causes for separation in UV adhesive bonded optical assemblies *Proceedings of SPIE*, San Diego, CA, USA (1993), Vol. 1999, 59–62.
25. A. J. Hudson, S. C. Martin, M. Hubert, and J. K. Spelt Optical measurements of shrinkage in UV-cured adhesives, *Journal of Electronic Packaging*, **124**(4), 352, 2002.
26. K. W. Lam, K. C. Hung, H. P. Chan, T. O. Tsun, and Y. C. Chan, The effect of CTE

mismatch on planar lightwave circuit packaging under thermal cycle, *Proceedings of 3rd Annual IEEE Photonic Device and Systems Packaging Symposium*, San Francisco, August 10–14, 70–74, 2003.

27. M. A. Uddin, W. F. Ho and H. P. Chan, Effect of surface structure on the interfacial adhesion of polymeric adhesive in the fabrication of polymer photonic devices, *Journal of Material Science – Materials in Electronics*, **18**(6), 655–663, 2007.
28. K.W. Lam, H. P. Chan, M. A. Uddin, Failure analysis of adhesive bonded planar lightwave circuit (PLC) based optical splitter packages *Optoelectronics and Advanced Materials – Rapid Communications (OAM-RC)*, **3**(10), 998–1004, 2009.
29. N. Marcuvitz, *Waveguide Handbook*, The Institution of Electrical Engineers, London, 1986, pp. 2–10.
30. M. Prasciolu, D. Cojoc, S. Cabrini, L. Businaro, C. Liberale, V. Degiorgio, E. M. Di Fabrizio, Fiber-to-rectangular waveguide optical coupling by means of diffractive elements, *Proceedings of SPIE–The International Society for Optical Engineering*, 5227, 132–138, 2002.
31. R. F. Feuerstein, Waveguide technologies, Chapter 7 in *Optoelectronic Packaging*, ed. A. Mickelson, N. Bassavahally, and Y. C. Lee: John Wiley and Sons, pp. 103–118, 1997.
32. Venkata A. Bhagavatula, A review of passive device fabrication and packaging, Chapter 8 in *Optoelectronic Packaging*, ed. A. Mickelson, N. Bassavahally, and Y. C. Lee: John Wiley and Sons, pp. 120–135, 1997.
33. Ray Chen, Polymer offers fabrication and economic advantages for photonic integrated circuits, *SPIE's OE Magazine*, 24–26, 2002.
34. C. P. Wong, *Polymers for Electronic and Photonic Applications*, Academic Press: Boston, Chapter 1, 1993.
35. M. A. Uddin, Polymeric Adhesive Material for Optoelectronic Device and Packaging, PhD Thesis, City University of Hong Kong, Hong Kong, 2004.
36. X. Li, Y. C. Han and L. J. An, Surface morphology evolution of thin triblock copolymer films during spin coating, *Langmuir*, **18**(13), 5293–5298, 2002.
37. M. A. Uddin, M. O. Alam, Y. C. Chan and H. P. Chan, Adhesion strength and contact resistance of flip chip on flex (FCOF) packages – effect of curing degree of anisotropic conductive film, *Microelectronics Reliability*, **44**(3), 505–514, 2004.
38. M. A. Uddin, H. P. Chan, C. K. Chow and Y. C. Chan, Effect of spin coating on the curing rate of epoxy adhesive for the fabrication of polymer optical waveguide, *Journal of Electronic Materials*, **33**(3), 224–228, 2004.
39. H. J. Lee, M. H. Lee, M. C. Oh, J. H. Ahn, S. G. Han, Crosslinkable polymers for optical waveguide devices. II. Fluorinated ether ketone oligomers bearing ethynyl group at the chain end, *Journal of Polymer Science – Part A: Polymer Chemistry*, **37**(14), 2355–2361, 1999.
40. J. F. Mano, D. Koniarova and R. L. Reis, Thermal properties of thermoplastic starch/synthetic polymer blends with potential biomedical applicability, *Journal of Material Science – Materials in Medicine*, **14**, 127–135, 2003.
41. S. Tomaru, K. Enbutsu, M. Hikita, M. Amano, S. Tohno and S. Imamura, Polymeric optical waveguide with high thermal stability and its application for optical interconnection, *International Conference on Integrated Optics and Optical Fiber Communication. OFC/IOOC '99*. 21–26 February, *Technical Digest*, **2**, 277–279, 1999.
42. H. P. Chan, M. A. Uddin, and C. K. Chow, Effect of spin coating on the adhesion strength of epoxy adhesive for the fabrication of polymer optical waveguide,

Proceedings of 54th Electronic Component and Technology Conference (ECTC), San Siego, CA, USA, 2004.

43. M. A. Uddin, W. F. Ho, C. K. Chow and H. P. Chan, Interfacial adhesion of spin coated polymeric thin film for the fabrication of polymer optical devices, *Journal of Electronic Materials*, **35**(7), 1558–1565, 2006.
44. K. B. Yoon, C.-G. Choi and S.-P. Han, *Japan. J. Appl Phy.*, **43**(6A), 3450, 2004.
45. Yong Chen *et al.*, *Electrophoresis*, **22**(2), 187, 2001.
46. Krchnavek *et al.*, *J. Appl. Phys.* **66**(11), 5156, 1989.
47. M. B. J. Diemeer, *et al.*, *Electron. Lett.*, **26**(6), 379, 1990.
48. K. K Tung, W. H. Wong, E. Y. B. Pun, *Appl. Phys. A*, 80, 621, 2005.
49. K. S. Chiang, K. P. Lor, Q. Liu and H. P. Chan, *Proc. of SPIE*, **6351**, 63511J-1, 2006.
50. G. Van Steenberge, N. Hendrickx, E. Bosman, J. van Erps, H. Thienpont, and P. van Daele, *IEEE Photon. Technol. Lett*, **18**(9), 1106, 2006.
51. A. Neyer, T. Knoche and L. Muller, *Electron. Lett.*, **29**(4), 399, 1993.
52. M. A. Khalil, G. Vitrant, P. Raimond, P. A. Chollet, F. Kajzar, *Appl. Phys. Lett.*, **77**(23), 3713, 2000.
53. M. A. Uddin, H. P. Chan, and C. K. Chow, *Chem. Mater.*, **16**, 4806, 2004.
54. L. M. Manske, D. B. Graves, W. G. Oldham, *Appl. Phys. Lett.*, **56**(23), 2348,1990.
55. S. F. Kistler and P. M. Schweizer, *Liquid Film Coating: Scientific Principles and their Technological Implications*, London: Chapman & Hall (1997).
56. K. J. Skrobis, D. D. Denton, A. V. Skrobis, *Polymer Engineering and Science*, **30**(3), 193, 1990.
57. S. K. Kim, J. Y. Yoo, H. K. Oh, *J. Vacuum Science and Technol. B*, **20**(6), 2206, 2002.
58. D. P. Birnie, *J. Materials Research*, **16**(4), 1145, 2001.
59. G. A. Luurtsema, *Spin Coating for Rectangular Substrate*, MSc thesis, University of California, Berkeley, (1997).
60. C. S. B. Ruiz, L. D. B. Machado, J. E. Volponi and E. S. Pino, *Nuclear Instruments and Methods in Physics Research, Section B: Beam Interactions with Materials and Atoms*, **208**, 309, 2003.
61. F. Y. Wang, C. C. M. Ma, W. J. Wu, *J. Material Science*, 36(4), 943, 2001.
62. S. P. Bhuniya, S. Maiti, *European Polymer Journal*, **38**(1), 195, 2002.
63. W. Schnabel, *Polymer Degradation: Principles and Practical Applications*, Hanser International, Munich 1981.
64. T. Dyakonov, P. J. Mann, Y. Chen, W. T. K. Stevenson, *Polymer Degradation and Stability*, **54**(1), 67, 1996.
65. H. Lee and K. Neville, *Handbook of Epoxy Resins*, McGraw-Hill, New York, 1967.
66. M. Reiche, U. Gosele, M. Wiegand, *Crystal Research and Technology*, **35**(6–7), 807, 2000.
67. Vigil G. *et al.*, *J. Colloid and Interface Science*, **165**(2), 367, 1994.
68. Good, R. J., Chaudhury M. K. *et al.*, in *Theory of Adhesive Forces Across Interfaces*, L-H Lee (Editor), Plenum Press, NewYork, 1991.
69. Q. Yao and J. Qu, *J. Electronic Packaging*, **124**(2), 127, 2002.

Index